Prevention of Hazardous Fires and Explosions

NATO Science Series

A Series presenting the results of activities sponsored by the NATO Science Committee. The Series is published by IOS Press and Kluwer Academic Publishers, in conjunction with the NATO Scientific Affairs Division.

A. **Life Sciences**	IOS Press
B. **Physics**	Kluwer Academic Publishers
C. **Mathematical and Physical Sciences**	Kluwer Academic Publishers
D. **Behavioural and Social Sciences**	Kluwer Academic Publishers
E. **Applied Sciences**	Kluwer Academic Publishers
F. **Computer and Systems Sciences**	IOS Press
1. **Disarmament Technologies**	Kluwer Academic Publishers
2. **Environmental Security**	Kluwer Academic Publishers
3. **High Technology**	Kluwer Academic Publishers
4. **Science and Technology Policy**	IOS Press
5. **Computer Networking**	IOS Press

NATO-PCO-DATABASE

The NATO Science Series continues the series of books published formerly in the NATO ASI Series. An electronic index to the NATO ASI Series provides full bibliographical references (with keywords and/or abstracts) to more than 50000 contributions from internatonal scientists published in all sections of the NATO ASI Series.
Access to the NATO-PCO-DATA BASE is possible via CD-ROM "NATO-PCO-DATA BASE" with user-friendly retrieval software in English, French and German (WTV GmbH and DATAWARE Technologies Inc. 1989).

The CD-ROM of the NATO ASI Series can be ordered from: PCO, Overijse, Belgium

Series 1: Disarmament Technologies – Vol. 26

Prevention of Hazardous Fires and Explosions

The Transfer to Civil Applications of Military Experiences

edited by

V. E. Zarko

Institute of Chemical Kinetics and Combustion SB RAS,
Novosibirsk, Russia

V. Weiser

Fraunhofer Institut Chemische Technologie ICT,
Pfinztal, Germany

N. Eisenreich

Fraunhofer Institut Chemische Technologie ICT,
Pfinztal, Germany

and

A. A. Vasil'ev

Lavrentyev Institute of Hydrodynamics SB RAS,
Novosibirsk, Russia

Kluwer Academic Publishers

Dordrecht / Boston / London

Published in cooperation with NATO Scientific Affairs Division

Proceedings of the NATO Advanced Research Workshop on
Prevention on Hazardous Fires and Explosions
The Transfer to Civil Applications of Military Experiences

A C.I.P. Catalogue record for this book is available from the Library of Congress.

ISBN 0-7923-5768-X (HB)
ISBN 0-7923-5769-8 (PB)

Published by Kluwer Academic Publishers,
P.O. Box 17, 3300 AA Dordrecht, The Netherlands.

Sold and distributed in North, Central and South America
by Kluwer Academic Publishers,
101 Philip Drive, Norwell, MA 02061, U.S.A.

In all other countries, sold and distributed
by Kluwer Academic Publishers,
P.O. Box 322, 3300 AH Dordrecht, The Netherlands.

Printed on acid-free paper

Printed in the Netherlands.

TABLE OF CONTENTS

PREFACE

Military research and development not only aims at destruction but also at protection of men, equipment and facilities against effects of weapons. Therefore concepts have been developed that improve safety for stationary and mobile facilities against action of pressure waves (detonation of warheads during conventional attacks), thermal radiation (nuclear bombs) and fires which induced precautions of fire safety to protect ammunition depots and crew's quarters. Effective fast fire extinguishing systems were designed for tank compartments and motors. To destroy closed buildings and remove land mines the gas and dust explosions (FAE = Fuel/Air Explosives) have been used. In the process of production of ammunition and explosives a number of safety requirements has to be met in order to eliminate the ignition sources and to avoid the total damage in the case of hazardous situations like explosions. On this way military and related industries have accumulated vast knowledge and sophistic experience which is very valuable in various civil applications like chemical industries and energy production.

All this expertise requires interdisciplinary specialists with comprehensive knowledge of complex relations in physics, chemistry and engineering sciences. The knowledge is based on theoretical and experimental research work the origin of which dates back to many centuries and often had been classified and is therefore unknown to civil researchers.

The aim of the workshop was the exchange of highly academic knowledge and results of scientific research on the topics of industrial safety technology. The main focus concerned effects, protection, suppression and avoidance of fire and explosion hazards. The researchers who primarily worked on military subjects met qualified experts in civil safety aspects of fires and explosions to discuss their experience and to work out new concepts concerning co-operation, especially application of military know-how for civil use.

In this context contributions from Europe, the former Soviet Unions and the USA to the following objectives are presented:

- Fire risk in plants and storage
- Ignition and flame spread
- Fires of solvents and industrial chemicals
- Gas- and dust-explosions
- Damage from fires and blasts
- Fire modelling
- Advanced extinguishing technologies

ACKNOWLEDGEMENTS

Major support for this useful workshop was provided by the North Atlantic Treaty Organization. So we are very grateful to the NATO for all sponsorship. Special thanks are due to Mrs. Nancy Schulte of the NATO Scientific Affairs Division for all her patience and help and to Prof. Hiltmar Schubert who also accompanied all phases of the workshop's organisation with willing suggestions.

The production of this volume was a collaborative work, and there are many to whom the editors like to thank. Specially the good co-operation between Novosibirsk and Pfinztal was exemplary.

Special thanks are due to all lectures and authors for their cooperation in preparing their manuscripts, excellent presentation and dedication to the meeting.

REVIEW ON EXPLOSION EVENTS
- A COMPARISON OF MILITARY AND CIVIL EXPERIENCES

C.O. LEIBER *, R.M.DOHERTY **
*Wehrwissenschaftliches Institut für Werk-, Explosiv- und Betriebsstoffe (WIWEB), D-53913 Swisttal, Germany - **
*Naval Surface Warfare Center, Indian Head, Maryland, USA - ***

Abstract

Some comparisons and distinctions between commercial explosives and those of military use result in the statement that these are very different in type, but quite similar with respect to explosive events. Whereas most military explosives are formulated based on expectations derived from a simple classical theoretical description of detonation, one finds in civil explosives a larger spectrum of explosive effects. Therefore, due to the large quantities produced and consumed, commercial explosives offer opportunities both for gaining knowledge and also for observing special occurrences. It is further shown here that the simple classical theory caused one of the largest explosive catastrophes in this century, which we cannot predict (and therefore prevent), even today, by appropriate testing. Regulations in both the civil and the military fields are more of a pragmatic nature than based on scientific facts. By using a more generalized detonation model, one can also describe physical explosions with a common set of basic mechanisms. An outline is given of several types of such physical explosions, which are not used for civil or military purposes, but do occur in nature, as well as in daily life.

Introduction
It is a great honor to be able to give some comments on dangerous hydrodynamic events in the Mecca of hydrodynamics, Novosibirsk.
The intended message of this contribution is:

- With respect to explosion safety, we cannot learn all that we need to know from military experiences; rather we are in need of a back-conversion from civil to military safety.
- Also with regard to the science of explosives, military aspects only partly illuminate (sometimes in a misleading way) some obvious aspects, thereby ignoring basic aspects of explosives behavior.
- Comparing western and eastern work, one finds that the western hemisphere was leading in technical and legalistic solutions that can be applied in a straightforward

V.E. Zarko et al. (eds.), Prevention of Hazardous Fires and Explosions, 1–15.

manner, whereas the eastern (Russia) part brought an overwhelming insight into more basic aspects.

- Explosion is not restricted to energetic substances such as those used in the military, but is a property of any dynamical excited two-phase system.

A further goal of this lecture is to explain why there is a big difference between performance aspects and safety aspects of explosives. Whereas technology solutions to real applications problems can be straightforward solutions, which apply in 95% or more of the cases, real safety science requires a solid understanding of the more basic aspects of the problem in order to address questions of what can happen with a probability of 1%, and much less.

Spectacular explosives accidents have happened all over the world in all centuries, both in civil and military use. Some shaded ones in the following table contributed significantly to safety science.

Year	Location	Kind and amount	Consequence estimate
1546, 1570	Regensburg, Venice [1)]	only notice	?
1687	Athens [1)]	black powder	Parthenon partly destroyed.
1700-1800	Switzerland, Netherlands [1)]	several black powder explosions	military, pictures exist.
1864 - 1894	Heleneborg - Krümmel [1)]	Some civil factories of A. Nobel	considerable
1917	Halifax, CA [1, 2, 3, 4)]	Ship accident, about 3.000 t military explosives involved	2000-5000 killed, 9000-10000 injured.
1920	Stolberg, [1, 2)]	AN, all induced by blasting	civil explosion.
1921	Kriewald, [1, 2)]	25 t AN,	19 died, 23 injured.
1921	Oppau [1, 2, 5)]	4500/750 t AN + sulfate	561 died, >1991 injured.
1941	Smederewo, Serbia [1)]	Ammunition store	> 2000 killed.
1942	Tessenderloo, Belgium [1)]	150 t AN, the same as in Oppau happened	100 killed.
1944	Bombay [4, 6)]	1400 t military expl., fire	1500 killed, 3000 injured.
1944	Seeadler Harbor, Manus Island [4)]	1000 t ammunition	350 killed.
1944	Port Chicago, USA [4, 7)]	2100 t ammunition	320 killed, 390 injured.
1947	Texas City (2 x) [4, 8)]	2300/962 t AN fire⇒explosion	450 died, 4000 injured.
1947	Brest, F [4, 9)]	3300 t AN, fire⇒explosion	21 died.
1953	Black Sea, USSR [1)]	4000 t AN, fire⇒explosion	
1964	Japan [1)]	2200 t AN, fire⇒explosion	
1956	Cali, Columbia [1)]	8 dynamite trucks exploded opposite the railway station.	1500 killed, 1 squaremile of the city destroyed.
1967	<John Forrestal> [4)]	fire⇒explosion	134 killed, 162 injured
1899 -	many Chlorate explosions [10, 11)]	fire⇒explosion also mechanical impacts	many victims, serious damages.

EARLY EXPERIENCES WITH EXPLOSIVES

While explosions have occurred with black powder over the centuries, the discovery and industrialization of nitrocotton by SCHÖNBEIN, and nitroglycerine by SOBRERO and NOBEL in the second half of the last century was a bloody success story. Nearly all possible failures leading to catastrophes with these materials have been experienced. It was learned, that heat, shock, friction, chemical compatibility, phase transitions, all kind of sparks and so on can be causal for explosions. At the time of this development, the influence of viscosity and the heterogeneity of the explosive on its sensitivity had been recognized (blasting gelatin) but the general applicability of these *principles* was not too widespread.

Nitrocotton powder, poudre B, and picric acid were used in military applications after their commercial development, and military use was based on civil practice. Considering the consumption of military and commercial explosives over 100 years, this was a good deal. It is less widely known, that the amount of commercial explosives consumed exceeds at least by an order or magnitude that of military explosives (excluding the time of the world wars). This is also the reason why there is considerable feedback of experiences regarding the safety of commercial explosives, which is usually not present to such an extent for military articles. With respect to explosion phenomena and irregularities, we get the most input from the commercial side, therefore.

Whereas commercial explosives - in present day terms - have always been highly non-ideal explosives, explosives used by the military behave more ideally. With respect to reliable high performance designs (shaped charges, explosively formed projectiles, and metal accelerating munitions) this is necessary. Consequently, the original explosive is often a precision material. In contrast to this, non-ideal explosives exhibit lower detonation velocities than theoretically possible, and can be crude mixtures. But it was learned that also in the military field non-ideal explosives can increase some performance aspects.

Further differences are that military explosives usually are designed for a long storage, even under extreme conditions, for good survivability when subjected to any threats, and reliability. Depending on the application, very sophisticated safety measures may be in use. Such requirements are not addressed to such an extent for a cheap mass product, which can be manufactured beneath a bore hole. Such differences are also reflected in the associated paper work.

In summary, it must be stated that commercial and military explosives are completely different types, so that a comparison is only possible with many caveats and premises. Nevertheless the explosive events, risks, and their causes are quite similar.

CATASTROPHIC INTERACTION BETWEEN THEORY AND SAFETY

Explosion accidents in which the cause can be found in classical threats occur with military as well as industrial explosives (see above). But there are also accidents for which the cause is less obvious, or is not explainable in current terms. For any assessment of the cause, an insight into explosives behavior, usually provided by a model or theory, is necessary.

The physical phenomenon of the detonation of gases was discovered in 1882 by MALLARD & LE CHATELIER, and BERTHELOT & VIEILLE. In 1893 SCHUSTER [12] made a discussion note to DIXON's lecture on gas detonation. This discussion note just contained in principle the current detonation theory. It is remarkable, that SCHUSTER also pointed out the weakness of this theory! But this theory, known as the piston model of detonation, was so successful for gases that it was also applied for condensed explosives. A plane piston in motion compresses and heats up the material until by thermal means a decomposition occurs, which drives this piston, and plane detonation waves result. This was the origin of the thermally driven detonation.

In accordance with this theory, supported by 'appropriate' testing, at the beginning of this century it became common practice, approved by authorities, that heaps of ammonium nitrate (AN) were loosened by blasting procedures. After the first accidents, the ammonium nitrate was desensitized (Oppau-salt), until the first large industrial catastrophe of this century resulted, see Figure 1. Even after this catastrophe by classical testing - also applied nowadays - this risk could not be evaluated. In spite of the fact that POPPENBERG, who was not officially involved in the accident investigation, detected phenomena, which we attribute today to Low Velocity Detonation phenomena, the officially adopted reason was that, most probably, the double salt (ammonium nitrate/ammonium sulfate) demixed. But a regulation was issued: Do not blast ammonium nitrate containing mixtures.

Figure 1. Oppau, Germany, 21.09.1921. Result of blasting in 4500 t AN/ammonium sulfate. The mass detonated was only 750 t. Safe blasting performed between 15.000 to 30.000 times until the catastrophe resulted!

Even after the Oppau experience, no instrumentation was developed that could predict such a hazard. More catastrophes of the same (Tessenderloo, Belgium, 1942) or similar kind resulted. Examples are Texas City (1947), Brest (1947), Black Sea, Japan for ammonium nitrate initiated by external fires. None of these accident investigations resulted in a means of evaluating the risk associated with these materials. But we must keep in mind that the most common response of ammonium nitrate to an external fire is burning too.

Other explosion hazards have been observed around the world with **non** UN-class 1 substances. Some of these have occurred spontaneously (by marshaling – i.e., handling in railway yards), but usually they are caused by external fires. Examples are nitromethane explosions, 1958 [13)], MMAN-explosions [14)] (marshaling), and explosions of pure chlorates, perchlorates and even bleaching powders by external fires. These have been known since 1899. The latter substances do not show any indication of an explosion hazard by classical testing. One can speculate, therefore, on what is the point of such tests, which cannot even predict the cratering risk if a drum falls from the table to the floor.

Coming back to military safety. The most insensitive explosives and articles are assigned to UN-class 1.5 and 1.6, defined by appropriate testing [15)]. It is stipulated that the probability of any accidental explosion is negligible. Therefore reduced safety distances, based on the expectation of mass fire rather than mass detonation, are thought to be adequate. Serious consideration should be given to whether such a conclusion is really acceptable. For the following reasons it really is not. Originally the UN-class 1.5 had been created for ANFO- and slurry-explosives, which are used as explosives but in hazard classification tests show no apparent explosive properties. Nevertheless the comparison of explosion hazards shows that there are a greater number of explosion hazards on 1.5-substances compared to explosives like TNT and so on [16)]! (In order to remain honest, much larger quantities commercial explosives are produced than classical 1.1-explosives.)

LESSONS LEARNED DURING THE LAST DECADES

The classical piston model of detonation 'allows' only 1 detonation velocity, which is called High Velocity Detonation (HVD). Early in the last century for nitroglycerine, and since the 50s of this century for porous and high density explosives, it was observed that there exist also Low Velocity Detonations. All kinds of chlorates, perchlorates and hypochlorites exhibit such phenomena. In our institute for the first time, LVD of pure alkaline chlorates in charges of the order of up to 500 g was observed, the first positive explosive response in testing [17)]!

In the accidents mentioned above, double explosions have been reported very regularly, wherein, after the first event and within the same heap, after a time interval of the order of second(s), a second usually more powerful event occurred. For pyrotechnic mixtures such occurrences have been demonstrated also for small charges in our laboratory. The reason is quite obvious. The capability of any LVD depends on the hydrostatic pressure. Even pressures of the order of some 10 bar prevent a LVD. A first reaction can, therefore, pressurize a vessel or a heap, and the explosive reaction dies away. If this pressure is released, onset of LVD can occur (again) [18)].

It has also been observed that insensitive explosives usually exhibit a large critical diameter, larger than the usual dimensions for classical sensitivity testing. Therefore testing in an appropriately large scale resolved the explosives sensitivities. It is commonly believed that critical diameters, sensitivity, and power of an explosive run parallel to one another, but this is not generally true. There are insensitive explosives with low critical diameters, exhibiting performances comparable to Comp B [19)]. A further fact is that sensitivity depends more on the systems properties than on the properties of the substance.

In initiating trains many irregularities can happen due to Low Velocity Detonation phenomena. Initiation is no matter of a simple GO or NO GO [15)].

LOISON [20)], 1952, from accident investigations concerning safety in mines found that 'empty' tubes, covered with combustibles, together with the surrounding air can 'detonate'. A shock formation in the tube is possible, not restricted by its diameter. 'Empty' explosive tubes or lines can transfer shock stimulations over long distances, and, under certain circumstances, also initiate the condensed explosive in a volume. Technical use of this is made by NONEL-cords. This type of shock wave generation is outside of the classical piston-view. But this mechanism explains very many curious events in tubes, including some without chemical reactions [21)].

From military experience of accident investigations it was learned that the current electrostatic sensitivity tests of materials in a test with a prescribed geometry show little relevance. More aspects of field breakthrough must be taken into account. Accidents with Teflon-based flares, propellants and the sometimes super sensitivity of primary explosives are evidence of this. It has also been found not unlikely that triboelectric phenomena can result in risks [22)]. A carbon dioxide extinguishing experiment on a NATO-fuel tank in BITBURG, 1954, blew up 29 German safety authorities standing on the tank roof, and injured 9, teaching them too late the unrecognized risk of electrostatics [23)].

Both in gases and dusts there exists a low concentration limit of possible explosions. Experience in mine safety provided the evidence, that by the combination of such two (non-explosive) phases, an explosion can be obtained, which can be, under certain circumstances, more violent than that of one phase in proper concentration. Such systems have been named „hybrid systems“ This effect is already recognized for military applications.

REGULATIONS

Explosives safety has been evaluated by accidents that have occurred. All these experiences have been condensed into rules and laws, describing what is proper and what is not. Since the railways were the first motor of progress, these early experiences have been internationally written into legal rules for transportation, which still apply today. These rules have also been adopted successively by other responsible agencies, which added their own points of view. Contrary to the classical explosives regulations, which prescribed the means of achieving safety goals, the modern approach in nuclear and chemical safety regulations is to set general legal goals for safety and leave the means of achieving them to be determined.Nevertheless all regulations have not only a real scientific and/or technical basis, but also many other combined interests, maybe from practicability or from economic interests. Therefore the regulations also differ, sometimes greatly, depending on the interests of the user. An example is the tables of distances.

An early answer for safe distances was given by the New Jersey tables of distances with respect to storage and manufacturing. Whereas the old New Jersey and German distances, based on actual accidents, used larger distances, the later NATO distances were reduced according to a reasonable risk with respect to defense [24]. The K-factors are defined as $K = D\left[in\ m\right] / \sqrt[3]{m\left[in\ kg\right]}$, where D is the distance in m, and m is the applicable weight of explosive (TNT) in kg. Whereas the NATO-criteria expect a safe distance at K = 22.2 for a normal house, and K = 55.5 for hospitals, and other most sensitive buildings, we know from real accident experiences (Port Chicago, 1944) that the limit for injured persons can be K = 157, and that of glass fracture $K \approx 300$. One reason for this is that in the military, field damage criteria are in use. But it cannot be concluded that at distances beyond „a guaranteed damage level" safety is present. Since tables of distances cannot be easily evaluated in a civil world, the NATO-table of distances was adopted in many civil regulations as a *scientific* result.

Regulations can also have an adverse effect on safety. The density of regulations by various responsible agencies has increased to such an extent, that it has now become easy to attribute an accident that occurs to the violation of a rule, and more detailed explorations are often not felt to be necessary - to the detriment of real safety!

Another misleading input both from theory (piston model of detonation) and the regulations is the definition of an explosion as the consequence of the gas production rate by a chemical decomposition. This definition erroneously suggests that only chemically reactive substances can be sources of explosion hazards. Due to this definition, more general explosion hazards have been completely excluded from any safety considerations. But there are many examples of explosion hazards that do not involve chemical decomposition.

CASE HISTORIES OF NON-CHEMICAL BASED EXPLOSIONS (PHYSICAL EXPLOSIONS)

Nature

A miraculously devastating explosion was observed in 1908 in TUNGUSKA, in the neighborhood of Novosibirsk, and was also observed in 1994 in the Shoemaker-Levy impacts on Jupiter, apparently large scale piston driven explosions!

Another type of explosions is the volcanic explosion. The most powerful events in recent times were in 1815 with the TAMBORA eruption of about 150 - 180 km³ erupted mass, and KRAKATOA in 1883 with 'only' about 18 km³ erupted mass. These explosive eruptions are models for the NUCLEAR WINTER. The Tambora eruption resulted in snowfall in the summer in the southern part of Germany, and famine all over Europe. But it was also the reason why QUEEN CATHARINA of Württemberg, a rich daughter of CZAR ALEXANDER of Russia established many social institutions, which are still alive today. It is a characteristic of such events, that the number of immediate victims can be of the order of 12000, but in time this number increased to much more than 80.000 due to famine in the immediate neighborhood, and much more all over the world. Many social changes - positive and negatives - resulted!

Another kind of volcanic activity is a forerunner of industrial explosions. If hot magma interacts with water, powerful steam explosions result in the formation of crater lakes (German word: Maare). In the spring 1977 the formation of two such crater lakes within 11 days had been observed in Alaska (Ukinrek crater lakes).

Other kinds of natural explosions by explosive degassing (Nios Lake, 1986, Cameroon, CO_2), (Kivu Lake, Monoun Lake, 1984, Tanganjika Lake, all Africa), and the Ocracoke in the Gulf of Mexico, CH_4) will be addressed below. All these eruptions caused many victims by suffocation.

Industry

Classical examples of industrial physical explosions are cavitating hydraulic JOUKOWSKI shocks, the fuel/liquid interactions (FLI), which are more commonly known as vapor explosions. These appear in many types of industry, such as manufacturing of cellulose, oil industry, foundries, power stations, explosions of hot cinders, in chemical processing, by fire extinguishing, and most often in the kitchen. The most catastrophic case is in nuclear reactors.

Explosion probabilities could not be determined until nuclear reactor safety claimed the numerical evaluation of the probability of such a risk for the first time. What was the procedure? They defined that any vapor explosion is caused by a melt fragmentation only [25]. This they also brought into an official definition of a physical explosion [26]. They attributed an estimated probability of an explosion by a given core meltdown. Then, as a solution, the risk of a core meltdown is quantitatively *down*calculated. It is therefore a real matter of safety science to verify or disprove such a master piece. The following case histories demonstrate that this criterion is not adequate! Explosions occur also without any fragmentation!

Hydraulic Transients

Hydraulic transients are well known as Joukowski shocks, but if these are cavitating, a powerful pressure augmentation can take place. An improper release of water in the TARBELA dam, 1974, Pakistan, resulted in a powerful explosion.

Vapor Explosions

Quebec foundry accident [27]

45 kg molten steel of 1560°C dropped into 295 l water. By calculation only 16 l water evaporated, but an explosion ordered to a TNT-equivalency of 5.4 kg.Cratering and explosive devastation up to 53 m distant brick walls where effective, more than 6000 windows had been broken.

It is unlikely that this is the result of evaporated water driving a pressure piston. Even if

1 cm^3 water is evaporated in its own volume, at 1200°C only 7.325 bar can result [+)], far from any cratering capability.

Reynolds Metal Co., McCook, USA, 1958 [28)]

Moist or possibly wet scrap aluminum was inserted in a melting furnace. An explosion caused 6 victims, and 40 injured. The damage ordered to 1 million US$.

This accident shows, that the explosion damage does not correlate with the mass of the evaporable water.

„Boiling Liquid expanding vapor Explosions (BLEVE)" [see Reviews 29-31)]

Liquid (pressurized) gas explosions of any kind (LNG, O_2, NH_3, CH_4, C_2H_6..., vinyl chloride, and so on) demonstrate best, that a fragmentation process is not required for any explosive event. A foaming up (by superheat in the case of a depressurization) and/or evaporation/condensation- or sorption/desorption-resonances activated by mechanical means (shaking of a Champagne bottle!) or thermal transfer processes seem to be the cause. Since RAYLEIGH we know the pressure in the condensed phase of a collapsing cavity is of the order of 20 kbar. But current accident investigations reveal that the highly increased gas pressure ruptures the vessel at the site of any (assumed or real) embrittlement, and induces fragments. The evaporated gas should propel the relatively large fragments, which travel an increased distance by virtue of the aerodynamic lift.

This view is invalidated by comparing the fragment distances and fireball-diameters and -durations (in the case of combustible gases) with exploding monergol tanks, which are approximately the same. The order of the pressure of exploding monergol tanks is around 20 kbar at the low end (LVD), and more than 100 kbar at the upper end.

Explosion of liquid carbon dioxide tanks

It was therefore „luck" that in the past several explosions occurred with liquid pressurized carbon dioxide vessels (15 bar, -30°C), well below the superheat limit. These accidents are of value for the following reasons: (1) The superheat theory is invalidated. (2) Furthermore, according to the Mollier diagram of state, with isenthalpic depressurization to atmospheric pressure, at 5.2 bar 50 weight % of the carbon dioxide condenses to „dry ice" (i.e. solid CO_2). Therefore any hypothetical propelling gas phase disappears drastically. By simple shock wave considerations one gets from the maximum detected fragment width of 350 m in this case a condensed phase pressure of the order of 7 kbar, which is attributable to an explosive event. Figure 2 demonstrates the explosive result of this tank car explosion (of liquid CO_2!).

Without any chemical reaction, such types of explosions can occur spontaneously (Brooklyn, 1971, liquid oxygen accident), by non or even weak mechanical impacts like marshaling, see Fig. 2, or by stronger impacts like derailment, see Fig. 3, and finally by external fires as for example the Crescent City, and Challenger [32)], 1986, events demonstrate, see Figure 4. Sometimes there are observed indications of build-up

---

[+)] Kindly Dr. Volk, ICT, calculated these values.

to such an explosion: unusual noises in the tank, or repeated safety valve clearances, but this is no reliable rule!

The 1948 dimethyl ether accident in Ludwigshafen [33)], see Figure 5, demonstrates also possible political risks. A witness saw that the tank car dismantled, and after the sudden and violent dispersion of the contents a room explosion followed, causing widespread devastation, 207 victims, 3800 injured people, damages at 3200 buildings, and 9450 flats. After World War II it was rumored - even from explosives authorities! - that Germany again was working on miraculous explosives. It was lucky that an international committee, with such prominent members as Straßmann (nuclear researcher), Professor Richard, Nancy, and mining engineer Stahl, Washington, testified that the explosion was not the result of a super explosive. But they did not recognize the reason, and speculated on a thermal overfill, and a *concatenation of misfortunes*, ignoring thereby very similar accidents before (1943). Other accidents of the same type followed therefore.

Shocks in Silos

For powdered solid materials silo-shocking appears to be the equivalent of the bubble resonance explosions of liquids. This can be produced by a sudden breakdown of so-called „silo bridges“ if a silo is being cleared. An explosive silo rupture can result, predominantly if the grains are very uniform in size. In performing the Low Velocity

Detonation studies on neat chlorates, experiments were also done on salt (sodium chloride) of very uniform grainsize, and sand of varying grain sizes. The result was that the shock driven compaction of salt dented a lead ingot, whereas sand did not [17)].

Figure 2. Explosion of a liquid pressurized tank car with carbon dioxide in Haltern, Germany, 2. 9. 1976. (With kind permission of Dipl.-Ing. Freudenthal, Minden)

Figure 3. Vinyl chloride tank explosion, 01.06.1996 Schönebeck caused by derailment. The picture shows the ruptured tank car. The top end of the tank was found at a distance of 55 m. (Kind permission of Prof. Grabski, Institut der Feuerwehr Sachsen-Anhalt, Heyrothsberge.)

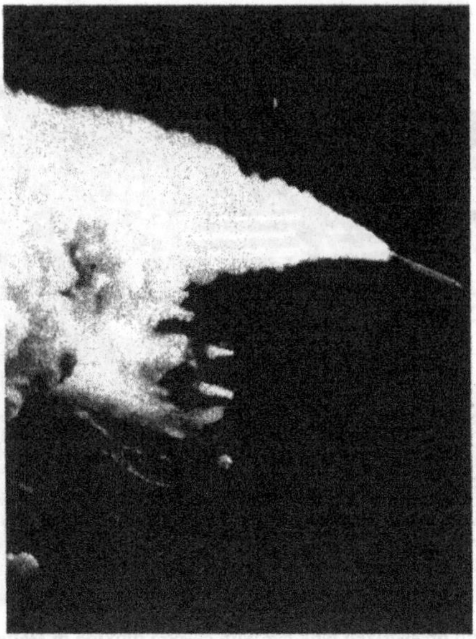

Figure 4. Start of Challenger explosion, 28.01.1986: The rocket motor fumes by a defect heated up the liquid hydrogen tank, which leaked 73.137 s after start. Flash at 73.191 s with rupture of oxygen tank, second flash 73.213 s, maximum size at 73.327 s with full explosion. (Permission of US NASA, Linthicum Heights)

Figure 5. 1948 Ludwigshafen dimethyl ether-accident. In the foreground the dismantled tank car is visible (the dismantling had been observed by a witness!), and the consequent large devastation of the area, typical for room explosions. (By kind permission.)

Examples of bubble resonance explosions

LNG-tanks	Cleveland, Ohio (1944), La Specia,
Liquid gases of all kind • External fire • After extinguished external fire • By mechanical impacts • By mechanical rupture of the vessel, • Apparently spontaneous	**LNG, O_2, NH_3, CH_4, CO_2, propane..., vinyl chloride....** Propane, Crescent City, 1970 Butyl alcohol, Litchfield, 1967 Vinyl chloride, Schönebeck, 1996 NH_4, Crete, 1969, CO_2, Haltern, 1976
Oils	Shell-Pernis (1968) (mechanism first evaluated!) [34] Triest, Oil tank sabotage, 1972. Tacoa, Venezuela, 1982, many victims: Burning oil erupted like from a volcano, and distributed fire simultaneously over a large area [35].
Tectonic CO_2 in a lake Tectonic CH_4 in a lake or sea	Nios Lake great amounts of explosively deliberated carbon dioxide produced about 2000 victims, and many animals died by suffocation [36].

Rolling over

Liquids of different densities (perhaps caused by different temperatures of the same liquid, or different gas saturation, or completely different and immiscible liquids) can be layered. Such a situation can also result from an external fire, which is extinguished! This is not a stable point. The liquids can suddenly mix by (an external) mechanical stimulus, or in the case of different temperatures by temperature conductivity. It was less well-known that above a critical difference in temperature, even after long times, a spontaneous onset of fluid convection (BENARD-convection) results. These can increase by several orders of magnitude the temperature conductivity, so that a whole system can be equilibrated within short times. The result is that volatile products suddenly evaporate, and by temperature balancing dissolved gases in the liquid degas. An explosive foaming up, creating damages like explosions, becomes possible. But all degrees between quasi-static and highly dynamic events are usual. Examples are given in the above table.

Spontaneous degassing [36]

Tectonic gases can dissolve in any kind of water, where their solubility depends on the static pressure. In the deep water the gas concentration is highly increased compared to the surface layers, and the density reduced therefore, so that a (gradually) layering results. By any mechanical or thermal instabilities mass transfers can be started, which can lead to a spontaneous even explosive degassing.

The Nios Lake contained about 4 m^3 of dissolved CO_2 per m^3 water. It was estimated, that 150 10^6 m^3 CO_2 explosively degassed, and still 250 10^6 m^3 remained in the water.

Another effect of degassing is that, by sparkling and foaming up, the original liquid's density decreases drastically, so that a normal ship (built for normal water buoyancy) sinks into this foam.

CONCLUSIONS

A great number of victims of explosive accidents have accumulated over the centuries. The most spectacular numbers resulted from unforeseen or unexpected explosions. These appear therefore as ***low probability - high risk*** explosive phenomena, which are completely outside of scientific considerations. With respect to the innocent victims, a call for more activities - potentially on an international basis - on potentially dangerous hydrodynamic events is appropriate. Novosibirsk would be an excellent starting place for improving safety in the military and nuclear field by returning to more fundamental considerations and undertaking appropriate basic work!

REFERENCES

1. Biasutti,G. S.: History of Accidents in the Explosives Industry, Vevey, Selbstverlag. Josef Pointner: Im Schattenreich der Gefahren, Int. Publikationen, GmbH, , Wien, 1994, 496 p.
2. Assheton, R.: History of Explosions on which the American Table of Distances was based, Bureau for the safe Transportation of Explosives and other dangerous Articles, The Institute of the Makers of Explosives, 1930.
3. Spencer, Len: (UK Health & Safety Executive): Explosive Lessons, Hazardous Cargo Bull., November 1980, p. 20/21.
4. Korotkin, I. M.: Seeunfälle und Katastrophen von Kriegschiffen, 5. ed., Brandenburgisches Verlagshaus, Berlin, 1991.
5. Kast, H.: Die Explosion von Oppau am 21. September 1921 und die Tätigkeit der Chemisch-Technischen Reichsanstalt, Sonderbeilage zur Z. für das ges. Schieß- und Sprengstoffwes. 20 $(1925)_{11,\,12}$, and 21 $(1926)_{1\,,9}$. Bericht des 34. Untersuchungsausschusses zur Untersuchung der Ursache des Unglücks in Oppau, Z. für das ges. Schieß- und Sprengstoffwes. 19 $(1924)_3$, p. 42/46 and 60/63.
6. Spencer, Len: (UK Health & Safety Executive): An act of self-mutilation, Hazardous Cargo Bull., April 1981, p. 25/26.
7. The Port Chicago, California, Ship Explosion of 17 July 1944: Technical Paper No 6, Army-Navy Explosives Safety Board, Washington, DC, 1948.)
8. Amistead, G.: Report to John G. Simmonds & Co Inc., Oil Insurance Underwriters, New York City on THE SHIP EXPLOSIONS AT TEXAS CITY, TEXAS ON APRIL 16 AND 17, 1947 AND THEIR RESULTS, 1947. - Wheaton, Elisabeth Lee: Texas City remembers, The Naylor Comp., San Antonio, Texas, 1948, 109 p.
9. Ville de Brest Archives Municipales: III - Rapport Annuel sur l'Activité du corps des Sapeurs Pompiers. Année 1947. Spencer, Len, (UK Health & Safety Executive): Learning the hard way, Hazardous Cargo Bull., February 1981, p. 22/23.
10. Leiber, C.O.: IHE - 2000, Wunsch und Wirklichkeit, 11. Sprengstoffgespräch, 5. 10. 1987, Mariahütte, in japanese: J. Industr. Expl. Soc., Japan 49 $(1988)_5$ p. 300/304, (Kōgyō Kayaku 49 $(1988)_5$ p. 300/304.).
11. Steidinger, M.: Die Gefahrklassifizierung von Alkali- und Erdalkalichloraten, Amts- und Mitteilungsblatt der Bundesanstalt für Materialforschung und -Prüfung (BAM) 17 $(1987)_3$, p. 493/504.
12. Schuster, A.: Note to H. B. Dixon, Bakerian Lecture: On the Rate of Explosion in Gases, Phil. Trans. Roy. Soc. London A 84 (1893), p. 152/154.
13. Ex Parte 213, Accident near Mount Pulaski, Ill., Interstate Commerce Commission No 305. Egly, R. S.: Analysis of Nitromethane Accidents, in Symposium on Safety and Handling of Nitromethane in Military Applications, Monroe, Louisiana, 14-15 November, 1984.
14. Railroad Accident Rept., Burlington Northern Inc., Monomethylamine nitrate explosion, Wenatchee, Wash., August 6, 1974, NTSB-RAR-76-1 (1976).

15. Recommendations on the Transport of Dangerous Goods (Tests and Criteria), Rev. 1, United Nations, February 1989

16. Prinz, J.: Communication from 18. 11. 1991.

17. Leiber, C. O.: Detonation Model with Spherical Sources H: Low Velocity Detonation of solid Explosives - a summary, Proc. 18th Int. Pyrotechnics Seminar (IPS), Breckenridge, COL, 1992, p. 563/591. (CA 118(24): 237083x), J. Ind. Expl. Soc., Japan, 48 $(1987)_4$, p. 258/271

18. Leiber, C. O.: On Double Explosions, Proc. 18th IPS, Breckenridge, COL, 13./17. 7. 1992, p. 593/606. (CA: 119(4): 31112t)

19. Lamy, P., C. O. Leiber, A. Cumming and M. Zimmer: Air Senior National Representative Long Term Technology Project on Insensitive High Explosives (IHEs) - Studies of High Energy Insensitive High Explosives, 27. Int. Jahrestagung ICT, 25./28. Juni 1996, Karlsruhe. Energetic Materials - Technology, Manufacturing and Processing, ICT, 1996, p. 1-1/1-14. Collaboration on insensitive high explosives. J. Defence Science **1** $(1996)_4$, p. 539/546.

20. Loison, R.: Propagation d'une déflagration dans un tube recouvert d'une pellicule d'huile, C. R. 234 (1952), p. 512/513.

21. Leiber, C.O., P. Steinbeiß and A. Wagner: On Apparent Irregularities of Pressure Profiles in Shock Tubes, EUROPYRO93, 5e Congrès International de Pyrotechnie du Group de Travail, Strasbourg, 6./11. June 1993, p. 171/178.

22. Hazard Studies for Solid Propellant Rocket Motors (Etudes des Risque pour les Moteurs-Fusées à Propergoles Solides), AGARDograph No. 316, AGARD-AG-316, September 1990, 191 p.

23. Nabert, K. and G. Schön: Folgerungen aus den Untersuchungen über die Ursache der Explosions-katastrophe bei Bitburg, Erdöl und Kohle **8** (1955), p. 809/810.

24. NATO/AC 258 documents.

25. Mayinger, F.: Wie sind Dampfexplosionen im Lichte neuerer Erkenntnisse zu beurteilen?, atomwirtschaft, Februar 1982, p. 74/81.

26. Lafrenz, B.: Physikalische Explosionen, Fb 771, Bundesanstalt für Arbeitsschutz, 1997.

27. Lipsett, S. G.: Explosions from Molten Materials and Water, Fire Technology **2** (1966), p. 118/126.

28. Epstein, L. F.: Metal-Water Reactions: VII. Reactor Safety Aspects of Metal-Water Reactions, Vallecitos Atomic Lab., GE Co, Pleasanton, CA, GEAP-3335 (1960).

29. Leiber, C. O.: Explosionen von Flüssigkeits-Tanks. Empirische Ergebnisse - Typische Unfälle, J. Occ. Acc. **3** (1980), p. 21/43.

30. Strehlow, R. A. und W. E. Baker: The Characterization and Evaluation of Accidental Explosions, NASA CR 134779, 1975.

31. Gugan, K.: Unconfined Vapour Cloud Explosions, Inst. Chemical Engineers, 1978.

32. Report to the President by the Presidential Commission on the Space Shuttle Challenger Accident, 6. 6. 1986, Washington.

33. Schnell, B.: Erster zusammenfassender Bericht über die Untersuchungen zur Aufklärung der Ursache der Explosionskatastrophe vom 28. 7. 1948 bei der B.A.S.F. - Stand vom 4. 9. 1948 -, Referat vor dem Parlamentarischen Untersuchungsausschuß des Landtags Rheinland/Pfalz. Storch: Bericht über die Explosionskatastrophe in der Badischen Anilin- und Sodafabrik in Ludwigshafen am 28. Juli 1948, Arbeitsschutz **1** (1949), p. 34/35.

34. Brief vom Minister für Soziales und Volksgesundheit No 1 vom 26.3.1968 an den Vorsitzenden des Parlaments (der Niederlande).

35. Boilover im Kraftwerk Tacoa, Magazin der Feuerwehr 8 (1983)10, p. 468/474.

36. Tietze, K.: Gefangene Gase in geschichteten Seen, Frankfurter Allgemeine Zeitung, 10. 9. 1986, Nr. 209, p. 31. - Personal communications (as primary investigator).

EXPLOSIONS CAUSED BY FIRES AT HIGH EXPLOSIVES PRODUCTION

B.N.KONDRIKOV
Mendeleev University of Chemical Technology
9 Miusskaya Square, 125047 Moscow, Russia

1. Introduction

Manufacturing, storage, transportation and practical application of energetic materials, particularly high explosives (HE) are tightly connected with the danger of accidental explosion. The real cause of the explosion can be different, e.g. impact, friction, self-heating, influence of a shock wave, radiation, electrical discharge, etc. However, the first stage or the first link of the chain of events (called lately Bowden's chain [1]) leading to energy evolution which results in destruction and demolition of the environment is always combustion of HE. Correspondingly, every accidental explosion in some sense might be considered as a consequence of a fire. However, the aim of this paper is to analyze only those accidents where explosion can be considered as a result of more or less prolonged influence of a fire on the relatively big amounts of HE.

The paper consists of three main sections.

– Description of the several, most impressive, or perhaps most typical accidents connected with fires;

– Discussion of situations at which the extended combustion of a big amount of explosive did not lead to explosion;

– Analysis of the ways which may result in the accidental explosion as a consequence of fire.

2. Explosions following fires

Several heavy accidents had taken place at fires in the course of HE production, storage and transportation during last two decades.

December 1984. The city of Tshapaevsk, Kuybyshev Region. A magazine containing approximately 100 tons (2.5 thousand bags) of dry flaked TNT had caught a fire as a result of heating in the badly designed unit of the conveyer belt. The quietly burning bag came into space of the magazine from the transport gallery and undoubtedly might be extinguished by the more or less experienced worker because burning of a single bag of TNT offers no danger of explosion in its own right. Unfortunately, the loader was inexperienced. He got fear and run away. The team of the

V.E. Zarko et al. (eds.), Prevention of Hazardous Fires and Explosions, 17–28.

firemen came to the place of the accident very quickly and immediately began to fight fire. The attempts, however, had no success and the fire became even more strong. Moreover, the hazardous claps and booms became heard. The head of the team, a young lieutenant ordered the subordinates to retreat. The people rushed out of the place of the growing fire and in about a quarter of minute the awful explosion occurred.

The magazine itself and a freight car containing about 40 tons of TNT, placed close to the magazine, eventually disappeared. The parts of the metal constructions heated by the prolonged fire got a second magazine, near the first one, where 200 tons of granulated TNT were stored. Its roof was destroyed and TNT was ignited. All of the fire-fighters were quite near the second magazine at this very moment. Quite naturally however nobody of them tried to extinguish this fire and the TNT burned out quietly, without explosion, in about 24 hours. The small spots of flame were still being seen between the magazine floor slabs through about 3 days after the event. Two freight cars containing TNT situated within the distance of about 0.5 km from both magazines were also inflamed by the hot construction pieces. Again nobody fought the fires and cars burned out without explosion.

The test experiment aimed to elucidate the way of the fire development were carried out with the single bag of TNT on the proving ground immediately after the accident. It was performed at about -20^{0}C out of doors. It was practically impossible to ignite TNT at this temperature by means of a match (or several matches) or by the burning splinter, or a chip of wood, TNT even hardly melted. The paper bag turned out be able to burn slightly better. It burnt for several minutes, then combustion of the upper layer of TNT started. TNT combustion lasts about 40 minutes with an abundant evolution of the black smoke. A dim reddish flame near the surface of the explosive was so weak that it seemed that its quenching is rather simple task.

August 1985, Pavlograd, Ukraine. The freight car loaded with TNT, caught a fire inside the territory of a big ammunition plant in the process of unloading into the truck. The reason was that a truck driver did not stop the engine. Accidentally, the car was not equipped by the spark arrester (actually it did not have even the noise suppresser). Several teams of firemen took part in fighting the blaze. Fortunately, when the claps and the booms had began they also recognized that time came to retreat (the loaders felt it much earlier, they just jumped into the track with the engine working as only they noticed the first traces of flame). Only one of the firemen who retreated not fast enough was killed. After the explosion had occurred about seventy local fires had arisen in the territory of the plant initiated by thrown away red-hot parts of the car. Two of the fires were related to the manufactories for Alumotol (15/85) and Grammonit (TNT/granulated AN) production. The technological schemes of both materials manufacturing included the long steel pipes containing HE. Both manufactories exploded after 7-10 minutes of burning. All the workers escaped after the first explosion, when the freight car burst and no one was killed or injured.

It is relevant to note that the production of Alumotol was later restored, in spite of the author's protests, on the basis of the same technological scheme that includes the long steel pipes from the tank containing atomized aluminum in molten TNT suspension to the granulators. It had been processing only about a year after the restoring. The inflammation has been occurred in the aluminum unloading unit resulting in fire following by an explosion.

This example demonstrates very well that along with the high danger of the attempts to extinguish a well developed fire with the sprays of water, there exist a great hazard of deflagration to detonation transition in the long rigid pipes. This is a well known fact though. Cutting, welding, "burning up" of the tubes containing some amount of an explosive substance is followed by explosion probably in two cases of three. Upon occasional ignition of an explosive substance inside a tube as a result of friction or impact, the accidental explosions, sometimes with tragic consequences, are reproduced quite often.

Presumably this was a cause of an accidental explosion of a building where Alumotol production was planning to be arranged in the end of 50th in Russia. Eight employees were killed including the very experienced engineer, one of my teachers, Irina P. Benzeman.

At least one of the accidental explosions at RDX production in the middle of 60th in the town of Rubezhnoye, Ukraine, presumably was also a consequence of the attempt to clean up the pipe piling by the thick RDX suspension.

It may not be excluded that explosion of 1990 in the city Biysk, Altai Territory, of a commercial explosive manufactured by the method of granulation into cold industrial oil of a suspension of aluminum in the ammonium nitrate melt, was a result of the same attempt to clean up the piled pipe.

Even ammonium nitrate melt may explode inside the long durable pipe at ignition. The accident of this sort was registered during AN production in the town of Tchirtchick, Uzbekistan in the very beginning of the 80th. The real cause of self-ignition and explosion was probably an overdose of sulfuric acid which usually is added, in the negligible concentrations though, to the AN melt as a remedy for the small quantities of ammonium nitrite to destroy. The melt of AN detonated in the feeding tube and then in the supply tank on the top of the granulation tower (about 70 meters height). The photograph of a plant taken from a helicopter looked like a picture made after the nuclear explosion. Fortunately, the explosion occurred during a holiday, and a number of suffered people was relatively small.

The explosives more powerful than TNT, and moreover than AN, sometimes need existence of special, and not always easily elucidated, combination of the external conditions for DDT or DET to proceed as a consequence of fire. The accidents in Sverdlovsk, Arzamas and Dzerzhinsk of 1988 are the examples of such combination of rather complicated conditions for the fire initiation and development leading to explosion

September 1988, Sverdlovsk. The freight car containing 60 tons of RDX (about 1500 bags) had been moving in the head of a train. A locomotive was in the tail. A locomotive operator noticed suddenly that another train, loaded with coal, was moving across the pass of the travel of his train. He switched on the system of the emergency braking but it did not help him to stop his train, and the first freight car, loaded with RDX, cuts its way in the car of coal, reared and tumbled down. RDX was poured out and ignited. It had been burning for about two minutes, and then exploded. Nobody was killed but the destruction was tremendously great.

Much more serious accident took place in 1988 in the city of Arzamas, Gorky Region. It was one of the most heavy catastrophes in the history of HE production and application. It took place at the outskirts of the town during transportation of HE. Three 60-tons freight cars one of which was loaded with the bags of RDX and HMX (partly desensitized, partly neat, dry "white powder"), another contained ammonite, the third besides ammonite was filled in with the boxes of some pressed items for the gas and oil wells, were included in the train just behind the locomotive. The train had reached Arzamas at about 9.15 a.m., June 4. It was hot summer Sunday. A lot of people were around this place, in their houses, in cars, at the barriers of the railway level crossing, in the vegetable gardens, at the low ground. Only few of them were happy to survive. The witnesses described in detail all the visible circumstances of the case.

It was discovered by investigation that at the same place, 15-20 m apart of the railway level crossing, the underground gas pipe, 108 mm id., crossed embankment of the railway. It was used for transport of the natural gas at moderate, about 3 atm, pressure to some place in suburb. The crossing was constructed about 15 years before the events described, and (as it was stated at the investigation of the case) with the essential violations of technical regulations. The tube of 280 mm I.D. used as a jacket of the basic gas pipe under the embankment was broken during the construction and the upper part of the destroyed jacket tube was only about 0.6-0.8 m under the railway bed surface instead of 1.5 m as it was demanded by the regulations. One of the joints of the basic pipe was welded also with the great infringement of the rules, and some time before the accident it was already destroyed. One may assume that the natural gas flowed out the slit, moved along the pipe walls in the gap between the pipe and jacket and through the slot in the upper part of the jacket and the porous bed of the railroad embankment went out in surroundings. The concentration of the gas in the atmosphere near this place was high enough to allow the inhabitants of the area to feel it. Some of the complaints of the people were collected in an inquire volume.

Of course the very important moment of the case was the usage of the soft bags (the paper bag inside the jute bag) as the containers for the extremely dangerous high explosives without any water or some other kind of desensitization employment. Pouring out of the explosive from the bag obviously could not be prevented. Actually the crystals of RDX were clearly detected by the experts along the way of its railway transportation.

Presumably one, maybe more, of the bags in the case investigated were torn, and some amount of the explosive was poured out on the surface of the stack of bags and on the floor of the car. If gas under the floor in space between the car and the railway bed was occasionally ignited the flame could penetrate through the slits under the door into the car and to inflame the explosive powder. Flame spread over the surface of the layer on the stuck, and on the floor. Then it appeared in the gap between the door and the wall in the bottom part of the car. Witnesses clearly noted this moment. The flame moved along the slit to the upper part of the car. About 5-10 sec later the shining ball of blaze grew up over the carriage roof, and the awful explosion thundered. Detonation of about 140 tons, in TNT equivalent, happened. All the houses in the radius of several hundred meters were blown off, the cars near the level crossing were fully destroyed, the broken parts of the freight cars were found at the distance of about 1 km from the site of explosion. About a hundred people were killed immediately. The total sum of losses was estimated to be approximately $ 100 million.

1988, the city Dzerzhinsk, Gorky Region. Explosion which seemed to be absolutely inexplicable at first, took place in the ammunition and HE plant during RDX drying process. This case, as fate has willed it, was assumed to be very similar in essence to the Arzamas catastrophe. The main points were, on the one hand, the neat RDX being involved into the accident, and, on the other hand, the very short time, several seconds, between the moments of ignition and explosion.

The process of drying was being implemented in a dryer of the fluidized bed. The wet RDX was cut from the surface of the vacuum drying drum, and passed through a vibro-trough into inlet of the dryer. A hot air produced by a blower was supplied through the mesh floor of the dryer at such a speed that particles of RDX were involved into permanent moving and like a liquid flew from the first section of the dryer to the fourth one where RDX became dry. The vibro-trough situated on the steel plate, 20 mm thick, and the massive concrete base had the trapezoidal cross-section along the main plane. The narrow side of it was closed by a welded lid, the broad part was open. The flow of the wet RDX along the trough was able in principle to fill up the narrow crossection confined by the lid.

The most probable consequence of events which had led to explosion was assumed to be as follows. The trough got filled up with the compressed RDX powder. A worker was sent to discharge it. The only tool by means of which it could be done was the brass crowbar that was being employed usually to break ice in front of the building in winter time. After the worker had cleaned up about three fourth of the trough surface he probably ignites the RDX layer that led to fire transmitting to the dryer. The worker threw the crowbar and started to run along the isle to the exit. When he was near one of the PMMA windows in the dryer walls, some 5-7 meters from the place he had been working, the explosion occurred.

Almost all of the events appeared to be imprinted with the photography exactness at the steel base-plate under the trough. Along the main axis of the trough, near one of its walls remote from the isle, it was a stripe on the plate, about 20 cm width, of the traces of impacts of the small fragments formed upon destruction of the bottom sheet of the trough (stainless steel, 3 mm thick) at detonation. About 20 relatively big fragments made the holes, up to 3 cm diameter, in the plate, and penetrated deeply into the concrete base. Using the standard formulae of the theory of explosion one may calculate easily that thickness of a layer of powdered RDX necessary to make the holes should be about 10 cm, almost precisely the height of the trough walls.

It should be also noted that many fragments of PMMA were found, at medical investigation, in the body of a worker after explosion. Obviously, in the moment of explosion, he really was in front of one of the PMMA windows in the wall of the dryer. It probably took him several seconds to understand what is happened, to throw the crowbar and to get over these 5 to 7 meters that separated him from the place where explosion occurred.

Some four hundred kilograms of RDX exploded. The building was fully destroyed. Only the substructure of it was remained. The bend brass crowbar was found near the place of the explosion.

It must be mentioned that 10 years before the accident described, at the same manufacturing plant, or to be more exact, at the plant predecessor, an explosion had taken place, also at the presence of a worker. That time the experts explained it by formation of the slot in the trough walls, as a result of the bad welding. It was ordered to check the welds by X-rays. In 1988 the welds of the trough under consideration were thoroughly checked some time before the explosion.

In the 80th, in the town of Rubezhnoye, Ukraine, two explosions took place at inflammation of the wet RDX, in a container (about 300 kg) and in a stack of PVC bags (about 1600 kg). In the first case ignition was supposed to be a result of mechanical stimuli, in the second case, of the thermal decomposition and self-ignition.

In all instances, namely in Arzamas, in Dzerzhinsk, and perhaps in Sverdlovsk the main link of the chain of events leading to explosion was probably formation of a cloud of RDX particles in the hot gases. Detonation of the aerosol may initiate the detonation process in the bed of granulated RDX.

3. Fires with no explosion

Fires upon production or in course of transportation or storage of high explosives proceed very often without explosion or detonation.

Many cases are known of fires with TNT, ammonites, and ammonium nitrate in the magazines or freight cars when the big amounts of the explosive substance burned out with no explosion. Moreover it was stated definitely in the course of the very expensive experiments that RDX and HMX may burn in the amount of up to 10 tons without detonation or explosion. The experiments were performed to determine the DDT conditions for RDX and HMX burning in the freight cars. Explosives were being ignited in the truck containers with mass of explosive in every run from 100 to 10,000 kg. Ignition of HE performed inside the stack of bags, and outside it, directly by the hot wire, or through a some amount of black powder. In no one case, about twenty or thirty trials, the DDT conditions were established.

An explosion proceeded, however, when the special freight car constructed to move explosives across the country was tested. The walls of the car were made of the steel plates 10 mm thick, the roof was fasten by a hundred of bolts, the strong doors were closed tightly. Burning of about 9 tons of RDX in such conditions resulted in detonation. The fragments of the car walls of the awful killing power were thrown away at a distance of about 1.5 km.

An extensive study of the DDT conditions at burning of conical piles of RDX, or stacks of bags was performed [4]. In most of the experiments detonation was not observed.

In the 70-80th the numerous inflammations of the hexamethylenetetramine (HMTA) were observed at the manufacturing of RDX. The method of RDX production in Russia consists in injection of concentrated nitric acid and HMTA into a reactor containing suspension of RDX, HMTA and half-products of RDX formation in HNO_3. The thick layer of RDX is formed usually on the wall of the reactor and on the shaft of the mixer blades. The RDX layer burned out, at the inflammations, but no explosion was actually noted. These inflammations inside (and sometimes outside) the reactor in the HE manufacturing plant, where matches or cigarettes in the pocket of a worker is the criminal offense, were counted eventually by hundreds.

There were no DDT cases also in the experiments called to check the possibility of explosion stipulated by inflammation of HMTA in the reactor. The experiments were carried out at the ground floor near the city of Dzerzhinsk in summer of 1978. Two cylindrical vessels about 2m in diameter and 1.2 m high made of the stainless steel were connected with the tube about 1 m long, 100 mm id. The installation should represent a model of the system of two reactors at the conditions of RDX production. The walls of the cylinders were being covered from inside by the layer of the thick suspension of RDX in the concentrated nitric acid. The same suspension had covered the walls of the connecting tube. The overall amount of RDX in the system was 80 kg, i.e. two bags, one and a half bag in the form of the suspension, and a half in the form of the conical pile at the bottom of one or both vessels, in front of inlet of the connecting tube. Ignition was implemented by the wire inserted into the connecting tube and heated by the electric current after the steel lids of the vessels had been placed and fitted by the bolts.

There were no explosions in five consequent experiments of this kind. Burning proceeded very intensively, laud noise, big pale of fire, bolts were broken, lids were thrown away, but no detonation in any of the tests. In the last experiment the detonating cap was placed into the pile of RDX at the bottom of one of the vessels . After the usual period of intensive burning, the moment of the absolute silence was noted followed immediately by the awful explosion. The cylinders, the lids, the bolts and the connecting tube were broken into pieces, a fragment of the lid about $1m^2$ size had flown over the high pines of the forest around the test site and fell down in the marsh about 1.5 km away. The paradoxical result of the series of runs consists in the fact that probability of detonation at burning of RDX layer on the walls of the reactors and the connecting tubes is negligible small.

4. Analysis of possible causes of detonation following fire

Burning of high (secondary) explosives is stable, on definition, in relation to transition into explosion. Primary explosives detonate at ignition, secondary explosives need a blasting cap filled in with the primary explosives to detonate reliably and effectively. Meanwhile the abundant, and preferably mournful, experience of production and transportation, storage and application of HE clearly shows that they, and not only RDX, PETN, and HMX, but also TNT, ammonites and even neat ammonium nitrate, are able to detonate at a fire. The main ways of initiation of the high explosives at fires may be formulated as follows.

1. Burning in closed or semi-closed volume at the definite strength of a confinement and/or at the corresponding relationship between the burning surface and the free cross-section of the system. Burning at the complicated geometry of the space where it is implemented.

2. Preliminary heating, demolition, or cracking of a charge that stipulates flame penetration into the cracks, slits and pores and deflagration to detonation transition

3. Formation of sufficiently big amounts of a liquid material during prolonged heating. Melting HE, depolymerization of the polymeric constituents of HE or propellant, foaming, and Landau effect manifestation.

4. Gasification and dispersion of HE as a result of self-heating, radiation and turbulence influence resulting in the cloud of vapors and/or aerosol formation.

5. Chemical transformation in the system under influence of preliminary heating which may result in formation of substances of higher sensitivity.

The volume of this paper does not allow us to review these points in considerable detail. Therefore only the brief description of the modern state of the problem is presented.

1. This problem is analyzed in Refs. [2] and [3]. The main part of the experiments were carried out in steel tubes of the tensile strength up to 0.5 GPa [5-15]. Experimental data corresponding to the strength of the confinement influence are given in Ref.[16]. They have shown that the relatively weak explosives, TNT and ammonites, exhibit DDT in the tubes filled in with explosive, at the certain relations between length, diameter, and strength of the tube. Pre-detonation distance, L, increases slightly in the strong tubes, with tube diameter, and much more significantly in weak tubes, near the limit of the tensile strength necessary for DDT. This limit is approximately 70-100 MPa for TNT and twice as low, 43-47 MPa, for the ammonite. The flaked TNT behaves nearly the same way as powdered TNT does. Obviously, the flakes are crushed during the preliminary stages of DDT and detonation occurs in a powdered TNT.

DDT of the powerful HE, like RDX, HMX, Nitroglycol, PETN happens very easily in the tubes filled in only partly, having some space over the explosive layer to provide a flow of the hot gas along the burning surface. Detonation was observed at HMX and RDX burning in glass tubes, 8 to 12 mm diameter, 20 to 40 cm long, with the overall amount of explosive of the order of several grams [17].

Solution of differential equation

$$UU'' + (U')^2 - kU^\alpha = 0 \tag{1}$$

allows to obtain a criterion of DDT in these conditions [17,18]

$$K_l = x(kU_0)^{1/2}, \qquad K_l^* \approx 2 \tag{2}$$

where x is the length of HE layer in the tube (when both ends of the tube are open it is a half of the length), U is the velocity of gas in the free space of the tube, $U_0 \approx 20$ - 30 m/s is the critical velocity of the flow at which burning of the wall layer converts into burning of the particles in the free space of the tube,

$$k = d_p^2 u_m bA/(\sigma_c \rho_g)$$

where d_p is the mean diameter of the burning particles, u_m is their mass burning rate, A is the width of the burning layer in the direction perpendicular to the tube axis, σ_c is the surface of the free cross-section, ρ_g is the density of the gas, b is the coefficient in a formula

$$m_1 = bU^{\alpha},$$

where m is the velocity of a process of the particles carrying away from the walls of channel ($m^{-2}\,s^{-1}$), α is equal 2 to 3. When deriving formula (2), $\alpha = 3$ was taken. The criterion K_1 is valid for diameter of a tube from 8 to 100 mm and their length from 0.2 to 7 m [17-18].

2. The processes considered here are connected with those of the previous section. HE may exhibit DDT mostly upon conditions of a strong confinement. The initial stage in all cases is penetration of flame into cracks, slits and pores and transition from the steady state to convective burning (conductive to convective burning transition, CCBT). The process of penetration a gas into pores is described by the well known equation

$$\partial^2 p^2/\partial x^2 = (2m\mu k)(\partial p/\partial t) \qquad (3)$$

where μ is the coefficient of viscosity of the gas, m and k are porosity and gas permeability of the porous media, correspondingly. Developing A.Margolin's ideas to solve equation (4) the corresponding criterion has a form [3]

$$K_2 = An_1 \xi^{1/2} \tau^{-1/4}, \qquad K_2^* \approx 3, \qquad (4)$$

$$An_1 = d_p/\delta_1, \qquad \xi = P\rho_o C_g \Delta T_g/(\rho_1 Q_i), \qquad \tau = [\mu(1-m)/\Delta p](\alpha_1/u_1^2),$$

where $\delta_1 = \alpha_1/u_1$ is the thickness of the Mikhelson layer at m=0, α_1 is the coefficient of thermal diffusivity in the condensed substance, u_1 is the burn rate, Δp is the difference of pressure inside and outside the pores, $P = p/p_0$ is the pressure of gas outside the pores, ρ_0 is the density of the gas at $p = p_0$ and $T = T_0$, c_p is the specific heat of gas, ΔT_g is the mean difference of temperatures between gas and a wall of the pore, ρ_1 is the density of HE, Q_i is the heat necessary for ignition.

3. There is abundant experimental material concerning liquid HE combustion and intensity of the combustion at conditions of the thermal explosion [20, 21]. The main cause of explosion of the liquids at self-ignition conditions was discovered by K.Andreev [20]: DET proceeds when u_m reaches a value of u_m^* corresponding to the critical Landau number La*

$$La=[g\sigma\rho\rho_g/u_1]^{-1/4}, \qquad La^{*}\sim 1,$$

where σ and ρ are the coefficient of the surface tension and the density of the liquid, respectively.

It would be relevant to note that the Landau and Zel'dovich numbers may interact each other. The Zel'dovich number equals

$$Ze = \beta(T_s-T_0), \qquad Z_e^{*}\sim 1$$

where β is the coefficient of the burn rate temperature sensitivity, T_s is the temperature of the burning surface and T_0 is the initial temperature. The substances which are not able to burn at low T_0 because of the high β and correspondingly high Ze number, burn very fast and are able to explode at a high T_0 due to high β and correspondingly high values of u_m and La (NG, PETN, TNR, NM [20]). And versus versa, HE able to burn steadily at room temperature, (RDX, HMX, Tetryl) usually burn stable at the conditions of self- ignition.

4. Explosion of the gases, vapors, dust, or aerosols is one of the main sources of DDT at fires involving HE. In spite of the great number of works devoted to investigation of turbulent burning, detonation and DDT in these systems (we will not enumerate them here) a simple criteria or any other approach useful for are still not formulated. The accidents described in Section 2 and the results of the experiments represented in Section 3 allow us however to calculate the critical Reynolds number for the cases, and to check reliability of the criteria K_1 for them. Let us assume the mean value of kinematic viscosity coefficient for the aerosol being equal $2.10^{-4}m^2s^{-1}$, the effective diameter of the flame over the car is about 5 m, and the gas velocity in the case of the intense turbulent burning is about 50-100 m/s. Thus the critical Reynolds number is about 10^6. It is close to the value estimated by Zel'dovich and Rozlovsky some fifty years ago in a closed bomb experiments to study DDT in gas proceeding due to the Landau instability effect. On the other side, at d_p = 0.1 mm, u_m = 1 kg/m²s, a = 2.10^5 s²m⁻⁵, U_0 = 30 m/s [17,18], σ_c = 40 m², ρ_g = 0.2 kg/m³, , x = 7 m (a half of the freight car length), we have from the formula (2) $K_1 \sim 2$, at the very reasonable d_c value, of about
5 m.

5. This is a hypothetical mechanism in some sense. Formation of products of low stability and higher sensitivity was suggested in many cases. Catastrophic explosions at TNT production of 1985 and 1988 in Tchapaevsk seemed to be connected with formation of some diazo-derivatives as a result of prolonged influence of the nitrogen oxides on the layer of aromatic compounds on the lid and walls of the apparatus (these were not the direct result of a fire but in many detail they were close to the cases considered here, G.M. Shutov, private communication). The possibility of the excessive yield of methylene glycoldinitrate, at some conditions of RDX production, was discussed after the explosion of RDX manufacturing plant in 1977. During the prolonged heating of TATB at high temperature some quantity of a furoxane can be formed. In most of the cases the sensitive products formation, their composition and

real role as the source of explosion are open for discussion. Even thoroughly investigated by F.P. Bowden and coworkers problem of the spontaneous explosions of the mother liquor at lead azide manufacturing [22] is not solved yet. One may say that all of these are very complicated, sophisticated and challenging tasks.

5. References

1. Kondrikov, B.N. (1996) Methodology of Examination of the Ways of the Accidental Explosions Prevention at EM Production and Application, Energetic Materials- Technology, Manufacturing and Processing, 27th Int. Ann. Conf. of ICT, Karlsruhe, 35-1-35-11
2. Belyaev, A.F., Bobolev, V.K., Korotkov, A.I., Sulimov, A.A., Tchuyko, S.V. (1973) Burning to Explosion Transition in Condensed Systems, Ìoscow, Nauka, 292 pp.
3. Kondrikov, B.N. (1997) Hydrodynamic Instability of Granular Explosives Burning, Russ. Chem. Journ., 61, 54-62
4. Ermolaev, B.S., Foteenkov, V.A., Khasainov, B.A., Sulimov, A.A., Malinin, S.E. (1990) Critical Conditions of DDT in Granulated Explosives, Comb., Expl., Shock Waves, 26, N 5, 102-110.
5. Macek, A. (1959) Transition from Deflagration to Detonation in Cast Explosives , J. Chem. Phys., 31, N 1, 162167.
6. Griffiths, N., Groocock, J.M. (1960) The Burning to Detonation of Solid Explosives ,
7. J. Chem. Soc., London, 11, part IV, 4154-4161.
8. Sokolov, A.V., Aksenov, Yu.N. (1963) Initiation and Development of Detonation in RDX, in. Vzryvnoe Delo, N 52/9, Moscow., Gosgortekhisdat.
9. Price, D., Wehner, J.F. (1965) The Transition from Burning to Detonation in Cast Explosives , Comb. and Flame, 9, N 1, 73-80.
10. Obmenin, A.V., Korotkov, A.I., Sulimov, A.A., Dubovitsky, V.F. (1969) Investigation of the Character of Propagation of Pre-detonation Regimes in Porous Explosives , Comb., Expl., Shock Waves, 5, N 4, 461-470 (Russ.).
11. Afonina, L.V., Babaytsev, I.V., Kondrikov, B.N. (1970) Method of estimation of HE Ability to DDT , in. Vzryvnoe Delo, N 65/25, M.., Nedra, 149-158.
12. Ermolaev, B.S., Sulimov, A.A., Okunev, V. A., Khrapovsky, V.E. (1988) Mechanism of DDT in Porous Explosives , Comb., Expl., Shock Waves, 24, N 1, 65-68 (Russ.).
13. Bernecker, R.R., Price, D. (1974) Studies in the Transition from Deflagration to Detonation in Granular Explosives - I. Experimental Arrangement and Behavior of Explosives Which Fail to Exhibit Detonation , Comb. and Flame, 22, N 1, 111-118.
14. Bernecker, R.R., Price, D., Erkman, J.O., Clairmont, A.R. (1976) Deflagration to Detonation Transition Behavior of Tetryl, Proc. 6th Symp. on Detonation, 381-390.
15. Ermolaev, B.S., Novozhilov, B.V., Posvyansky, V.S., Sulimov, A.A. (1985) Results of the Numerical Modeling of Convective Burning of the Granular Explosive Systems at Growing Pressure, Comb., Expl., Shock Waves, 21, N 3, 3-15.
16. Strakovsky, L.G., Kuz'min, S.B. (1989) Peculiarities of the Transitional Processes in Pressed Desensitized RDX…, IX USSR Symposium on Combustion and Explosion, Detonation, Chernogolovka,. 30-33.
17. Zav'yalov, V.S., Kondrikov, B.N. (1980) Ignition and Explosion of Granular Explosives in the Long Metal Pipes, in: Problems of Theory of Condensed Explosive System, MICT Proceedings, M., 112, 82-90.
18. Kondrikov, B.N., Karpov, A.S. (1992) DDT of Charges with the Longitudinal Cylinder Channel, Comb., Expl., Shock Waves, 28, N 3, 58-65.
19. Kondrikov, B.N, Ryabikin, Yu.D., Smirnov, S.P., Tchekalina, L.K., Alymova Ya.V. (1992) DDT of the Layer of Granular HE in the Big Pipes, Comb., Expl., Shock Waves, 28, N 5, 66-71.
20. Margolin, A.D. (1961) Stability of the Porous Substances Combustion, Proc. of the USSR Acad. Sci. (Doklady), 140, N 4, 867-869.

21. Andreev, K.K. (1957) Thermal Decomposition and Burning of Explosives, 1st ed., Moscow, Gosenergoizdat, 311 pp, 2nd ed., Moscow, 'Nauka", 1966, 346 pp.
22. Kondrikov, B.N. (1963) Intensity of Self-Ignition of Explosives , in: K.K.Andreev, A.F.Belyaev, A.I.Gol'binder, A.G.Gorst (Eds), Theory of Explosives, M., Oborongis, 515-528
23. Bowden, F.P., Ioffe, A.D. (1952) Fast Reactions in Solids, Cambridge, University Press.

PREDICTION OF LARGE SCALE FIRE BEHAVIOR USING NUTERIAL FLAMMABILITY PROPERTIES

Michael A. Delichatsios
Northeastern University and
Renewable Resources Associates Inc, 1 75 East Street
Laxington NM 02173, E-mail: madelicha@aol.com

1. Abstract

A methodology is presented to obtain flammability properties of solid materials from small-scale flammability measurements. An example is given for the flammability properties of a fire retardant polypropylene. Flammability properties include properties related to heating and ignition, properties related to pyrolysis histories, properties related to gaseous phase combustion and properties related to lateral or downward (creeping) flame spread. The deduced properties and fire modeling are shown to be sufficient to predict fire growth in large-scale fires by using fire spread models for complicated geometries or similarity for simple geometries. The magnitudes of the fire and heat, fluxes allow the quantification of thermal and non-thermal damage to people and equipment.

Flammability measurements were conducted in a flammability apparatus similar to Cone Calorimeter apparatus for half an inch thick polypropylene. These measurements include ignition time measurements at various imposed heat fluxes, and pyrolysis history measurements at a heat flux of 50 kW/m2 for two ambient conditions a) in normal air including combustion and b) in an inert atmosphere. The methodology shows that solid properties can be determined from a single imposed heat flux test in an inert atmosphere and gaseous properties from a similar test in normal atmosphere with combustion. Unambiguous interpretation of the measurements in the current flammability apparatus requires a treatment of a sample to establish thermally thick conditions either by using high imposed heat fluxes and / or by stacking samples together.

FLAMMABILITY PROPERTIES OF POLYMERIC MATERIALS

2. Introduction

This paper demonstrates how to obtain flammability properties of solid materials using certain standard measurements in so-called flammability apparatus such as the Cone Calorimeter. The so defined properties are essential for determining fire spread and fire growth in upward or radiation dominated fire growth situations. As we have presented

V.E. Zarko et al. (eds.), Prevention of Hazardous Fires and Explosions, 29–33.

in a previous paper [1], these measurements have to be supplemented with additional measurements in other type apparatus [l] needed to characterize creeping (lateral) fire spread and material sootiness. A new flammability apparatus has been designed to accommodate all measurements for determining flammability properties for the various requirements outlined earlier. The new apparatus, for which a patent disclosure has been submitted, has the capability of a) applying heat fluxes up to 300 kW/m2, b) performing downward (creeping) flame spread to characterize lateral fire spread properties and c) measuring smoke-point flames from solid materials to characterize smoke and radiation.

The focus in this paper is, however, to present a methodology for obtaining flammability properties of non-charring polymeric materials from small scale flammability testing in a standard flammability apparatus such as the Cone Calorimeter. We also outline how to use these properties to predict fire growth in large scale burning of single and parallel panels. In the present paper we concentrate in the description of the methodology for obtaining flammability properties for fire retardant polypropylene. Polypropylene is a thermoplastic material that first melts and then pyrolyzes.

FLAMMABILITY PROPERTIES AND THEIR MEASUREMENT

3. What are the flammability properties

Flammability properties are material properties that can determine the transient pyrolysis of the solid and the combustion of pyrolysing gases. Through appropriate models these properties can be used to predict fire growth and heat release rates in a given geometry and scale.
From the many material properties related to flammability, it is practical to select the properties that are essential to predict large scale fire behavior [1,2,3,4]. The essential properties for a charring and non-charring material are defined categories as[1,2]:

3.1. PROPERTIES FOR THE SOLID PHASE AS OBTAINED FROM A CONE CALORIMETER TYPE APPARATUS

The solid material flammability properties are determined as those associated with a thermal pyrolysis modeling of (pilot-assisted) ignition and solid degradation. In thermal pyrolysis, ignition occurs and pyrolysis starts when the surface temperature reaches a material characteristic temperature T_p. Experiments [l, 2, 3, 4, 5] show that this description is appropriate and adequate for materials sustaining endothermic pyrolysis for a range of heat fluxes imposed in a fire environment (20 to 120 kW/m^2). For these conditions, analysis and numerical calculations [6,7] demonstrate that thermal modeling of pyrolysis is an excellent approximation of TGA measured Arrhenius type pyrolysis rates, because the activation energy in these rates is large.
Thermal pyrolysis modeling allows the determination of flammability properties using measurements only in a flammability apparatus without having to simultaneously run TGA measurements.

There are seven essential properties needed to predict ignition and pyrolysis rates: k_v conductivity of virgin material, c_v specific heat of virgin material ρ_v density of virgin material T_p pyrolysis/ignition temperature, L heat of pyrolysis, k_c conductivity of char, ρ_c density of char.

3.2. PROPERTIES FOR THE GASEOUS (PYROLYSING GASES) PHASE AS OBTAINED FROM A CONE CALORIMETER TYPE APPARATUS

The minimum requirements to predict the combustion of pyrolysing gases related to flame spread are: the fuel stoichiometry, the efficiency of combustion in a turbulent diffusion flame environment and the soot concentration profiles that control flame radiation. The first two properties determine the flame extent (e.g. flame heights) and soot concentration profiles determine radiation heat fluxes which are the dominant heat transfer mechanism in fire spread and fire growth. Soot concentration profiles can be obtained if the sooting tendency of the material is known [8]. The sooting tendency can be deduced from laminar smoke-point heights of the pyrolysis gases from the material (see part 4.) which is also correlated to the efficiency of combustion and the smoke yield of the material. A standard flammability apparatus can also provide yields of products of combustion (e.g. CO yield). These measurements are not considered in this work The essential combustion properties for fire spread predictions are:

ΔH_c effective heat of combustion
χ_A efficiency of combustion: $\Delta H_c = \chi_A \, \Delta H_T$
Y_s smoke yield

Where ΔH_T is the total (theoretical, bob calorimeter) heat of combustion.
The efficiency of combustion and smoke yield are related to the material laminar smoke-point so that they can be added into an existing data base to consolidate and extend such correlations [8]. If smoke-point measurements are not available, these measurements, χ_A, Y_s can be used to deduce smoke-point heights and then, soot concentrations [9].

3.3. PROPERTIES FOR LATERAL, HORIZONTAL AND DOWNWARD FLAME SPREAD

An additional parameter for this case is the convective energy near the flame front. It can be determined in a new flammability apparatus by running a downward flame spread test, or in the LIFI' apparatus [10,11]. This parameter is particularly important if no external or flame heat fluxes are dominant in driving lateral, horizontal or downward flame spread.

3.4. PROPERTIES FOR SOOT FORMATION: SMOKE-POINT MEASUREMENTS FOR SOLID FUELS

A laminar flame is produced by pyrolysing the solid material and guiding the pyrolysis gases through a small nozzle [9]. The maximum soot concentration in the axis and the smoke-point height are measured. These measurements can be used to deduce soot formation rates for a given solid material in order to determine smoke production and radiation in large scale fires.

4. References

1. M. A. Delichatsios "Basic Polymer Material Properties for Flame Spread" Journal of Fire Sciences, 11, 287, 1993.
2. M. A. Delichatsios, Th. P. Panogiotou and F. Kiley, "The use of Time to Ignition Data for Characterizing the Thermal Inertia and the Minimum (Critical) Energy for Ignition or Pyrolysis," Combustion and Flame 84, 223, 1991.
3. Y. Chen, M. A. Delichatsios and V. Motevalli, "Material Pyrolysis Properties, Part I: An Integral Model for One-Dimensional Transient Pyrolysis of Charring and Non-Charring Materials," Combustion Science and Technology, 88, 309-328, 1993.
4. Y. Chen, V. Motevalli, M. A. Delichatsios, "Material Pyrolysis Properties Part II :Methodology for the Derivation of Pyrolysis Properties for Charring Materials", Combustion Science and Technology 104, 404-425, 1995.
5. J. Villermaux, B. Antoine, J. Lede, F. Soulignac, "A New Model for Thermal Volatilization of Solid Particles Undergoing Fast Pyrolysis" Chem. Eng. Sci. 41, 151, 1986.
6. M. Kindelan, F. A. Williams "Theory for Endothermic Gasification of a Solid by a Constant Energy Flux" Combust. Sci. Technol. 10, 1, 1975.
7. J. Lede and J. Villermaux "Comportement Thermique et Chimique de Particules Solides Subissant une Reaction de Decomposition Endothermique sous l'action d'un Flux de Chaleur Externe" The Can. J. of Chem. Eng., 71, 209, 1993.
8. M. A. Delichatsios, L. Orloff and M. M. Delichatsios "The effects of Fuel Sooting tendency and the Flow on Flame Radiation in Luminous Turbulent Jet Flames" Combust. Sci. and Technol. 84, 199, 1992.
9. M.A Delichatsios, "Smoke yields from turbulent Buoyant Jet Flames", Fire Safety Journal, 20,2,99,1993.
10. M. Harkleroad, J. Quintiere and W. Walton " Radiative Ignition and Opposed Flow Flame Spread measurements on Materials" DOT/FAA/CT-83/28, 1983.
11. M. A. Delichatsios, "Creeping Flame Spread, Energy Balance and Applications to Practical Materials" Twenty Sixth (Symposium International) / The Combustion Institute, 1405,1996.
12. Mary Delichatsios, "Ignition Times for Thermally Thick and Intermediate Conditions in Flat and Cylindrical Geometries" Eastern Section Meeting of The Combustion Institute, Hartford Connecticut, October 1997.
13. A Tewarson "Generation of Heat and Chemical Compounds in Fires" p. 3-53 in Fire Protection Engineering SFPE Handbook, Second Edition, 1995
14. M. A Delichatsios and Y. Chen "Asymptotic, Approximate and Numerical Solutions for the Heat-up and Pyrolysis of Materials including reradiation Losses" Combustion and Flame, 92, 192, 1993.
15. M. A. Delichatsios, M. M. Delichatsios, "Upward Flame Spread and Critical Conditions for PE/PVC Cables in a Tray Compartment", Fire Safety Science Proceedings of the Fourth International Symposium, 433-444, 1994.
16. M. M. Delichatsios, P. Wu, M. A. Delichatsios, G. D. Lougheed, G. P. Crampton, C. Qian, H. Ishada, K Saito, "Effect of External Radiant Heat Flux on Upward Fire Spread Measurements on Plywood and Numerical Predictions," Fire Safety Science Proceedings of the Fourth International Symposium, 421-432, 1994.

17. M. A. Delichatsios, Y. Chen, "Flame Spread on Charring Materials: Numerical Predictions and Critical Conditions", Fire Safety Science - Proceedings of the Fourth International Symposium, 457-468, 1994.
18. M. M. Delichatsios, M. K. Mathews and M. A. Delichatsios "Upward Fire Spread Simulation Code: Version 1: Non-charring Fuels", FMRC Technical Report J. 1. OROJ2.BU, November 1990. Also published in the Proceedings of the Third International Symposium of Fire Safety Science, pp. 207-216, Elsevier, NY, 1991.
19. M. A Delichatsios, M. M. Delichatsios, Y. Chen, Y. Hasemi, "Similarity Solutions for Upward Flame Spread and Applications to Non-Charring Materials such as PMMA", Winter Annual ASME Meeting, Anaheim, California, October 1992. Also published in Combust. and Flame, 102, 357-370, 1995.
20. F. R. Menter "Eddy Viscosity Transport Equations and Their Relation to k-ε Model" Trans. Of the ASME, J. of Fluids Eng., 119, 876, 1997.

USE OF MODERN COMPOSITE MATERIALS OF THE CHEMICAL HEAT ACCUMULATOR TYPE FOR FIRE PROTECTION AND FIRE EXTINGUISHING

V.N.PARMON, O.P.KRIVORUCHKO, Yu.I.ARISTOV
Boreskov Institute of Catalysis SB RAS,
Novosibirsk 630090, Russia

1. Introduction

For the last two decades, a huge amount of work has been done on developing efficient systems for absorbing excessive heat. This activity was mainly addressed to conventional applications for accumulation and recovery of low- and middle temperature industrial heat wastes as well as for utilisation of renewable sources of thermal energy. Here we extend this approach to non-traditional application for fire protection and extinguishing.

The Boreskov Institute of Catalysis (Novosibirsk) has recently invented Chemical Heat Accumulators (now the name Selective Water Sorbents (SWS) is more widely used) which employ chemical and physical dehydration of water molecules tightly coupled to an appropriate lattice material [1-11]. These storage media are able to absorb large amount of heat (more than 2000 kJ kg^{-1} for dry material) that could make them promising for fire protection applications. These materials are now under extensive experimental studies and have been successfully tested in some commercial applications like air conditioning, seasonable energy storage, sorption heat pumps, refrigeration machines, efficient fire-retarding powders, etc.

In this communication, basic physicochemical properties of the SWS materials as well as examples of their application to fire protection and fire extinguishing are presented and discussed.

2. Physicochemical Processes Associated with the SWS Materials

Among physical processes efficient for heat absorption, evaporation and sublimation have the best specific parameters. Indeed, evaporation of liquid water is associated with consumption of 40.6 kJ mol^{-1} (or 2260 kJ kg^{-1}) being extremely promising for heat absorption. Some chemical processes demonstrate even better specific parameters than "pure" physical ones. At the moment, one of the most simple and attractive chemical reactions for heat absorption seems to be the chemical analogue of evaporation - a

V.E. Zarko et al. (eds.), Prevention of Hazardous Fires and Explosions, 34–48.

dehydration process. At present, three types of systems permitting a reversible hydration-dehydration attract much interest. These are crystalline hydrates, hydroxides and porous oxides (zeolites, silicas, etc.).

The ability of crystalline hydrates of simple inorganic salts to retain water in large amounts (about 100 % in respect to the weight of the anhydrous salt or even more) is well known. Their dehydration is associated with breaking chemical bonds of a moderate strength, and requires sufficiently large energy supply. Table 1 presents data on the enthalpy and typical dehydration temperatures for selected crystalline hydrates (under ambient partial pressure of water vapour). Despite the promising energy and temperature characteristics practical application of the common crystalline hydrates is still not very spread because of their undesirable fluidisation at large water sorption. The formed aqueous solutions contains even more water than initial hydrates (on a dry salt basis) and have extremely attractive specific parameters, however, they cannot be used in actual practice because of low contact surface and high corrosivity.

Inorganic hydroxides have the energy storing parameters no worse than crystalline hydrates, being, however, very corrosive and unstable upon interaction with CO_2-containing air.

TABLE 1. Enthalpy and dehydration temperature for some simple crystalline hydrates [1,12-14]

No	Formula of crystalline hydrate and the amount of water molecules capable of release	Enthalpy of dehydration, $kJ\ mol^{-1}\ H_2O$	Specific heat of dehydration, $kJ\ kg^{-1}$*	Decomposition temperature, °C**
1	$Na_2SO_4 \cdot 10H_2O$ (-10)	51.4	1595	32.4
2	$Na_2SO_3 \cdot 7H_2O$ (-7)	51.7	1436	87
3	$CaCl_2 \cdot 6H_2O$ (-4)	53.4	1061	30
4	$SrCl_2 \cdot 6H_2O$ (-4)	56.7	797	60
5	$MgCl_2 \cdot 6H_2O$ (-6)	70.0	1948	120-280
6	$CuCl_2 \cdot 2H_2O$ (-2)	59.6	1420	200
7	$CoSO_4 \cdot 7H_2O$ (-7)	55.6	1385	300
8	$MgSO_4 \cdot 7H_2O$ (-7)	58.5	1661	280
9	$Na_2B_4O_7 \cdot 10H_2O$ (-5)	54.2	710	25-50

* Calculated per unit mass of the hydrated state and for the amount of the releasing water molecules denoted in the parentheses; the final state of water is considered to be gaseous (see comments in [1]).
**see also comments in [1].

The third type, porous oxides, appear to be highly suitable for heat absorption due to water desorption process, too. Indeed, for these systems the heat of water desorption is higher than for those considered above. Moreover, they can be prepared as granules, pellets or monolyths thus facilitating drastically the reactive surface and mass transfer. Their main disadvantage is relatively low amount of water which may be bound under normal conditions (even for zeolites it commonly does not exceed 30-35 % of the dry sorbent weight) that results in the lower specific heat absorbing capacity.

The main idea in designing the composite SWS materials was to combine pronounced heat-absorbing properties of salt hydrates and high technological convenience of porous solids. This aim was achieved by filling a porous host matrix of various nature (silica gels, aluminas, metal sponges, etc.) with highly hygroscopic material of a chosen type (for instance, inorganic salts, their crystalline hydrates or liquid solutions). Being prepared in such a way, the SWS materials appears to be very promising for heat absorption in fire protection applications.

In this communication, we, first, briefly consider a general SWS concept, and then describe basic physicochemical properties of the SWSs important for fire protection. To be more specific, we present here just one particular SWS material based on the meso-porous silica gel KSKG as a host matrix and calcium chloride as an active filling. Finally, we shall discuss our experimental results on the use of SWS and SWS-like materials for heat retardation and fire extinguishing.

3. Physico-Chemical Properties of the SWS Materials

3.1. GENERAL CONCEPT

As it was mentioned above, a typical SWS material is
two-component system and consisting of a porous host matrix and an active heat absorbing substance A impregnated into the matrix pores. The substance (for example, an aqueous salt solution or a crystalline hydrate) takes part in a gas-solid (gas-liquid) reaction of dehydration type

$$\text{A (solid, liquid)} \underset{\text{energy storage}}{\overset{\text{energy release}}{\rightleftarrows}} \text{B (solid, liquid)} + H_2O \text{ (vapour).} \qquad (1)$$

During above process the heat is absorbed producing water vapour and a compound B with lower water content (for instance, lower crystalline hydrate or anhydrous salt). Since this process is highly endothermic, it is accompanied by absorption of large amount of heat. It is clear from Table 1 that by a proper choice of the energy-accumulating substance one can design energy absorbing materials adopted for any required temperature region.

Preliminary analysis showed that one of the most interesting SWS materials for efficient heat absorbing and, hence, for fire protection and retardation is based on composite "calcium chloride-silica gel" which is referred below as an "SWS-1L". For this material the reaction of the water removal (dehydration) can be written as

$$CaCl_2{\cdot}nH_2O \text{ (hydrate or solution)} \rightarrow CaCl_2{\cdot}(n\text{-}m)H_2O + mH_2O.$$

The water sorption and energy accumulating properties of the SWS-1L have been studied in many details in Refs. [3-8]. Among the properties measured so far are the "water vapor - SWS" sorption equilibrium at different temperatures and partial pressures of vapor, specific heat, thermal conductivity, melting-solidification diagram, NMR- and IR-spectra of sorbed water. As fire protection effects are mainly caused by water desorption process, we focus here on the brief description of thermodynamics of the water sorption on the SWS-1L.

3.2. "WATER VAPOUR - SWS-1L" SORPTION EQUILIBRIUM

3.2.1. Experimental

The KSKG silica gel (Reakhim, former USSR) with a BET specific surface area $S_{sp} = 350\ m^2/g$, pore volume $V_{pore} = 1.0\ cm^3/g$ and average pore diameter of 15 nm was prepared as a powder of 0.25-0.5 mm grain size and used without additional purification. The confined "calcium chloride - silica gel" system was prepared by filling pores of the silica gel with 40 wt.% aqueous solution of $CaCl_2$. Then the samples were dried at 200°C until their weight remained constant. Calcium chloride content in the anhydrous samples was measured to be 33.7 wt.%.

A sorption equilibrium between water vapor and dry SWS-1L was studied by a thermal balance method [3,5]. Relative equilibrium water content was calculated as $w = m(P_{H_2O},T)/m_0$, where m_0 is the dry sample weight. As the water sorption by the silica gel was found to be less than 6-7% of total sorption, the calculation of the equilibrium number of water molecules with respect to 1 molecule of calcium chloride is the most convenient way of the data presentation: $N(P_{H_2O},T) = (m(P_{H_2O},T)/\mu(H_2O))/(m(CaCl_2)/\mu(CaCl_2))$, where μ stands for a molecular weight.

3.2.2. Isobaric chart of water sorption on the SWS-1L material

Water sorption isobars on the SWS-1L are presented in Fig.1 as a $N(P_{H_2O},T)$ dependence on temperature at different partial water vapor pressures P_{H_2O} ranging from 8 to 133 mbar. It is seen that all the curves are similar but shifted and partially extended along the temperature axis. Each curve has a plateau corresponding to N = 2, the plateau being longer for low vapour pressures. The isobars display a rapid increase of sorption at temperatures close to the condensation temperature.

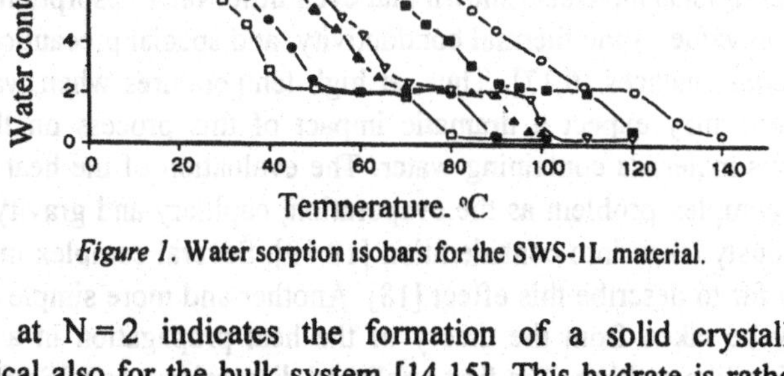

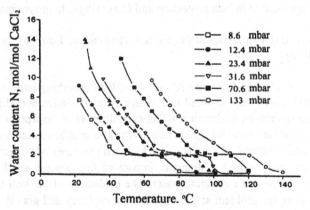

Figure 1. Water sorption isobars for the SWS-1L material.

The plateau at N = 2 indicates the formation of a solid crystalline hydrate $CaCl_2{\cdot}2H_2O$ typical also for the bulk system [14,15]. This hydrate is rather stable and possesses no transformation in 10-20°C temperature range showing a monovariant type of the sorption equilibrium typical for gas-solid processes. At high temperatures this hydrate undergoes decomposition to form a compound with an overall formula $CaCl_2 \cdot nH_2O$, where n = 0.34 ± 0.04 and then to anhydrous salt. At temperatures lower that the left boundary of the plateau, water sorption depends on both temperature and pressure, so that the equilibrium becomes bivariant, typical for liquid solutions of $CaCl_2$ [14].

Water sorption isosters on the SWS-1L are straight lines [3,5]

$$\mathrm{Ln}(P_{H_2O}) = A(w) + B(w)/T$$

with slope depending on the water content w and giving an isosteric water sorption heat $\Delta H_{is}(w) = B(w)\,R$, where R is the universal gas constant. For N > 2 isosteric sorption heat (43.9 ± 1.3 kJ/mol) appears to be close to the water evaporation heat from aqueous $CaCl_2$ solutions [16]. A significant increase in ΔH_{is} at $N \le 2$ (up to 60-63 kJ/mol) is caused by the formation of solid hydrates where water molecules are bound stronger.

The study of the sorption equilibrium allows to make estimates for specific water sorption capacity A_{sp} and specific energy storage capacity E_{sp} of SWS-1L. The former value of A_{sp} strongly depends on the relative pressure and can reach A_{sp}=0.7÷0.8 g of water sorbed by 1 g of the dry SWS-1L at $P/P_0 \approx 0.7$. Evidently, this value is far beyond that of common adsorbents such as silica gels, alumina, zeolites, etc., and close to that of crystalline hydrates. So large A_{sp} value causes an extremely encouraging specific energy absorbing capacity E_{sp} of the SWS-1L material, that is defined as the maximum energy required for a complete water desorption calculated per 1 g of a dry SWS. Taking into account the desorption heat ΔH_{is} and sorption capacity A_{sp}, the value of E_{sp} can be estimated as 2000 kJ/g that is quite consistent with the results of our direct DSC measurements.

Thus, the data on water sorption equilibrium presented above evidence the promising properties of the SWS-1L material for heat absorbing and can be used for analysing its application in heat protection and fire extinguishing systems.

4. Heat Protecting Properties of the SWSs: Abnormally Low Thermal Conductivity

As it is shown above the SWS materials are capable of absorbing large amount of water vapor, and their pores may be filled with water even at relative humidity of about 60-70%. The experiments have also shown that even little water desorption during the tests may affect true value of the thermal conductivity, and special precautions were done to avoid these consequences [6,17]. Thus, at high temperatures when water evaporation takes place one may expect a dramatic impact of this process on the heat transfer through porous materials containing water. The evaluation of the heat transfer in such systems is a complex problem as the evaporation, capillary and gravity effects have to be simultaneously taken into consideration [18,19]. Several complex models have been developed so far to describe this effect [18]. Another and more simple approach to this problem may be taken from the theory of the heat propagation in a medium with a phase transition like melting. The first published discussion was of Stefan [20] although some general results on the problem were first obtained by Neumann (see ref. [21,22]). All available exact solutions are comprehensively described in Ref. [22].

In this Chapter, we present experimental data on the heat transfer in horizontal SWS consolidated bed saturated with water and adjacent (by lower side) to

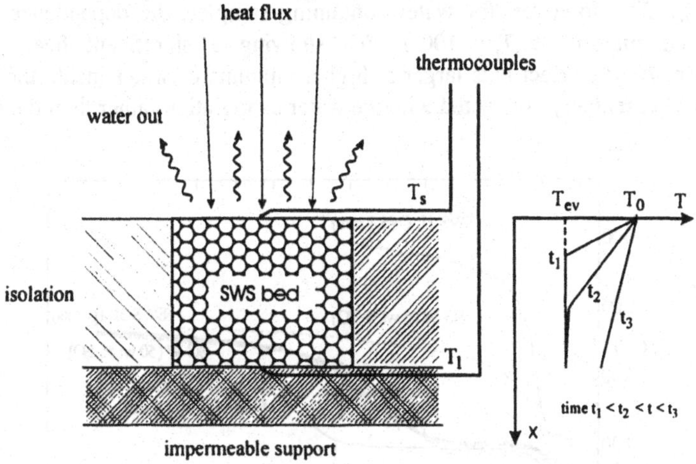

Figure 2. Scheme of the heat transfer experiments. Also shown are the temperature distributions in the sorbent bed at different moments of time.

impermeable surface. The layer is heated from the opposite permeable side by an intense radiant flux of that makes surface temperature T_s higher than the evaporation temperature T_e (Fig. 2). Thus, this study imitates the use of SWSs as a sort of heat protecting materials.

4.1. EXPERIMENTAL

Preparation of samples for this study followed the procedure included four stages: a) impregnation of a KSKG silica gel powder with water, $CaCl_2$ or $MgCl_2$ aqueous solutions (samples SG-W, SWS-1L and SWS-3L, respectively); b) moulding in cylindrical bricks of a 24.5 mm diameter and of 4 to 20 mm thickness; c) thermal treatment at 473K to evaporate water; d) saturation with water vapor in a desiccator until the desired water content w was reached.

The samples were irradiated with a 10 kW Xenon lamp of a DKSHRb-10000 type. The average radiant flux W_{in} was measured by a calorimetric method and was varied from 5 to 50 W/cm^2. The temperatures T_s of the upper (heated and permeable) and T_l of lower (adjacent to the impermeable support) surfaces were measured using K-type thermocouples (Fig.2). Thermal insulation was used to avoid heat losses from the lateral brick surface.

4.2. STUDIES OF THE HEAT RETARDING PROPERTIES OF THE SWSs

Under incident heat fluxes of 15-40 W/cm^2 the temperature T_s of the upper (heated) surface is found to be rapidly increasing up to limiting values of 500-1200°C that depends on W_{in}, while the temperature T_l of the protected (lower) surface remains much lower (Fig. 3). Moreover, for water containing samples, the dependence $T_l(t)$ has a pronounced plateau at T_p= 100-120°C, showing a significant heat propagation retardation. As the effect gets larger at higher amounts of water inside the samples, it seems to be essentially connected with the water evaporation rather than the salt

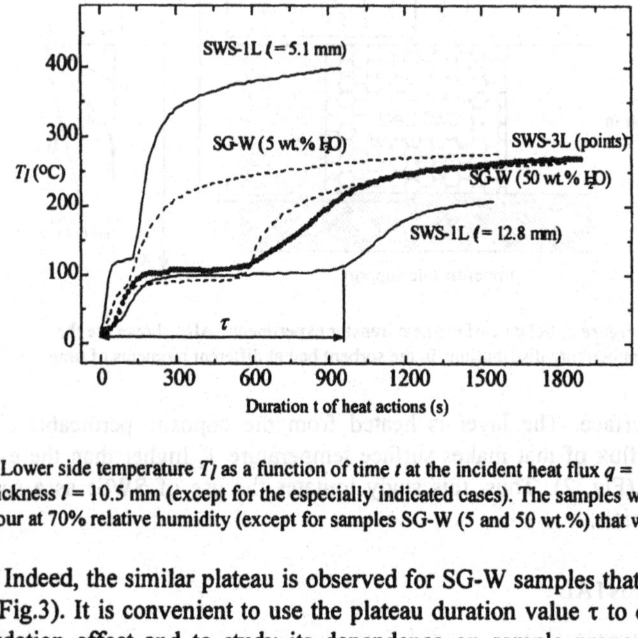

Figure 3. Lower side temperature T_l as a function of time t at the incident heat flux q = 15 ± 1 W/cm^2 and sample thickness l = 10.5 mm (except for the especially indicated cases). The samples were saturated with water vapour at 70% relative humidity (except for samples SG-W (5 and 50 wt.%) that were force wetted).

presence. Indeed, the similar plateau is observed for SG-W samples that do not contain any salt (Fig.3). It is convenient to use the plateau duration value τ to characterise the heat retardation effect and to study its dependence on sample properties (thickness, water content, salt nature), incident heat flux, etc.

As it has been mentioned above, the water content w affects strongly the heat propagation through the SWS consolidated layer, so that the delay time τ is nearly proportional to the mass of water m_{con} has to be evaporated. It is worth noting that water-free samples do not demonstrate the retardation effect at all (Fig.3). No substantial distinctions are observed between $CaCl_2$- and $MgCl_2$ containing samples.

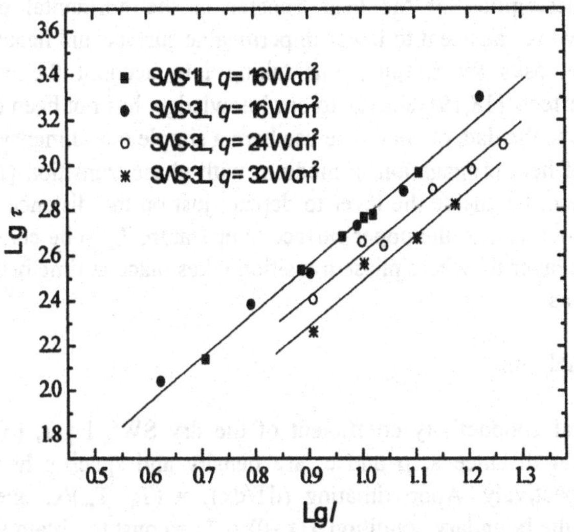

Figure 4. **Experimental dependencies of the delay time on the sample thickness at different values of the incident heat flux.**

Sample thickness l is another very efficient tool to change the delay time which is found to be about quadratic in l:

$$\tau \sim l^{\alpha}, \tag{2}$$

where $\alpha = 2.0 \pm 0.2$ (Fig.4). This equation is quite typical for kinetics of heat front propagation in a medium with phase transition like melting [20-22]. It is follows from relationship (2) that the value τ/l^2 does not depend on l and may be chosen as a convenient parameter to compare different samples and to study the influence of incident heat flux as well.

The surprising thing is that the intermediate heat fluxes (W = 10-40 W/cm²) make rather weak effect on the delay time, whereas at lower fluxes (<10 W/cm²) the τ-value increases drastically. This probably indicates that there is a threshold for water evaporation, that means the minimum heat flux necessary for total water removal. This is in line with the theory of evaporation in porous media [18] stating that no vapour is presented inside pores until heat flux exceeds some critical value W_{cr} depending on impregnated liquid and host-matrix properties. For fluxes lower than the critical one, the gravity and capillarity prevent vaporisation and increase evaporation time τ significantly.

For the salt-containing samples, the effect observed can be completely reproduced as soon as they restore their saturated state by moisture sorbing from the ambient air at relative humidity of 60-70%. At these conditions the salt-free silica gel can adsorb only 5-7 wt.% of water, so to recover its thermal protection properties it should be force wetted.

Precise analytic description of the heat transfer in the horizontal porous bed saturated with liquid, which adjacent to lower impermeable surface and heated from the upper permeable side, asks for simultaneous taking into account the evaporation, capillary and gravity effects [18,19], and, as to our knowledge, has not been done yet in the literature. To analyse the data obtained we use here a simple one-dimensional model based on the theory of heat propagation in medium with phase transition [20-22]. We assume a) the temperature T inside the layer to depend just on the distance x from the heated surface and time t, and b) the upper surface temperature T_0 to be constant. Then the heat balance in the layer dx where phase transition takes place at time between *t* and (t + dt) can be written as

$$\lambda\,(dT/dx)_x\,dt = \rho\,\Delta H_{ev}\,dx, \tag{3}$$

where λ is the thermal conductivity coefficient of the dry SWS layer, $(dT/dx)_x$ is a temperature gradient at distance x, ρ and c_p are density and specific heat of water containing layer, respectively. Approximating $(dT/dx)_x = (T_0 - T_{ev})/x$, one can easy integrate Eq. (3) with the boundary condition $T(x=0) = T_0 = \text{const}$ to obtain the time for finishing the phase transition (complete water evaporation) in the whole sample volume

$$\tau = \rho\,\Delta H_{ev}\,l^2 / 2\,\lambda\,(T_0 - T_{ev}). \tag{4}$$

This equation allows to describe the main features of the heat transfer in the system under study, because it represents well experimentally observed dependence $\tau(l)$ and shows a great increase in the protection time τ if the upper surface temperature T_0 tends to reach T_{ev}, that is indirectly presents discussed above threshold effect at incident fluxes close to the critical one.

An analytic dependence $\tau(W)$ can be evidently obtained from relationship (2) if the analytical relation between T_0 and W is known. If one approximates the dependence $W(T_0)$ as $W \approx \alpha T_0^{\gamma}$, where parameter γ summarises the relative contribution of reflection, convective and radiation heat losses from the upper sample surface (evidently, for radiation losses $\gamma = 4$), Eq.(4) can be rewritten as

$$\tau = \rho\,\Delta H_{ev}\,l^2 / 2\,\lambda\,\beta\,(W^{(1/\gamma)} - W_{cr}^{(1/\gamma)}), \tag{5}$$

where $\beta\ W_{cr}^{(1/\gamma)} = T_{ev}$. Plotting $\text{Log}(\tau/l^2)$ vs. Log(W) allows to obtain both the parametery $= 3.0 \pm 0.5$ and the critical heat flux $W_{cr} = 4 \pm 1$ W/cm^2 (Fig.5). So high γ-value demonstrates clearly that most of the incident energy is dissipating from the

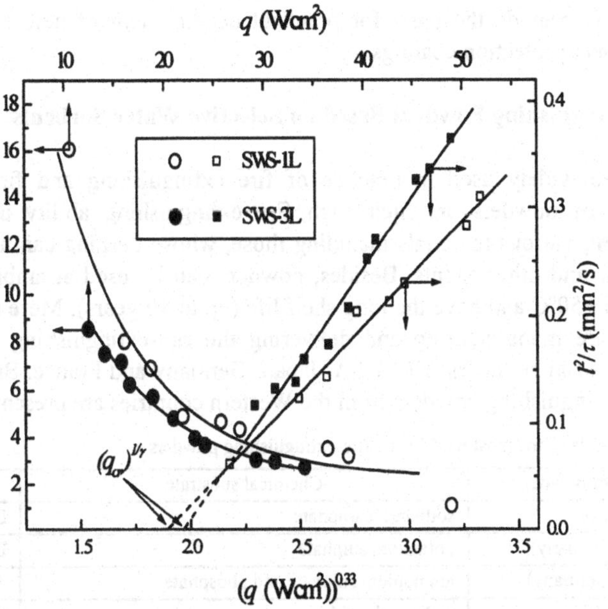

Figure 5. **Influence of the heat flux q on the parameter τ/l^2 and determination of the critical heat flux q_{cr} (straight lines).**

sample surface due to the radiation losses. It is reasonable since the T_0-value usually reaches 700-1200°C.

If compare Eq.(4) with the typical time of the heat front propagation in a homogeneous and isotropic solid without any phase transition $\tau = \rho c l^2/4\lambda$ (c is the specific heat capacity of a solid) [22] one can introduce the apparent thermal conductivity coefficient for porous layer saturated with liquid $\lambda_{app} = \lambda [c (T_0 - T_{ev})/\Delta H_{ev}] = (0.03\text{-}0.3) \lambda$. So, the water evaporation indeed may cause the retardation of the heat transfer and lead to apparent thermal conductivity 3-30 time lower (for $(T_0 - T_{ev}) = 1000$ and 100°C, respectively) than the actual one of dry material λ. As for porous solids $\lambda \approx 0.1\text{-}0.2$ W/m*K, the λ_{app}-values can be as low as 0.03-0.006 W/m*K. For comparison, typical high temperature insulating materials (laminated and corrugated asbestos, diatomaceous earth, etc.) possesses $\lambda > 0.05$ W/m*K at room temperature and $\lambda > 0.1\text{-}0.3$ W/m*K at 1000°C [23]. Of course, Eq.(2) predicts directly a rise of τ if the impregnating liquids with high evaporation heat and matrices with low specific heat and thermal conductivity are used.

The data obtained show that the SWS consolidated layers saturated with water have the ability of considerable heat front retardation and demonstrate abnormally low thermal conductivity at T = 100 - 1200°C. The delay time is strongly dependent on incident heat flux, sample thickness, water content and may change over a wide range. Simple one-dimensional model taking into account phase transition (water evaporation) inside the SWS has been developed and allows clear description of the results obtained.

These results provide the basis for practical development of new efficient and self-recovered heat protection coatings.

5. Fire-Extinguishing Powders Based on Selective Water Sorbents

Powders are widely used in practice of fire extinguishing and fire retarding. The advantages of powders are their high fire extinguishing ability and versatility in extinguishing various materials including those, whose burning can not be stopped by foam, water and other agents. Besides, powders can be used at ambient temperatures from -50 to +50°C and have the long shelf life (up to 10 years). More than 130 firms in the world are manufacturing and delivering the fire-extinguishing powders, mainly from 5 industrial countries: UK, USA, Japan, Germany and France. Brands of the most used fire-extinguishing powders from the Western countries are presented in Table 2.

TABLE 2. Brands of the most world-used fire-extinguishing powders

Brand of powder	Chemical substrate	Grade of fire
BCE (Germany)	sodium bicarbonate	B, C, E
CARATE (Germany)	potassium sulphate	B, C, E
TOTALIT (Germany)	ammonium sulphate and phosphate	A, B, C, E
FAVORITE (Germany)	sodium chloride	D
P-11-24 (France)	ammonium phosphate	A, B, C, E
MONNEX (UK)	melt of carbamide and potassium carbonate	B, C
BI-EX (Belgium)	ammonium phosphates, cryolites	A, B, C
PYROCHEM (USA)	monoammonium phosphate	A, B, C, E

Physicochemical properties of burning material are the most important parameters of fire, determining the conditions of fire extinguishing. Classification of fires is made also according to these properties. Table 3 shows the classification of fires and extinguishers presently accepted in Russia.

TABLE 3. Grades and characteristics of fires

Grade of fire	Characteristic of combustible medium or burning object	Recommended fire extinguishers
A	Conventional solid combustible materials (wood, coal, paper, textile)	All types of fire extinguishers, water, first of all
B	Combustible fluids and materials melting on heating (fuel oil, gasoline, alcohol, lacquers, oils)	Sprayed water, all types of foam, compounds based on alkyl halides, powders
C	Flue gases (hydrogen, methane, acetylene, etc.)	Inert diluents, halohydrocarbons, powders, water for cooling
D	Metals and their alloys	Powders
E	Charged equipment	Powders, CO_2, coolers

Existing and industrially produced fire-extinguishing compounds possess some specific properties which allow fire extinguishing due to the following factors:

(1) cooling the combustion zone below certain temperatures;

(2) intense retardation of chemical reactions rate in the flame through homogeneous inhibition by gaseous products of extinguishing powders decomposition and/or through heterogeneous chain breaking on the surface of powders;

(3) insulation of a combustion zone from oxygen or lowering the oxygen concentration through dilution of ambient medium by inert gases;

(4) creating the fire-retardant conditions, whereby the flame is propagated through thin channels.

The fire-extinguishing powders which exhibit a combining action on the process of burning (factors 1, 2, 3 and 4) possess the highest fire-extinguishing ability and provide extinguishing of such materials, which can not be extinguished by water or other agents. Each fire-extinguishing powder is characterised by a particular predominant factor. At the same time, factors 1 and 2 are the most important ones determining the fire-extinguishing power of a powder.

At the Boreskov Institute of Catalysis, a new class of versatile fire-extinguishing powders based on the CHA or SWS-like composites has been developed. These powders appeared to be suitable for extinguishing fires of the A, B, C, D and E grades [24] . The new fire-extinguishing powders are a multiphase composite material, the solidity of its each particle being provided by chemical interaction of constituent phases of a particle: complex hydroxide $[M(II)_XAl(III)_YM(IV)_Z(OH)_{(2X+3Y+4Z)}]$ as well as $[M(I)An_1]$ and $[M(I)An_2]$ phases, where M(I) and M(II) comprise mono- and bivalent cations of different non-transition metals, M(IV) is Si^{4+}, Ti^{4+}, Zr^{4+} or their mixture, while An_1 and An_2 are mono- and bicharged anions under different molar ratio of components and phases. Each particle of the proposed powder has the identical chemical and phase composition, every phase being active in fire-extinguishing.

Efficiency of the powders in extinguishing the fires of a certain grade may be raised by varying the powder composition, the basic technology of its production remaining unchanged. The granulometric composition of the powder is 30-70 μm, the content of this fraction is no lower than 95%. The surface area of the powder is no less than 0.5 m^2/g. Methods have been developed to vary the poured weight of the powder (the chemical composition retained) from 500 to 900 kg/m^3. This implies that the powders of the composite type with a reduced poured weight offer promise for fire retardation and extensive fire extinguishing. The powder particles with the size of 30-70 μm, when thrown into the flame, are crushed efficiently into micro-particles of 1-10 μm. This leads to a substantial growth of their geometrical surface area, where the heterogeneous inhibition of many radical reactions in flame proceeds with a high rate. Subsequently, endothermic processes of decomposition of the micro-particles result in lowering the flame temperature and in homogeneous inhibition of chain reactions by gaseous products of the solid phase decomposition. A high level of heat absorption in the flame is provided by the heat consumed on the heating of solid phase and by endothermic processes of decomposition of hydroxide and some salt phases comprising the particles of the composite fire-extinguishing powder.

One can see from Table 4 that the new fire-extinguishing powder OPK-1 is characterised by heat-absorbing ability, which is 1.5-3.0 times higher as compared with the known industrial powders. The starting materials for manufacturing the fire-extinguishing powder OPK are available in industrial scale being fairly inexpensive.

The technology of this powder production is based on the use of standard chemical equipment and is completely wasteless. The fire-extinguishing powder is also ecologically safe during its production, storage and use.

TABLE 4. Comparative characteristics on heat adsorption ability of the Russian fire-extinguishing powders with various composition

Brand of fire-extinguishing powder	Heat absorption due to physico-chemical transformations, kJ/kg	Total heat absorption on heating up to 1000°C, kJ/kg	Contribution to heat absorption through heating of the solid phase, %
PSB-3	750	890	15.0
P-1A	1420	1620	12.6
PYRANT-A	1450	1770	18.1
PF	1500	1700	11.6
OPK-1	1795	2630	31.7

The fire-extinguishing ability of the OPK powder has been examined in large series of tests on the bench-scale setup (the fires of the B and C grade). This powder has been tested successfully in full-scale fire of the B grade of 1 and 2 m^2 areas by CPNT «Pyrant» (Novosibirsk) on the facilities of the Russian Research Institute of Fireproof Defense VNIIPO (Moscow). The results of bench and full-scale testing of the powder are presented in Table 5. One can see that in the full-scale tests the new fire-extinguishing powder is 2.0-2.5 times more efficient than most of the Russian domestic powders and stands at the level of «Monnex» (UK), which is considered to be the best powder for extinguishing the fires of a B grade.

TABLE 5. Comparative testing of heat-extinguishing ability of powders with various composition for fires of B a grade

	Bench-scale testing	Full-scale fire testing	
Area of burning	0.01 m^2	1.0 m^2	2.0 m^2
Burnt substance	kerosene	gasoline	gasoline
Time of extinguishing	0.1 s	2 s	3 s
Consumption of powder	(g/100 cm^2)	(kg/m^2)	
OPK	0.1	0.40	
PYRANT (monoammonium phosphate)	0.3	0.80	
PSB-3 (sodium bicarbonate)	0.7	1.00	
MONNEX (melt of carbamide and potassium carbonate)	0.1	0.42	

6. Conclusions

The paper gives idea about new generation of materials for fire protection and extinguishing based on recently developed composites called Chemical Heat Accumulators or Selective Water Sorbents. These materials consist of a porous host matrix filled with a hygroscopic substance. When simple salts used as the hygroscopic substance, the materials allow to reach the heat storing capacity up to 2000 kJ*kg^{-1} and may be used for accumulation of low temperature heat (of ca. 20-50 °C) due to easy dehydration of the imbedded hydrates or aqueous solutions. The materials possess also improved properties for mass and heat transfer. All these make a serious advantage of these materials when using them for fire protection and extinguishing as well as for some other applications. A brief review of water sorption properties of these materials is presented and demonstrates their much promise as a base for fire protection coatings. The results of special experiments show that consolidated layers of new materials have an ability of considerable heat front retardation and manifest abnormally low apparent thermal conductivity λ_{app}= 0.03-0.01 W/m*K at T = 100-1200°C. The retardation effect is strongly dependent on incident heat flux, sample thickness, water content, etc. Simple one-dimensional model taking into account phase transition (water evaporation) inside the sorbent layer has been used for description of the results obtained. These results provide the basis for practical development of new efficient and self-recovered heat protection coatings.

Based on the same concept a new class of versatile fire-extinguishing powders suitable for extinguishing fires of the A, B, C, D and E grades has been developed. Efficiency of the powders in extinguishing the fires of a certain grade can be increased by varying the powder composition while the basic technology of its production remains unchanged. The moderate scale tests have shown that the new fire-extinguishing powder OPK-1 is characterised by heat-absorbing ability 1.5-3.0 times higher as compared with the known commercial powders. The technology of this powder production is based on the use of standard chemical equipment and is completely wasteless. The fire-extinguishing powder is ecologically safe during its production, storage and use.

Acknowledgements. A part of this work was supported by the Russian Foundation for Basic Research (Grant N97-03-33533).

7. References

1. Levitskij, E.A, Aristov, Yu.I., Tokarev M.M., Parmon, V.N. (1996) "Chemical Heat Accumulators" - a new approach to accumulating low potential heat, *Solar Energy Mater.Sol.Cells*, **44**, 219.
2. Levitskij, E.A., Parmon, V.N., Moroz, E.A., Bogdanchikova, N.V. (1991) New thermoaccumulating material, Patent application to PCT/SU 91/00173 of 26.08.91 and subsequent national patents.
3. Aristov, Yu.I., Tokarev, M.M., Cacciola, G., Restuccia, G. (1996) Selective water sorbents for multiple applications: 1. $CaCl_2$ confined in mesopores of the silica gel: sorption properties, *React. Kinet. Catal. Lett.*, **59**, 325.
4. Aristov, Yu.I., Tokarev, M.M., Restuccia, G., Cacciola, G. (1996) Selective water sorbents for multiple applications: 2. $CaCl_2$ confined in micropores of the silica gel: sorption properties, *React. Kinet. Catal. Lett..*, **59**, 335.
5. Aristov, Yu.I., Tokarev, M.M., Di Marko, G., Cacciola, G., Restuccia, G., Parmon, V.N. (1997) Properties of the system "calcium chloride - water" confined in pores of the silica gel: equilibria "gas - condensed state" and "melting - solidification", *Rus.J.Phys.Chem.*, **71**, 253.

6. Aristov, Yu. I., Tokarev, M.M., Cacciola, G., Restuccia, G. (1997) Properties of the system "calcium chloride - water" confined in pores of the silica gel: specific heat, thermal conductivity, *Rus.J.Phys.Chem.*, **71**, 391.
7. Aristov, Yu.I., Di Marco, G., Tokarev, M.M., Parmon, V.N. (1997) Selective water sorbents for multiple applications: 3. $CaCl_2$ solution confined in micro- and mesoporous silica gels: pore size effect on the "solidification-melting" diagram, *React.Kinet.Cat.Lett.*, **61**, 147.
8. Tokarev, M.M., Aristov, Yu.I. (1997) Selective water sorbents for multiple applications. 4. $CaCl_2$ confined in the silica gel pores: sorption/ desorption kinetics, *React.Kinet.Cat.Lett.*, **62**, 143.
9. Gordeeva, L.G., Resticcia, G., Cacciola, G., Aristov, Yu.I. (1998) Selective water sorbents for multiple applications: 5. LiBr confined in mesopores of silica gel: sorption properties, *React.Kinet.Cat.Lett.*, **63**, 81.
10. Gordeeva, L.G., Resticcia, G., Cacciola, G., Aristov, Yu.I. (1998) Properties of the system "LiBr-H_2O" dispersed in silica gel pores: vapour equilibrium, *Rus.J.Phys.Chem.*, **72**, 1236.
11. Mrowiec-Bialon, J., Jarzebskii, A.B., Lachowski, A., Malinovski, J., Aristov, Yu.I. (1997) Effective inorganic hybrid adsorbents of water vapor by the sol-gel method, *Chem.Mater.*, **9**, 2486.
12. Sharma, S.K., Jotshi, C.K., Kumar, S. (1990) Materials for solar energy accumulation, *Solar Energy*, **44**, 177.
13. Mozgovoj, A.G., Shpil'rain, E.E., Dibirov, M.A., Bochkov, M.M., Levina, L.N., Kenisarin, M.M. (1990) Thermophysical properties of heat-storing materials. Crystalline hydrates. in: *Overviews on thermophysical properties of substances*, No 2 (82). Moscow: Scientific-Informational Centrum on Thermophysical Properties of Pure Substances, 104 p. (in Russian).
14. Gmelins Handbuch der Anorganischen Chemie, Calcium Teil B - Lieferung 2. /Hauptredakteur E.H.Erich Pietsch. Verlag Chemie GmbH, 1957.
15. Kirk-Othmer Encyclopedia of Chemical Engineering, 4th Ed., v.4, Wiley, New York, 1992.
16. Gurvich, B.M., Karimov, R.R., Mezheritskii, S.M. (1986) Heat conductivity of $CaCl_2$ aqueous solutions, *Rus.J.Appl.Chem.*, **59**, 2692.
17. Aristov, Yu.I., Restuccia, G., Cacciola, G.· Tokarev, M.M. (1998) Selective water sorbents for multiple applications. 7. Heat conductivity of $CaCl_2$ - SiO_2 composites, *React.Kinet.Cat.Lett.* (in press).
18. Kaviany, A. (1991) *Principles of Heat Transfer in Porous Media*, Springer-Verlag, New York.
19. Luikov, A.V. (1966) *Heat and Mass Transfer in Capillary Porous Bodies*, Pergamon Press.
20. Stefan (1891) *Ann. Phys. u. Chem.(Wiedemann) N.F.*, **42**, 269.
21. Riesmann-Weber, H. (1912) *Die Partiellen Differentialgleichungen der Mathemat.Physik*, 5th Ed.,_2, 121.
22. Carslaw, H.S., Jaeger, J.C. (1988) *Conduction of Heat Transfer in Solids*, Oxford Science Publications, 2nd Ed., 282
23. *Handbook of Heat Transfer Fundamentals*, (1985) Eds. W.M. Rohsenow, J.P. Hartnett, E.N. Ganic, 2nd Edition, McGraw-Hill, New York.

TEST METHOD FOR RANKING THE FIRE PROPERTIES OF MATERIALS IN SPACE BASED FACILITIES

J. L. Cordova, Y. Zhou, C. C. Pfaff, and A. C. Fernandez-Pello*
Department of Mechanical Engineering
University of California at Berkeley
Berkeley, CA 94720

1. Abstract

The prevention of accidental fires in space-based facilities is a major safety concern. A new methodology for testing and ranking the flammability behavior of solid combustible materials is presented. Based on the principles of the LIFT (ASTM-E 1321-93), the proposed test method measures the piloted ignition delay and flame spread rate of a material as a function of a known external heat flux, but it also includes the forced flow velocity and oxygen concentration as test parameters. It is shown that the ignition delay, the critical heat flux for ignition, and the flame spread rate are a function of flow velocity. This method is considered more suitable for testing materials to be used in space-based facilities, since in the absence of buoyancy, forced convection is the dominant mode of mass and energy transport.

2. Introduction

The possibility of an accidental fire in a space-based facility is a primary safety concern of current space exploration programs. For this reason, it is of critical importance to characterize the flammability properties of materials that will be used in such facilities. At present, no standard testing methodology exists that specifically addresses the flammability properties of materials subjected to the conditions expected in space installations. Current tests are based on the common theory that materials in reduced-gravity burn less readily than in normal-gravity conditions. Although this is probably true in a quiescent microgravity environment (where diffusion is the primary transport mechanism), spacecraft environments generally have low velocity air currents that are produced by ventilation and

* Corresponding author

V.E. Zarko et al. (eds.), *Prevention of Hazardous Fires and Explosions*, 49–59.

heating systems (of the order of 0.1 m/s). Recent experiments of flame spread in microgravity [1] show that at these low velocities, flame spread is faster, and the limiting oxygen concentration for flame spread is lower than in normal-gravity. Furthermore, CO_2 removal and oxygen replenishing systems in the space facilities cause oxygen concentration fluctuations which may also have an effect on the burning of materials. Therefore, it is necessary to establish a flammability test that properly represents the fire hazards that are expected in space installations.

The Lateral Ignition and Flame Spread Test (LIFT) apparatus, ASTM-E 1321-93 [2], although not considered a general fire test method, provides important information about the ignition and flame spread characteristics of a material, including such flammability properties as critical radiant flux for ignition, ignition delay, ignition temperature, flame spread rate constants, and extinction limits. These fire properties are extracted from the *flammability diagrams* [3-5] of the material, which are constructed from the data obtained by means of the LIFT apparatus, and consist of curves of ignition delay and flame spread rate as a function of the value of an externally applied radiant flux. The LIFT apparatus, however, operates under the effect of buoyancy (i.e., normal-gravity), and consequently, cannot be used in microgravity.

The concepts underlying the operation of the LIFT have been used to develop the *Forced-flow Ignition and flame-Spread Test* (FIST) [6, 7]. This testing methodology is similar to the LIFT, but it is considered to reflect more accurately the specific conditions expected in a space based facility. Like the LIFT, the FIST provides information about the ignition delay, critical heat flux for ignition, and the flame spread rate of combustible solids as a function of an externally applied radiant flux. However, it also includes a controllable forced flow of oxidizer over the exposed fuel surface, thus allowing to perform the flammability tests under different flow conditions, and in microgravity environments. Since the theoretical basis of ignition and flame spread in forced and natural convective conditions is fundamentally the same, the LIFT and FIST flammability diagrams are similar, though the flammability properties obtained using the latter are dependent on the velocity and oxygen concentration of the flow.

This work sets out to fulfill two objectives. First, it applies the FIST methodology to a well characterized solid material, namely polymethylmethacrylate (PMMA), with the purpose of demonstrating that oxidizer flow characteristics have a significant effect on the flammability properties of materials. Second, it compares the flammability diagrams generated with the FIST methodology to those obtained with the LIFT [8]. This comparison is used to gain some insight into the implications that the flow-dependence of the material flammability properties has with regard to issues of fire safety in micro-gravity environments.

3. Experimental Apparatus and methodology

The flammability diagrams and properties of the fuel are obtained from a series of piloted ignition and flame spread tests. For both types of tests, the fuel sample is placed in the test section of a combustion wind tunnel. Like in the LIFT, the exposed surface of the fuel is subjected to an external heat flux. However, in contrast with the LIFT, a forced flow of a given velocity is imposed parallel to the irradiated fuel surface, simultaneously with the heating process. A schematic of the problem under consideration is shown in Fig. 1.

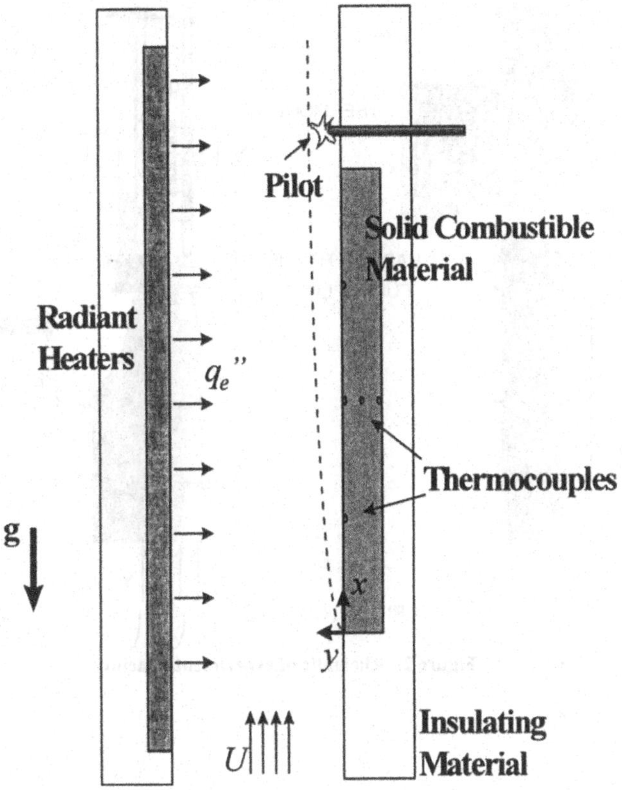

Figure 1: Basic problem configuration.

The experimental facility used for the FIST is shown schematically in Fig. 2, and it is described in detail in [7]. The apparatus used consists of a small scale combustion wind tunnel and supporting instrumentation. One of the walls of the tunnel test section is lined with a thick Marinite sheet, and the fuel sample is mounted flush with the Marinite surface. The side walls of the test section are made of Pyrex windows for optical access. The wall opposite to the fuel specimen is fitted with a computer controlled radiant panel, designed to impulsively irradiate the sample surface with a heat flux of predetermined intensity. The ignition pilot consists of an electrically heated Nichrome wire loop, located 10 mm downstream from the trailing edge of the fuel sample, and perpendicular to the fuel surface.

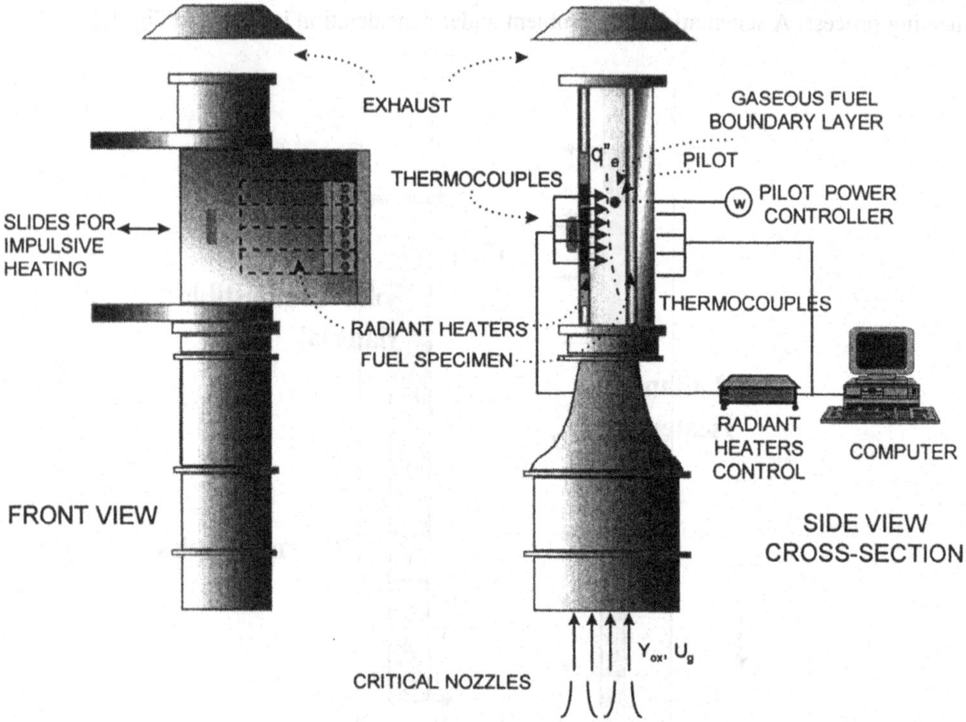

Figure 2: Schematic of experimental facility.

The oxidizer flow consists of house compressed air and bottled oxygen or nitrogen, each independently metered with critical nozzles. The gases are premixed to the desired oxygen concentration prior to entering the tunnel test section. The flow velocity is measured at the centerline of the test section with a LDV before heating takes place. The results presented in this work were all performed using air as oxidizer and PMMA (Rhom & Haas Plexiglas, Type G) as solid combustible.

3.1 IGNITION DELAY EXPERIMENTS

The PMMA samples used for the ignition delay experiments are 85 mm long, 50 mm wide, and 12.7 mm thick. For each test, the fuel specimen is centered with respect to the heater panel, as shown in Fig. 3, with the heaters extending 60 mm beyond the leading and trailing edges of the fuel. In this manner, the surface is irradiated with a nearly uniform and constant heat flux of a predefined value. A typical incident heat distribution used during the ignition tests is shown in Fig. 3. Experiments are conducted for cold flow velocities ranging from 0.2 to 2.5 m/s, and with radiant fluxes at the fuel surface ranging from 10.5 to 45 kW/m^2. The time required for the establishment of self-sustained burning, or *piloted ignition delay*, of the solid fuel is recorded, and the experiments are ended shortly after this is occurs by extinguishing the flames with a nitrogen flow.

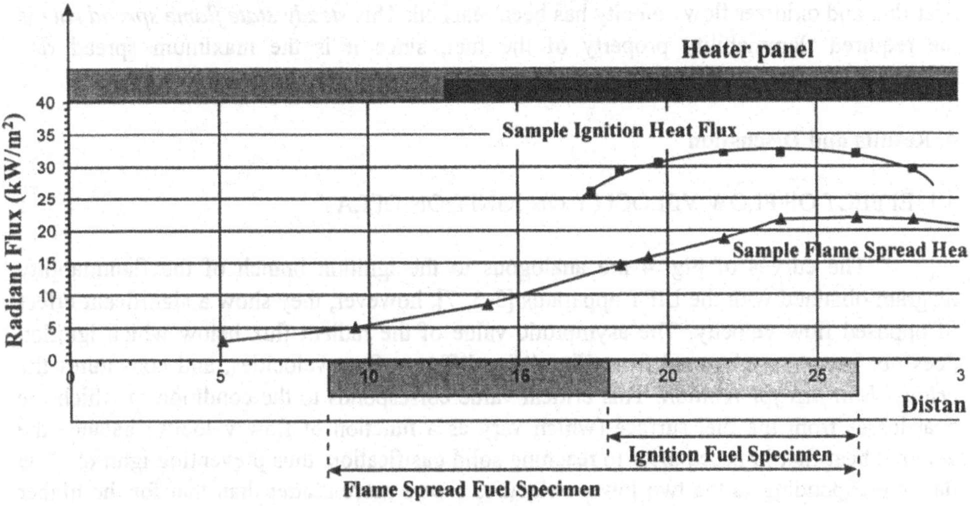

Figure 3: Sample heat flux distributions in test section.

3.2 OPPOSED FLOW FLAME SPREAD EXPERIMENTS

The PMMA samples used for the flame spread experiments are 175 mm long, 50 mm wide, and 12.7 mm thick. In this test, a decaying heat flux is imposed to determine the variation of the flame spread rate as a function of the incident heat flux. In order to produce the decaying flux, the samples are mounted off-center with respect to the heaters, with the leading edge of the fuel extending beyond the leading edge of the heater panel by 40 mm, but with its trailing edge and the pilot location identical to those of the ignition experiments. Figure 3 shows both the placement of the flame spread fuel sample, and a typical incident heat flux curve.

Air flows with velocities ranging from 1.0 to 2.5 m/s are made to flow over the combustible material surface, which is then impulsively heated by the radiant panel. The fuel is then ignited with the pilot, and the flame propagation is video-recorded. The flame location is then read from the video as a function of time. The flame spread rate is obtained by differentiating the position vs. time data. The flame spread data is then rendered as a function of incident heat flux, since this is a known function of position. For each flow velocity, the time of pilot activation is varied, from simultaneous with the initiation of fuel heating, through a range of predetermined pre-heat periods. In each trial, the preheat period is increased until the flame spread rate ceases to increase between successive tests. This indicates that the steady-state (maximum) fuel gasification corresponding to a particular heat flux and oxidizer flow velocity has been reached. This *steady-state flame spread rate* is the required flammability property of the fuel, since it is the maximum spread rate attainable for given conditions.

4. Results and Discussion

4.1. EFFECT OF FLOW VELOCITY ON IGNITION DELAY

The curves of Fig. 4 are analogous to the ignition branch of the flammability diagram obtained with the LIFT apparatus [3-5, 7], however, they show a significant effect of opposed flow velocity. The asymptotic value of the radiant flux below which ignition does not occur may be read from Fig. 4 for different flow velocities, and constitutes the *critical heat flux for ignition.* This critical value corresponds to the condition in which the heat losses from the fuel surface (which vary as a function of flow velocity) balance the external heat flux imposed prior to reaching solid gasification, thus preventing ignition. The data corresponding to the two lower velocities shows more scatter than that for the higher ones. This is due to the greater relative importance of natural convection effects for lower forced flow velocities.

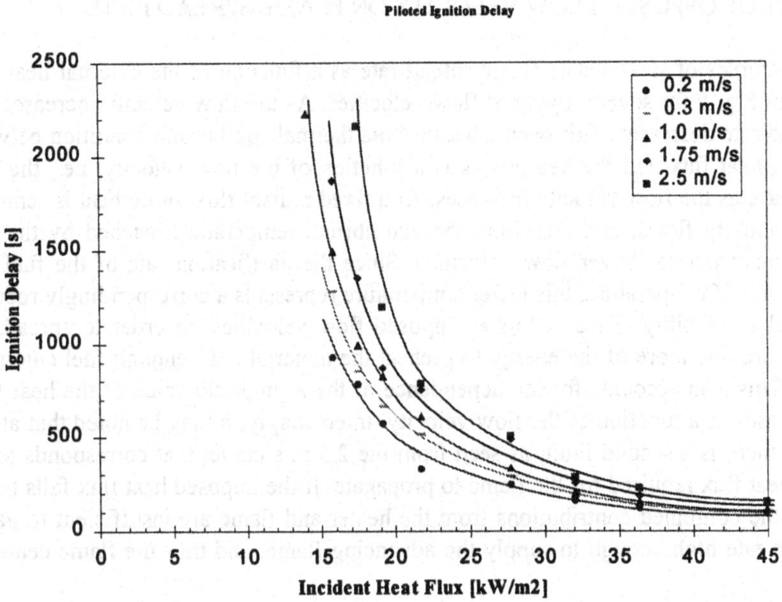

Figure 4. Ignition delay time as a function of incident heat flux for several flow velocities.

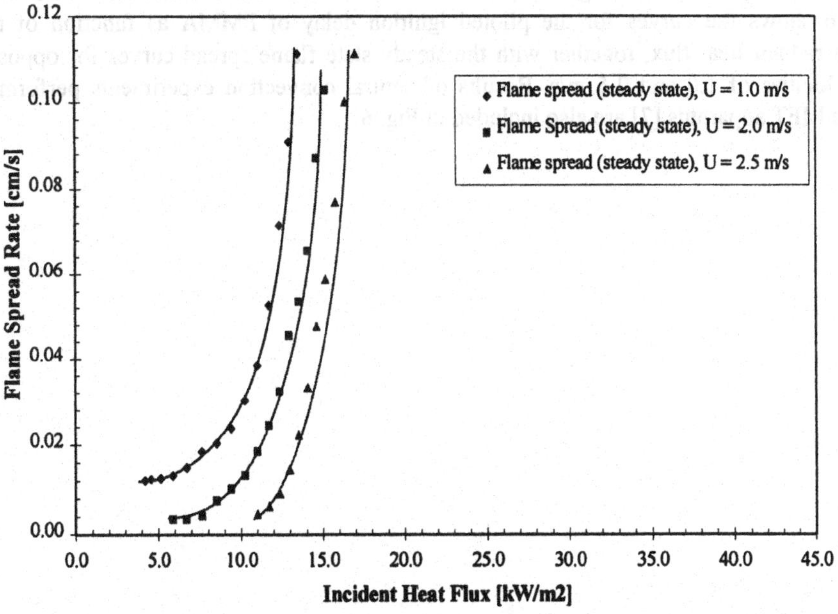

Figure 5: Flame spread rate as a function of heat flux for several opposed flow velocities.

4.2. EFFECT OF OPPOSED FLOW VELOCITY ON FLAME SPREAD RATE

Examples of steady-state flame spread rate as a function of the external heat flux are shown in Fig. 5 for several opposed flow velocities. As the flow velocity increases, the flame spread rate decreases. This occurs because the thermal equilibrium condition between the imposed heat flux and the heat losses is a function of the flow velocity, i.e., the heat losses increase as the flow velocity increases. At a fixed radiant flux, more heat is removed by higher velocity flows, and therefore, the equilibrium temperature reached by the fuel surface is lower than at lower flow velocities. Since the gasification rate of the fuel is a strong function of temperature, this lower temperature represents a correspondingly reduced gaseous fuel availability. Thus at higher opposed flow velocities, in order to spread, the flame must provide more of the energy to preheat the material until enough fuel enters the gas phase. This also accounts for the dependence of the asymptotic value of the heat flux, which increases as a function of the flow velocity. Interestingly, it may be noted that at low heat fluxes there is a second limit, as seen from the 2.5 m/s curve, that corresponds to the minimum heat flux required for the flame to propagate. If the imposed heat flux falls below such limit, the combined contributions from the heater and flame are insufficient to gasify the fuel at a rate high enough to supply the advancing flame, and thus the flame ceases to spread.

4.3. FLAMMABILITY DIAGRAMS FOR PMMA IN NORMAL-GRAVITY

The data collected from ignition and flame spread may be used to assemble what is known as the flammability diagrams [5] of PMMA for different external flow velocities. Figure 6 shows the curves for the piloted ignition delay of PMMA as function of the incident radiant heat flux, together with the steady state flame spread curves for opposed flow velocities of 1.0 and 2.5 m/s. Results of natural convection experiments performed with the LIFT apparatus [7] are also included in Fig. 6.

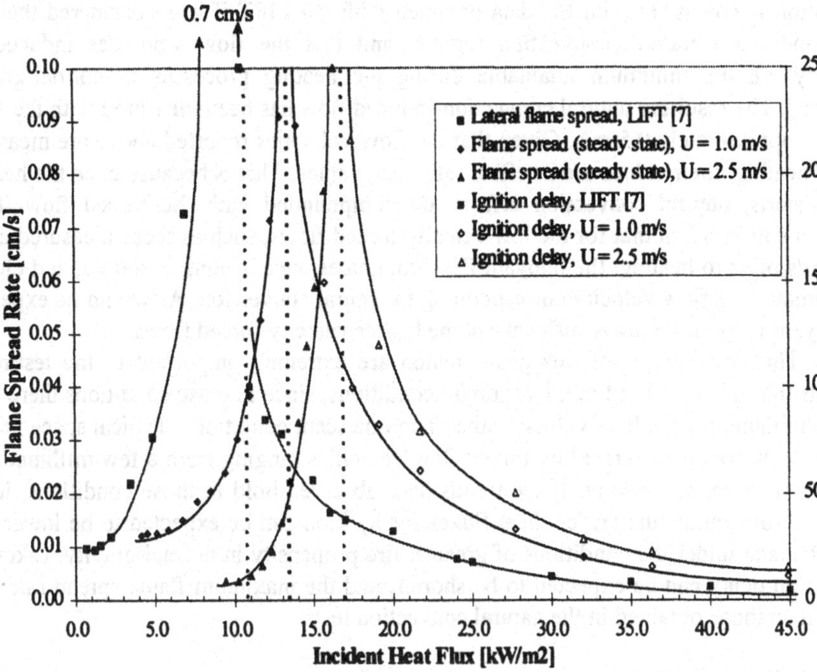

Figure 6: Flow velocity effect on the flammability diagram of PMMA

It may be seen that the asymptotic value of the heat flux corresponding to the steady-state flame spread rate is, for a given opposed flow velocity, the same as the critical heat flux for ignition. From the point of view of ignition, the critical heat flux is the minimum amount of heat required to overcome the heat losses, and to gasify the material at a high enough rate to produce an ignitable mixture. At a given flow velocity and prior to ignition, if thermal equilibrium is reached above this critical heat flux, gasified fuel production is already higher than the minimum required for the production of a flammable mixture. Thus, at such heat flux and higher, the flame rapidly flashes into a combustible gas region, at a rate that is only bounded by the premixed flame speed. The premixed flame speed is approximately 2 or 3 orders of magnitude greater than the flame spread rate over the solid fuel, and from the point of view of the present analysis approaches infinity. In contrast, for any heat flux below critical, the flame must supply extra heat to the fuel ahead of it, in order to raise its gasification rate, and therefore its propagation speed is limited by this process.

Figure 6 also shows that as the forced flow velocity is decreased, the steady state flame spread rate and the critical heat flux increase, while the ignition delay decreases. This observation is consistent with the data obtained with the LIFT if it is considered that this corresponds to a natural convection regime, and that the flow velocities induced by buoyancy are the minimum attainable during the heating processes in normal-gravity conditions. The resulting natural convection-induced flow has been measured with the LDV to be of approximately 0.5 m/s. Given that the flow velocities reported above are measured prior to heating, their values are in effect reference values. This is because once the heating process starts, natural convection effects are compounded with the forced flow. LDV measurements confirm that for the low velocity forced flows, such as those measured at 0.2 or 0.3 m/s prior to heating, the buoyant flow dominates once heating is started, and closely approximates the flow velocity corresponding to natural convection. As would be expected, the buoyant force has a lower influence at the higher velocity forced flows.

The implications of this phenomenon are extremely important to the testing of materials that are to be used in microgravity conditions, since in those conditions there is no minimum bound for the flow velocity other than quiescent conditions. Typical space facility environments are characterized by forced flow velocities ranging from a few millimeters to a few centimeters per second. If the trends here observed hold in those conditions, in the absence of buoyancy, the critical heat fluxes for ignition can be expected to be lower than those obtained under the conditions of greatest fire propensity in normal-gravity. Likewise, the ignition delay can be expected to be shorter, and the maximum flame spread rate even higher, than those obtained in the natural convection tests.

5. Concluding remarks

Comparison of the flammability diagrams of PMMA in the LIFT and FIST apparatuses in normal-gravity provides information about the effects of flow conditions on the flammability properties of materials. It is found that the critical radiant flux is a function of the air flow velocity, increasing as a function of the flow velocity. This is an important observation, since in the LIFT methodology, the value of the critical radiant flux for ignition is considered to be a constant material property, and consequently independent of the environmental conditions.

It is envisioned that the results here presented, together with those of ongoing investigations, may eventually lead to establishing the FIST methodology as standard means to more accurately rank the flammability of solid combustible materials used in microgravity applications.

6. Acknowledgments

The authors wish to thank Prof. J. L. Torero and Mr. R. T. Long for the LIFT data, and for their valuable discussions. Thanks are also due to Dr. H. Ross for his comments and enthusiastic support of the work. This work was funded by NASA under Grant No.NCC3-478.

7. References

[1] Olson, S. O., Mechanisms of microgravity flame spread over a thin solid fuel: Oxygen and opposed flow effects, *Combustion Science and Technology*, **76**, pp. 223-249, 1991.

[2] ASTM-E 1321-93, *Annual Book of ASTM Standards*, **04-07**, pp. 1055-1077, 1993.

[3] Quintiere, J. G., A simplified theory for generalizing results from a radiant panel rate of flame spread apparatus, *Fire and Materials*, **5**:2, pp. 52-60, 1981.

[4] Quintiere, J. G., Harkleroad, M., Walton, D., Measurement of material flame spread properties, *Combustion Science and Technology*, **32**, pp. 67-89, 1983.

[5] Quintiere, J. G. and Harkleroad, M., New concepts for measuring flame spread properties, NBSIR 84-2943, 1984.

[6] Cordova, J. L, Ceamanos, J., Fernandez-Pello, A. C., Long, R. T., Torero, J. L., Quintiere, J. G., Flow effects on the flammability diagrams of solid fuels, Fourth International Microgravity Combustion Workshop, NASA Conference Publication 10194, pp. 405-410, 1997.

[7] Cordova, J. L, Ceamanos, J., Fernandez-Pello, A. C., Piloted ignition of a radiatively heated solid combustible material in a forced oxidizer flow, AIAA/ASME Joint Thermophysics and Heat Transfer Conference,1, pp. 169-177, 1998.

[8] Long, R. T., An evaluation of the lateral ignition and flame spread test for material flammability assessment for microgravity environments, M. S. Thesis, University of Maryland at College Park, 1998.

FROM ROCKET EXHAUST PLUME TO FIRE HAZARDS

Methods to analyse radiative heat flux

V. WEISER, N. EISENREICH, W. ECKL
Fraunhofer-Institut für Chemische Technologie (ICT)
P.O.Box 1240, D-76318 Pfinztal (Berghausen), Germany

1. Introduction

Propellant flames and rockets exhausts consist of hot gases and particles and emit radiation ranging from the UV to the Infrared spectral region. The information inherent in the radiation has been used for various purposes in military research and technology:

1. When developing new solid rocket propellants the emission from flames was analysed to establish combustion mechanisms and to verify results of thermodynamic calculations. Especially, predicted adiabatic flame temperatures could be compared with experimental ones and the performance estimated which is closely related.
2. The observation of the signature of rocket exhausts allows early recognition of missiles, their identification and also affects guidance. Including after-burning these effects strongly influence the formulations of new propellants.

The investigation of the radiation of propellant flames and rocket exhausts lead to the development of sophisticated methods of measurement. The evaluation of the experimental results required adequate procedures to analyse the spectra with respect to the expected achievements.

In various branches of industry like chemical, pharmaceutical and food industry large quantities of fuel, flammable chemicals and explosive substances are produced and stored. To prevent disastrous impact on the environment if an accident occurs these materials have to be handled under safe conditions in buildings separated by safety distances. Beneath the pressure waves radiation is the main effect to influence the surroundings or even transmit a fire.

It is obvious to correlate the effects of propellant flames and rocket exhausts to those of fuel/air explosions and hazardous fires. Especially the experimental and theoretical methods to analyse radiation can be closely compared.

V.E. Zarko et al. (eds.), Prevention of Hazardous Fires and Explosions, 61–76.

2. Systems Investigated

The hot gases of flames, fires and explosions emit bands caused mainly by di-atomic molecules in the reactions zones and molecules like water and carbon dioxide which are final products of combustion. Particles like soot or aluminium oxide emit continuous spectra.

The emission spectra of solid rocket propellants show both characteristics. Composite propellants emit mainly broad band spectra. The emission spectra of double base or nitramine propellants are dominated by molecular bands. A similar variety can be found by pool fires of liquid fuels or solvents. Nitromethane and methanol burn with blue or white flames the spectra of which consist mainly of molecular bands. The yellow flames of fuels like diesel oil or octane emit a grey body spectrum with small contributions of bands.

These two classes of materials are compared with respect to their radiation. Emphasis is given on the correlation of results from a nitramine model formulation (compositions: RDX, RDX+5% Estane and RDX+10% Estane) [1, 2] and nitromethane [3, 4]. To get an insight into the combustion process propellant strands and a simple pool fire of the fuel was chosen. A metal pool with a diameter of 113 mm has been filled with fuel and placed on an laboratory jack allowing a detection grid to be scanned by the spectroscopic equipment. The propellant strands were investigated in an bomb equipped with quartz or CaF_2 windows which could be operated at pressures up to 10 MPa.

Flame temperatures, reaction zone profiles and state parameters (T_{rot}, T_{vib}) of intermediate combustion radicals have been determined by analysing band profiles of the emitting diatomic radicals OH, NH and CN, and by quantitative analysis of water emissions.

3. Emission Spectra

Emission spectra in the UV/Vis have been detected using a Trakor Northern grating spectrometer and a Jarrel Ash monochromator (focal length 275 mm, entrance slit 50 microns). To get rotationally resolved spectra, a grating with 2400 lines/mm has been chosen (wavelength resolution 20 wavenumbers). The detector is equipped with a 1024 element diode array. Minimum time resolution of the system is 10 ms. Considering assumed stationary conditions, experimental spectra were recorded with 100 ms per scan and averaged over 10 scans. If no rotational resolution was needed, a grating with 300 lines/mm has been used.

Spectra in the IR have been recorded applying a rotating filter wheel spectrometer developed at ICT [5]. The spectrometers consist of fast rotating wheels with interference filter segments. They continuously vary the transparency in the wavelength regions from 1200 to 2500 nm (InSb-detector) and 2450 to 14000 nm (InSb/HgCdTe-sandwich-detector).

It was developed at ICT. The intensity calibration was carried out by recording reference spectra of a black body radiator.

3.1. DATA ANALYSIS

The determination of rotational and vibrational temperatures of diatomic molecules from emission spectra is based on the calculation of line intensities and profiles. The intensity of a spectral line I_{em} is given by [6]:

$$I_{em} = N_n hc\nu_{nm} A_{nm} \tag{1}$$

where N_n is the number of atoms in the initial state, $hc\nu_{nm}$ the energy of each emitted light quantum of the transition and A_{nm} the Einstein transition probability of spontaneous emission. Considering thermal equilibrium, Born-Oppenheimer approximation and the explicit expression for Einstein coefficients equation (1) can be written as:

$$I_{em} = C \cdot q_{J'J''} \cdot S_{J'J''} \cdot \nu^4 \cdot \exp(-E_{vib} / kT_{vib} - E_{rot} / kT_{rot}) \tag{2}$$

where $q_{J'J''}$ and $S_{J'J''}$ are, respectively, the vibrational and rotational line strength, E_{vib} and E_{rot} are, respectively, the vibrational and rotational energy of the upper level, k is the Boltzmann constant and T_{vib} and T_{rot} are, respectively, the vibrational and rotational temperature. The constants and parameters for the different molecules and transitions are taken from the literature [6, 7]. At high radical concentrations, self absorption can not be neglected [8]. Assuming parallel beam geometry and a homogeneous layer the one dimensional solution of the equation of radiative transfer is [9]:

$$I = \frac{\varepsilon}{k}(1 - \exp(-kx)) \tag{3}$$

The temperatures are determined by comparing calculated line profiles to experimental data [10]. Therefore, a line profile must be taken into account. In case of our spectral resolution (about 20 cm^{-1}), the line profile is dominated by the slit function of the spectrometer and could be approximated by a Lorentzian.

Stable combustion products like water or carbon dioxide are emitting in the NIR spectral range and their vibrational spectra can be used to determine temperatures in colder flame regions, where no electronic excitation of diatomic molecules is observed. Therefore, additionally a code basing [11] on the data of the 'Handbook of Infrared Radiation of Combustion Gases' [12] has been developed allowing the band modelling of IR spectra. The used computer program calculates NIR/IR-spectra (1 - 10 μm) of non-homogenous gas mixtures of H_2O (with bands around 1.3, 1.8, 2.7 and 6.2 µm), CO_2 (with bands around 2.7 and 4.3 µm), CO, NO and HCl taking into consideration also emission of soot particles. It is based on the single line group model [12] and makes also use of tabulated data of H_2O and CO_2 in this reference. Because there are so many unknown parameters influencing the emission spectrum of an in-homogenous gas mixture, only a simplified model can be applied. Therefore, it was assumed that there is just one emitting layer of undefined thickness, constant temperature, constant concentration of the various gases and soot particles in thermal equilibrium. These assumptions lead to a reasonable number of parameters, which can be determined by fitting calculated spectra to experimental data. We have analysed only the NIR-spectrum from 1.7 - 2.2 µm dominated by the 1.8 µm band of water, since the 1.3 µm band of water is weak and shows a low signal/noise ratio. With this restriction, temperature and "concentration length" (concentration * length) of water and soot particles have been determined.

3.2. EMISSION OF DI-ATOMIC MOLECULES

An emission spectra of the investigated nitromethane flame is shown in fig 1. It is compared to the spectrum of a RDX/estane flame. In both cases the emission of OH, NH and CN and a continuous radiation and a NO emission is observable but the latter could not be rotationally resolved with the spectrometer available. Therefore, OH, NH and CN spectra were used for the analysis of the reaction zone. In the calculation of the molecular spectra, the rotational and vibrational distribution of the energy levels were separated. This allows the determination of a rotational and a vibrational temperature [13]. In thermal equilibrium, T_{vib} and T_{rot} are equal. If an additional energy is supplied to a molecule, e.g. chemical reaction energy, T_{vib} and T_{rot} show different values. Therefore, temperature determination can be used for checking chemical reaction kinetics if reaction branches exist or are discussed were the same molecule appears on the one hand chemically excited on the other hand only thermally excited [4].

The temperatures determined from the reaction zone of our nitromethane flame show this effect, too (see table 1). In case of CN, different values of T_{rot} and T_{vib} are found. Examples of calculated and experimental spectra are shown in figures 2 and 3.

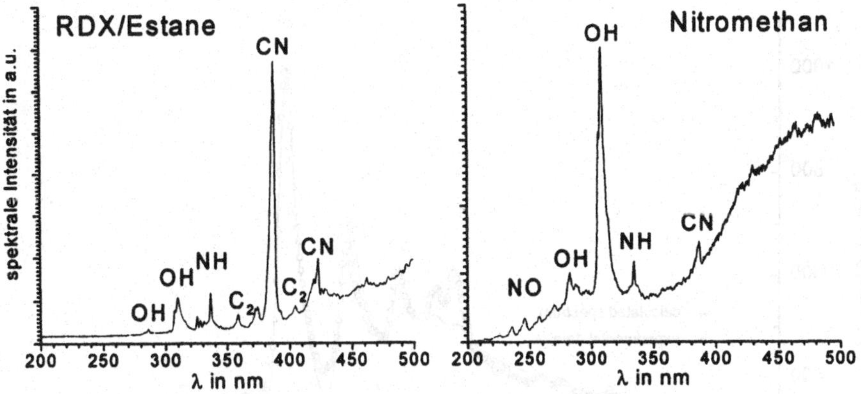

Figure 1. UV spectra of an burning RDX model propellant and a nitromethane flame

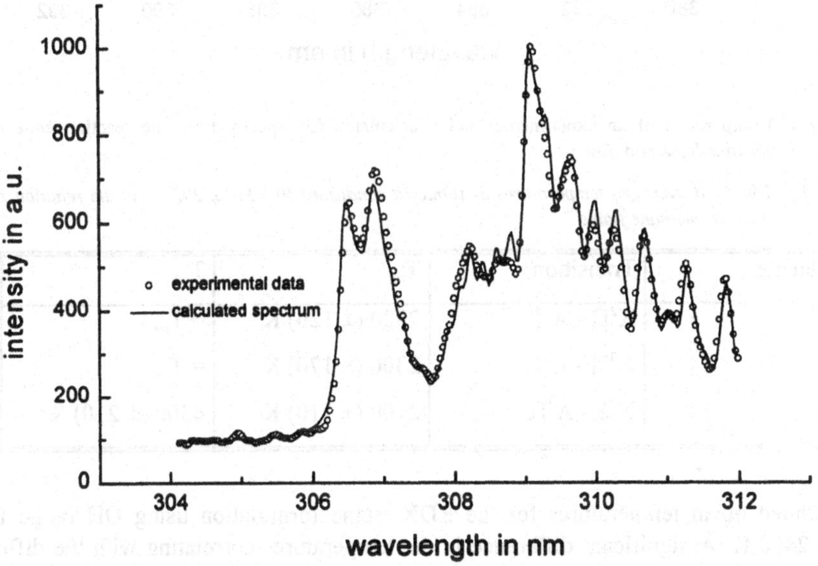

Figure 2. Comparison of an experimental and a calculated OH spectra from nitromethane

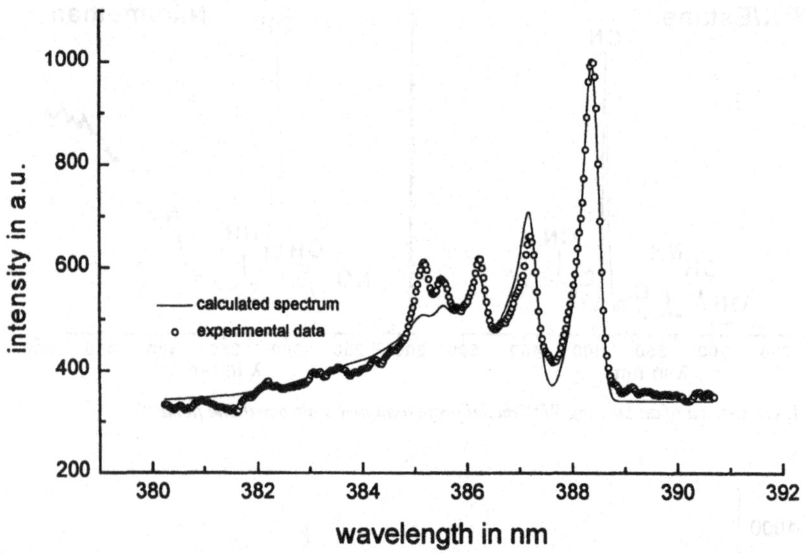

Figure 3. Comparision of an experimental and a calculated CN spectra from the reaction cone of the *nitromethane pool fire*

Table 1. Obtained averaged temperatures at standard conditions (0.1 MPa, 293 K) in the reaction zone of the nitromethane flame

Molecule	Transition	T_{rot}	T_{vib}
OH	$X^2\Pi$ - $A^2\Sigma$	2380 (± 120) K	= T_{rot}
NH	$X^3\Pi$ - $A^3\Pi$	2300 (± 170) K	= T_{rot}
CN	$X^2\Sigma$ - $A^2\Pi$	2100 (± 110) K	4300 (± 210) K

The obtained mean temperatures for the RDX/estane formulation using OH range from 2100 to 2400 K. A significant difference in the temperatures correlating with the different formulations could not be observed. The values are close to the adiabatic flame temperature of 2500 K calculated by the ICT code [14]. The ratio of the Q- and R-branch intensity is depending on the temperature. This stability of the ratio of both branches indicates that the OH emission is not strongly affected by chemical reactions throughout the flame zone which should be more pronounced in the reaction zone. Therefore, the rotational temperature of OH is reliable and can be estimated to be constant in the propellant flame.

This correlated to the fact that also the CN radical shows thermal excitation [1] which is in contrast to CN in the nitromethane flame or in a hydrocarbon/NO2 flame [15].

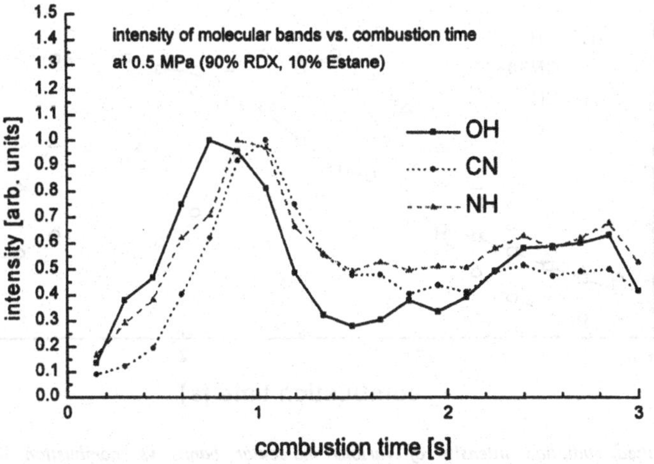

Figure 4. Emitted radiation intensity of various molecular bands vs. combustion time at 0.5 MPa (90/10 RDX/Estane formulation)

At higher pressures, no rotationally resolved spectra are observed. A continuous radiation dominates the flame emission, which can be fitted to the function of a grey body radiation (figure 6). This continuous radiation is stronger compared with pure RDX. The carbon content of the formulations is obviously converted partially to soot. The obtained mean temperatures were found to range from 2000 to 2200 K which are considerable lower than those obtained from molecular bands and the adiabatic flame temperature. It is assumed that radiation cools the flame or at least the particles strongly.

The emission profile of the important radicals along the flame axis is plotted in figures 4 and 5. Combustion time can be converted by use of the burning rate to height above the burning propellant surface.

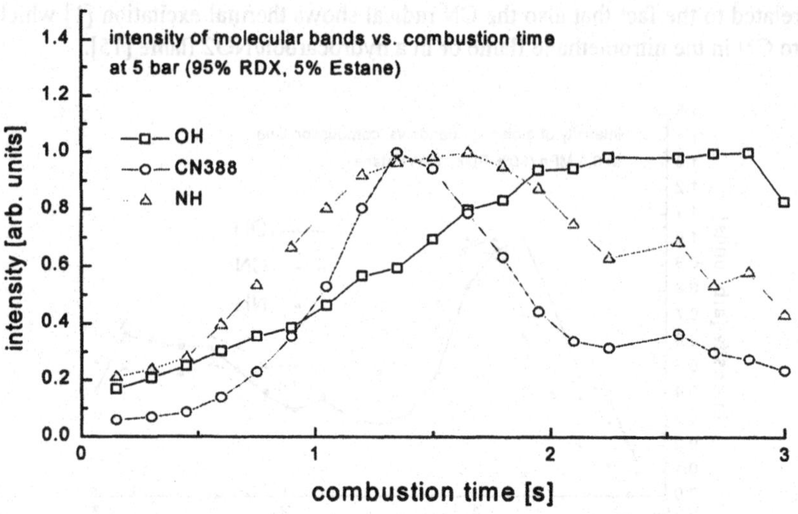

Figure 5. Emitted radiation intensity of various molecular bands vs. combustion time at 0.5 MPa (95/5 RDX/Estane formulation)

The emission of an iso-octane flame shows similar characteristics. A soot in a pool flame emits mainly a grey body radiation which is plotted in figure 7. The temperatures was found to be 1550 K.

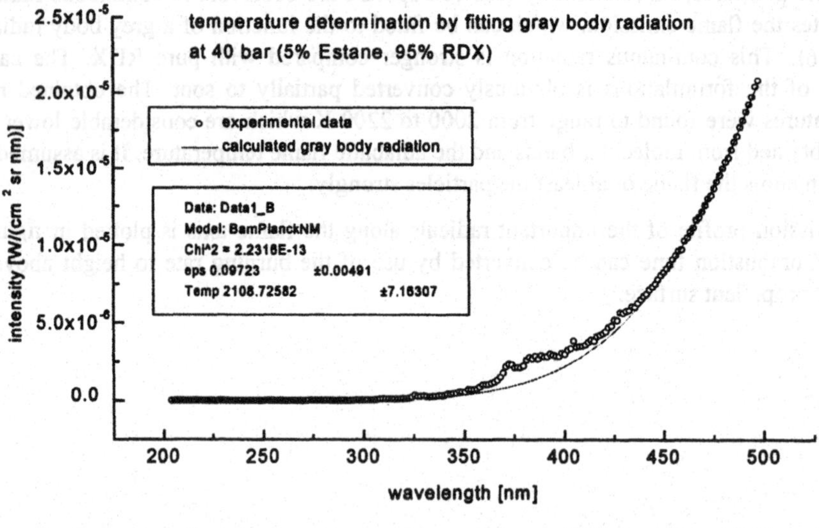

Figure 6. Temperature determination at increased pressure by grey body fit of the emitted intensity of the propellant flame

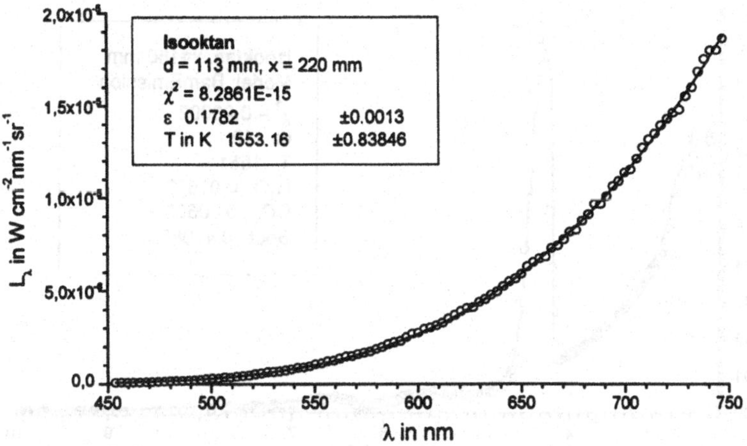

Figure 7. Continuous spectrum of an isooctane pool flame and its least squares fit by grey body radiation

3.3. BANDS IN NIR AND IR

Figure 8 shows a experimental and calculated spectrum of a water band emission of an iso-octane pool fire. The experimental intensity distribution could be reproduced very well. A series of spectra has been recorded with a time resolution of 20 ms at a centre position of the flame. The typical oscillating behaviour of pool fire flames [16] with a frequency of 5 - 10 Hz, depending on pool diameter) is observed in the variation of the concentration length. The temperatures are about 1850 K.

The bands in the IR are also similar which is demonstrated by the emission spectra of ADN and iso-popanol.

In the NIR spectral region mainly water bands are observed which can by modelled to for detimination of the temperature. In figure 10, an example is plotted.

The fluctuations of the bands is also observed in the infrared spectral region. Close to liquid pool the maximal intensity of the hydrocarbon bands and the CO_2 bands alternate which is plotted in figure 11.

These fluctuation of the pool fires are visualised in figure 12. In figure 13 the fluctuations in emitted radiation is shown.

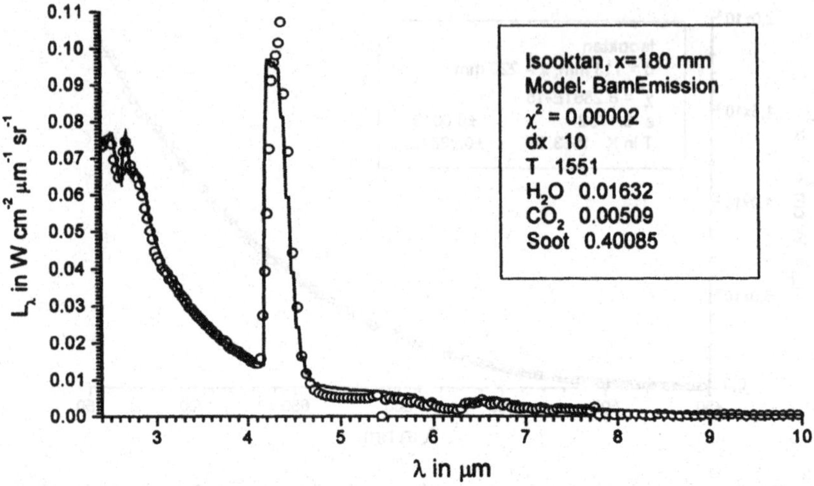

Figure 8. Least squares fit of a IR spectrum of an iso-octane pool flame modelling the bands of H2O, CO2 and of a grey body emitter

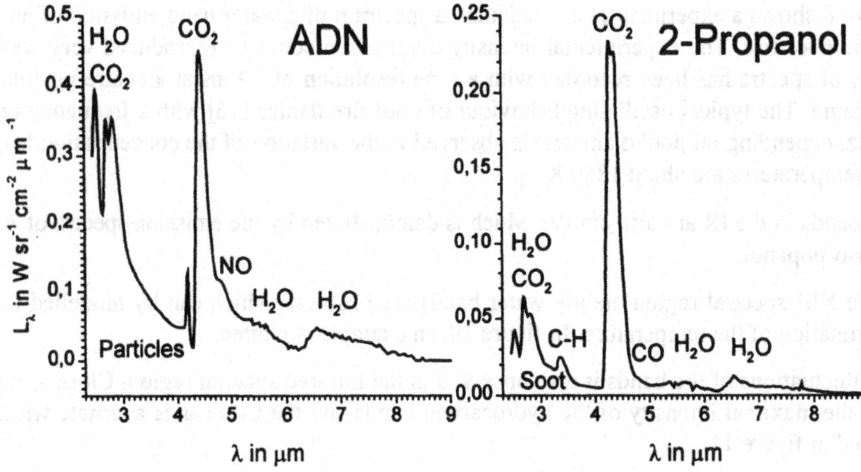

Figure 9. IR spectra of burning ADN propellant and 2-propanol flame

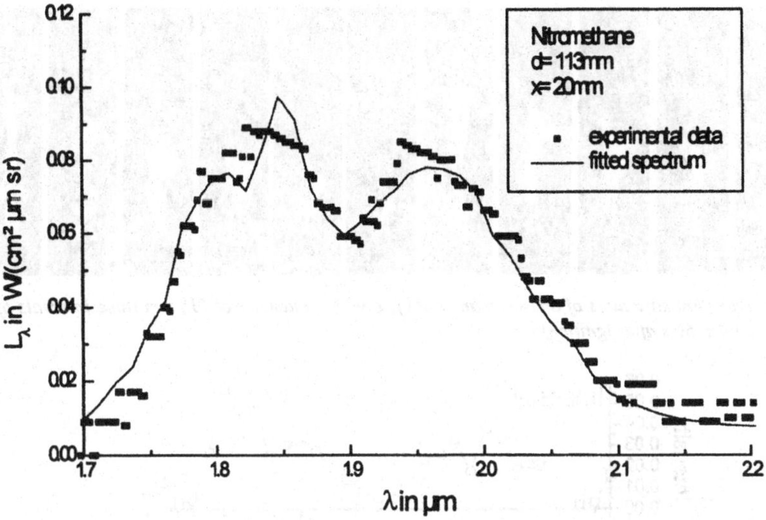

Figure 10. Water band temperature distribution in a nitromethane pool fire flame

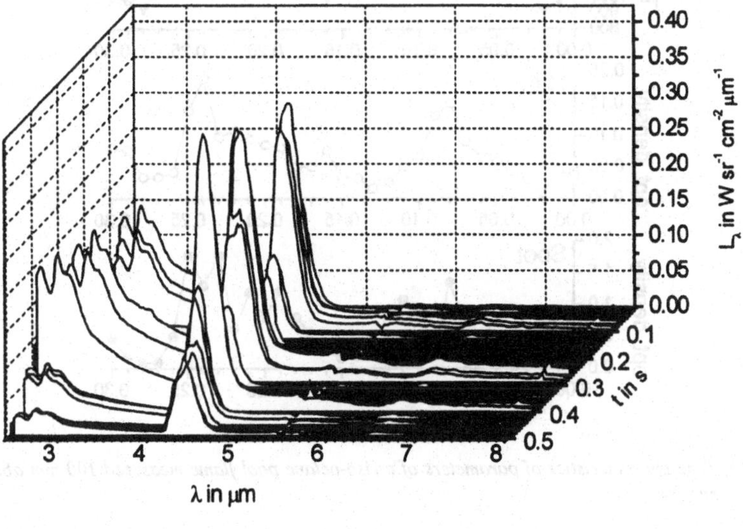

Figure 11. Sequence of spectra in the IR range measured 180 mm above a iso-octane pool surface

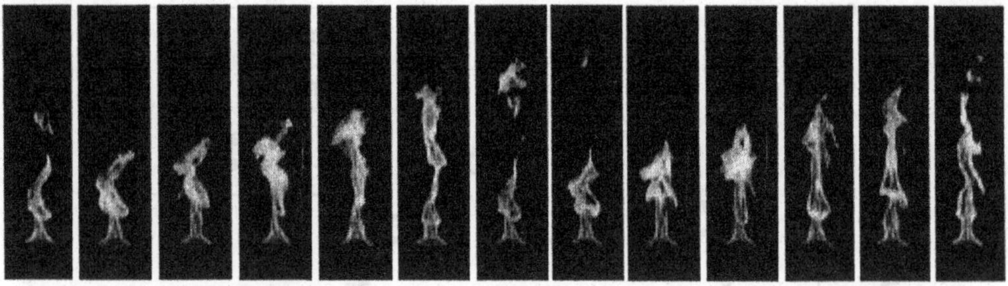

Figure 12. Transient structures of an iso-octane pool fire with a diameter of 113 mm (time intervals of frames 0.04 s, 60 s after ignition)

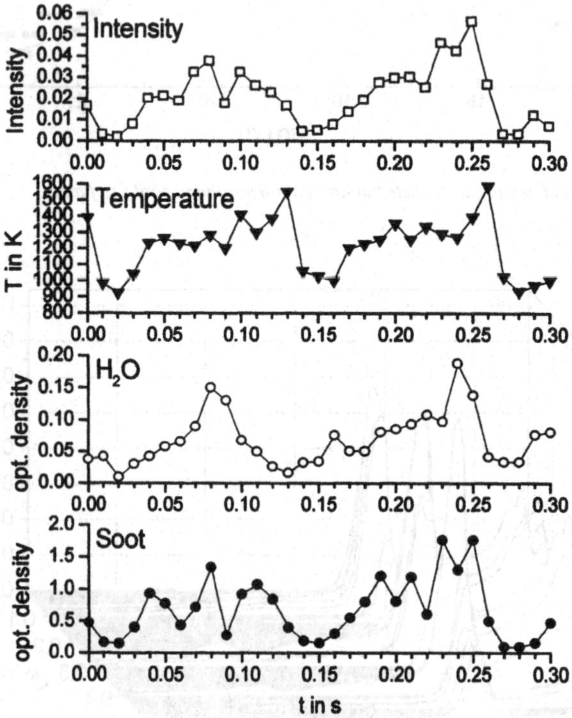

Figure 13. Time characteristics of parameters of an iso-octane pool flame measured 100 mm above the pool surface

4. Discussion

In a pool fire the liquid fuel is evaporated by energy transfer from the flame and burnt by entrained air. At diameters below 1 m the flame is strongly fluctuating at frequencies of some Hz. Nitromethane includes sufficient oxidiser for a self sustained flame leading to an adiabatic flame temperature of about 2400 K (calculated by ICT-code [14]). Air entrainment can rise the flame temperature up to about 2700 K. The reaction cone of the pool flame shows two torus of high intensities of OH emission. The outer torus is assumed to be mainly dominated by air supported combustion whereas the inner torus is mainly dominated by self sustained combustion. The temperature of hottest regions of the flame enclose 2500 and 3000 K indicating the expected air supported combustion. The averaged temperatures from molecular spectra except the vibration temperature of CN lie well below the adiabatic flame temperature which is explained by cooling due to subsequent air entrainment. The reaction product areas represented by the water bands show even lower temperatures by further cooling down.

The reaction mechanism seems to be similar to that of the flame of methane-nitrous oxide and oxygen [15] or to the flames of nitramines [2]. The initial step of the decomposition of nitromethane is assumed to be the bond scission of C-N forming CH_3 and NO_2 which than behaves similar to hdrcabon flame burning in air. The main emitters of these flames are the diatomic molecules NO, OH, NH and CN. These radicals are all included in the reaction mechanisms of the flame of $CH_4/NO_2/O_2$ which are more detailed discussed in the literature [15]. NO, NH and CN are intermediates in the reaction schemes whereas CN is involved in strongly exothermal reactions. Therefore CN molecule the radiative lifetime of which (see comparison in [15, 17]) is lowest is not observed in thermal equilibrium as indicated in tab. 1.

The radiation of the investigated pool fires is dominated by the emission of flame balls which are formed quasi-periodically above the flame base and ascend with acceleration. They are also the dominant part of the transient flame structure which emits the main part of energy. Therefore a simplified model of radiation of pool fires which could act on flammable surrounding objects can be reduced to the consideration of the radiation from ascending flame balls. The flame balls are approximated by spheres of diameters d_{FB}. The flame ball is formed one pool diameter above the pool surface, moves upwards accelerated by buoyancy forces (see Figure 12) and burns out more than 5 pool diameters above the pool surface. The radiation incident on a neighboured object is described by

$$q_T(y_T) = \varphi_{12}(y_T)\tau(y_T)E(T) \tag{4}$$

The atmospheric absorption is normally neglected ($\tau \approx 1$). The viewing factor φ_{12} is given by the solid angles of source and target [18] and is calculated according to Seeger [19]. It

reduces to an emitting disc with a diameter d_{FB} perpendicular to the straight line connecting flame ball and the centre of the object.

The radiation flux incident on a target area at a distance y_T depend on the geometry of the flame shape according to the flame model assumed. The total energy Q_R emitted by the surface A_{SEP} must be the same for all flame models compared. In the case of a grey body radiator:

$$Q_R = EA_{SEP} = \varepsilon\sigma T^4 A_{SEP} \tag{5}$$

If constant homogeneous flame temperatures T are assumed it is sufficient to adjust emissivity ε and surface to fulfil εA_{SEP} = const.

Figure 14 compares the radiation incident on a target area emitted from the static point source model, cylinder model of Seeger and the model of an ascending spherical flame ball. The data used were taken from the results of an iso-octane pool fire with 113 mm pool diameter.

The flame ball diameter d_{FB} is set equal to the pool diameter as plausible from the video frames like figure 12. From figure 13 results an emissivity $\varepsilon = 0.1$. The temperature was assumed to be T = 1600 K (comp. figure 8, 9, and 13). The total emitted power was then obtained by $Q_R = 0.1 \times \sigma \times 1600^4\,K^4 \times \pi \times 0.113^2\,m^2 = 1500$ W for all models. The viewing factors are derived according to Seeger [19].

Figure 15 shows the irradiance at a distance y_T on an area parallel to the flame axis at a height x=0 (height equal to the pool surface). The flame ball model predicts that more than twice the power can fall onto objects when compared with the cylinder model, especially in the case when the flame ball is close to flame base. If the flame ball is at higher positions the difference is lower and the cylinder model estimates quite well the irradiation on the average. Depending on the position of the flame ball the cylinder model predicts to low irradiation onto an endangered object.

The point source model overestimates the irradiation to objects close to the pool fire. For distances more than 4 d it agrees well with the maximum irradiation by the flame ball model. It is useful to estimate the worst case.

If safety distances have to be estimated it should be carefully proved what model to use. In cases where special materials have not to be exposed to radiation exceeding well defined limits the static models of surface radiators should be avoided.

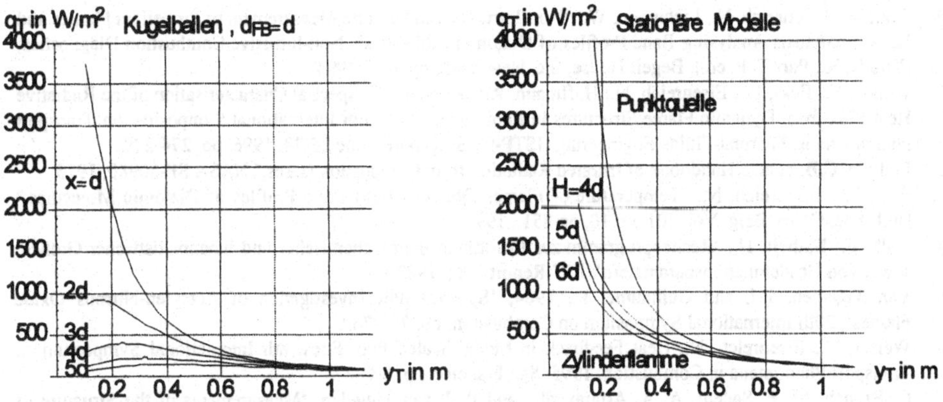

Figure 14. Irradiation onto objects at distancesyT by the dynamic flame ball model (left side) and the static point source and cylinder flame model (parameters are derived from aniso-octane pool flame of a diameter 113 mm)

5. Conclusions

All combustion processes emit radiation in the spectroscopic range from UV to IR. Military research developed fast scanning spectrometers and analysing tools to characterise rocket plumes and the burning zone of propellants. These methods can be applied successfully to investigate hazardous fires and explosions.

6. References

1. Eckl, W., Eisenreich, N., "Determination of the Temperature in a Solid Propellant Flame by Analysis of Emission Spectra," Propellants, Expl., Pyrotech., 17, 202 - 206, 1992.
2. Eckl, W.; Weiser, V.; Weindel, M.; Eisenreich, N.; Spectroscopic Investigation of Nitromethane Flames; Propellants, Explosives, Pyrotechnics 22; 180-183 (1997)
3. Eckl, W.; Eisenreich, N.; Weiser, V.; Spektroskopische Untersuchungen der UV/VIS Strahlungsemission der Flammen des Flüssigtreibstoffes Nitromethan; 25th International Annual Conference of ICT, 1994, Karlsruhe, pp. 58-(1-12);
4. Eckl, W.; Weiser, V.; Weindel, M.; Eisenreich, N.; Spectroscopic Investigation of Nitromethane Flames; Propellants, Explosives, Pyrotechnics 22; 180-183 (1997)
5. Baier, A.; Weiser, V.; Eisenreich, N.; Halbrock, A.; IR-Emissionsspektroskopie bei Verbrennungsvorgängen von Treibstoffen und Anzündmitteln; 27th International Annual Conference of ICT, 1996, Karlsruhe, pp. 84-(1-12);
6. Herzberg, 'Molecular Spectra and Molecular Structure I. Spectra of Diatomic Molecules', D. van Nostrand Company Inc., Princeton, New Jersey, 1950
7. Mavrodineanu, R. and Boiteaux, H., 1965, Flame Spectroscopy, J. Wiley & Sons, Inc, New York
8. Deimling, L.; Liehmann, W.; Eisenreich, Weindel, M.; N.; Eckl, W.; Radiation Emitted from RocketPlumes; Propellants, Explosives, Pyrotechnics 22; 152-155 (1997)
9. Lochte-Holtgreven, 1968, 'Plasma Diagnostics', North-Holland Publishing Company, Amsterdam

10. Eckl, W., Eisenreich, N., Liehmann, W., "Non-Intrusive Temperature Measurement of Propellant Flames and Rocket Exhausts Analysing Band Profiles of Diatomic Molecules", Non-Intrusive Combustion Diagnostics; (Kuo K. K., Parr T. P. ed.), Begell House, Inc. New York, pp. 673, 1994.
11. Weiser, V.; Eckl, W.; Eisenreich, N.; Hoffmann, A.; Weindel, M.; Spectral Characterisation of the Radiative Heat Flux from Dynamic Flame Structures in Pool Fires; The Ninth International Symposium on Transport Phenomena in Thermal-Fluids Engineering (ISTP-9), Singapore, June 25-28, 1996, pp. 274-279
12. Ludwig, C.B. et al., "Handbook of Infrared Radiation from Combustion Gases," NASA SP-30980, 1973
13. Eckl, W., Eisenreich, N., "Temperature of Flames Obtained from Band Profiles of Diatomic Molecules," Bull. Soc. Chim. Belg. Vol. 101, no 10, pp 851., 1992.
14. Volk, F.; Bathelt, H., 'Rechenprogramm zur Ermittlung thermochemischer und innenballistischer Größen, sowie von Gasdetonationsparametern', ICT-Report 3/82, 1982
15. Van Tiggelen, P.J. and Guillaume, P., 1984, 'Spectroscopic Investigation of Acetylen-Nitrous Oxide Flames', 20th International Symposium on Combustion, pp. 751-760
16. Weiser, V.; Eisenreich, N.; Heat Feedback in Model-Scaled Pool Fires; 8th International Symposium on Transport Phenomena in Combustion, 1995, San Francisco; p. 8-C-2
17. C. Branch, M. E. Sadeqi, A. A. Alfarayedhi, and P. J. van Tiggelen, 'Measurements of the Structure of Laminar, Premixed Flames of CH4/NO2/O2 and CH2O/NO2/O2 Mixtures', Combustion and Flames 83, pp. 228-239 (1991)
18. Siegel, R.; Howell, J. R.; Lohrengel, J.; Wärmeübertragung durch Strahlung (Teil 2); Springer Verlag, Berlin, 1991
19. Seeger, P.G.;Wärmeübertragung durch Strahlung und Konvektion bei Bränden in Flüssiggaslagern; VFDB 1/87 S.7-13

A REVIEW OF COMPUTIONAL FLUID DYNAMICS (CFD) MODELING OF GAS EXPLOSIONS

B.H. HJERTAGER, T. SOLBERG
Aalborg University Esbjerg (AAUE)
Niels Bohrs Vej 8, DK-6700 ESBJERG, DENMARK.
Telemark Technological R & D Centre (Tel-Tek)
Kjølnes Ring, N-3914 PORSGRUNN, NORWAY

The present paper reviews models and computer codes for detailed analysis of gas explosions. In particular computer models for gas explosion propagation in complex and highly congested geometries are described. The governing equations are formulated according to the quasi-continuum principle. The influence of obstacles is taken care of by specifying volume and area porosities and by including distributed resistances in the calculation domain. Examples of prediction capabilities are given.

1. Introduction

Background Gas explosion hazard assessment in flammable gas handling operations is crucial for obtaining an acceptable level of safety. In order to perform such assessments, good predictive tools are needed. These tools should take account of relevant parameters such as: geometrical design variables and gas cloud type and distribution. A theoretical model must therefore be tested against sufficient experimental data prior to becoming a useful tool. The experimental data should include variations in geometry as well as gas cloud composition and the model should give reasonable predictions without use of geometry or case dependent contants. The tools that seem best suited to take full account of the interaction between geometry and explosions are the models based on detailed numerical modeling - the socalled CFD (Computational Fluid Dynamics) models. In order for industry and approving authorities to gain confidence in such models, comprehensive comparisons between experiments and predictions must be shown and the degree of uncertainty must be quantified. In the last 10-15 years some attempts to show the capabilities of CFD models like the EXSIM code (Hjertager[1,2,3], Hjertager et al.[4], Sæter et al[5]), the REAGAS code (van den Berg et al.[6]), the FLACS code (van Wingerden et al.[7,8,9]) and the COBRA code (Catlin et al[10]) have appeared. In a recent CEC (Commision of the European Community) sponsored research programme named MERGE all these codes were further developed and tested in simplified 3D geometries (Popat et al.[11]). In fact the CEC has issued a set of Guidelines for model developers[12] and a Model evaluation report [13] as well as a Gas explosion model evaluation protocol[14] that will play an

V.E. Zarko et al. (eds.), Prevention of Hazardous Fires and Explosions, 77–91.

important role in unifying and quantifying the performance of explosion models. The present paper will review models that are available for analysis of gas explosion propagation in complex industrial geometries. Special attention will be given to the EXSIM code developed by the authors.

Complex geometry modelling All geometries found in industrial practice may contain a lot of geometrical details which can influence the process to be simulated. Examples of such geometries are heat exchangers with thousands of tubes and several baffles, and regenerators with many internal heat absorbing obstructions. In the present context the geometries found inside modules on offshore oil and gas producing platforms constitute relevant examples of the complex geometries at hand. There are at least two routes for describing such geometries. First, we may choose to model every detail by use of very fine geometrical resolution, or secondly we may describe the geometry by use of some suitable bulk parameters. A detailed description will always need large computer resources both with regard to memory and calculation speed. It is not feasible with present or even future computers to implement the detailed method for solving such problems. We therefore use the second line of approach, which incorporates the so-called Porosity/Distributed Resistance (PDR) formulation of the governing equations. This method was proposed by Patankar and Spalding[15] and has been applied to analysis of heat exchangers, regenerators and nuclear reactors. Sha et al[16,17] have extended the method to include more advanced turbulence modelling. Hjertager et al[4] have formulated and applied thc PDR technique to offshore platform modules.

Review of models and codes It has in the past been usual to predict the flame and pressure development in vented volumes or unconfined vapour clouds by modelling the burning velocity of the propagating flame. This may be succesful if we have a simple mode of flame propagation such as axial, cylindrical or spherical propagation in volumes without obstructions in the flow. If these are present, however, it is difficult to track the flame front throughout complex geometries. It has been apparent that in these situations it is more useful to model the propagation by calculating the rate of fuel combustion at different positions in the flammable volume. It is also important to have a model which is able to model both subsonic and supersonic flame propagation to enable a true prediction of what can happen in an accident scenario. One such model which in principle meets all these needs has been proposed by Hjertager[1,2,3,4] and Bakke and Hjertager[18,19,20]. The model has been tested against experimental data from various homogeneous stoichiometric fuel-air mixtures in both large- and small-scale geometries. Similar models for gas explosions have subsequently also been proposed by Kjäldman and Huhtanen[21], Marx et al.[22], Martin[23] and van den Berg[24]. All the above models are similar in nature. They use finite-domain approximations to the governing equations. Turbulence influences are taken account of by the k-ε model of Launder and Spalding[25] and the rate of combustion is modelled by variants of the 'eddy-dissipation' model of Magnussen and Hjertager[26]. The Bakke and Hjertager models are incorporated in two computer codes named FLACS (FLame ACceleration Simulator) and EXSIM (EXplosion SIMulator). The solution method used is the SIMPLE technique of Patankar and Spalding[27]. The model of Kjäldman and Huhtanen uses the general PHOENICS code of Spalding[28]. Whereas the model of Marx et al. uses the CONCHAS-SPRAY computer code which embodies the ICE-ALE solution technique. The model of Van den Berg is similar to the Hjertager model and is incorporated into a code named REAGAS. The model of Martin which is embodied in a computer code named FLARE uses the flux-corrected transport (FCT) of Boris and Book[29]. The recent COBRA code of Catlin et al[10] uses novel techniques both for numerical and combustion modelling of the explosion. The numerical technique involves using an adaptive numerical grid and an explicit solution method due to Gudunov[30]. The combustion model involves using empirical relations for the turbulent burning velocity as well as the turbulent flame thickness. The model of Connel et al[31] also uses an adaptive mesh technique but involves use of unstructured grids. This last feature makes the model very flexible for geometry-resolved calculations. The latter model uses the socalled flamelet model of Bray[32] for combustion.

2. Governing equations

The simulation of gas explosion in complex geometries involves using the PDR method mentioned above. The following section will give a brief review of the governing equations and submodels of subgrid turbulence and combustion used in this method.

Volume and area porosities The presence of geometrical details modifies the governing equations in two ways. First, only part of the total volume is available to flow and secondly, solid objects offer additional resistance to flow and additional turbulence production. If we consider a control volume we can define a volume fraction or porosity occupied by the fluid, β_v and area porosity in the x_i-directions β_i. All the volume and area fractions (porosities) may take values between 0.0, completely blocked, or, 1.0, completely open.

Conservation equation If we apply the conservation principle to an obstructed control volume we obtain the following equation for a conserved variable Φ:

$$\frac{\partial}{\partial t}(\beta_v \rho \Phi) + \frac{\partial}{\partial x_j}(\beta_j \rho U_j \Phi) = \frac{\partial}{\partial x_j}(\beta_j \Gamma_{\Phi,eff} \frac{\partial \Phi}{\partial x_j}) + S_\Phi + R_\Phi$$

Here U_j is the velocity component in the x_j coordinate direction; ρ is the density; $\Gamma_{\Phi,eff}$ is the effective turbulent diffusion coefficient for the variable Φ; S_Φ is the non-obstructed part of the source term and R_Φ is the additional source term caused by obstacles located inside the control volume.

Momentum equations: The resistance per unit volume in the x_i-direction, R_{Ui}, may be expressed as:

$$R_{U_i} = -C_D \frac{1}{2} \rho |U_i| U_i \; ; \quad C_D = f_i \bullet A_w$$

where C_D is a drag coefficient, A_w is the so-called wetted area or frontal area of the obstructions per unit volume. The friction factor f_i may depend on parameters like velocity, porosity, typical dimension or hydraulic diameter, pitch between obstacles, obstacle shape and orientation.

Energy: The first law of thermodynamics applied to an obstructed control volume will include an additional source term that is the heat transfer rate between the fluid and internal obstructions in the volume. Most gas explosion models ignores this heat transfer, mainly because of the short time scales involved in explosions.

Chemical species conservation: The conservation principle for a chemical species applied to an obstructed control volume will result in the following conservation equation for the mass fraction of a chemical species, Y_j:

$$\frac{\partial}{\partial t}(\beta_v \rho Y_j) + \frac{\partial}{\partial x_i}(\beta_i \rho U_i Y_j) = \frac{\partial}{\partial x_i}(\beta_i \Gamma_{Y_j,\mathrm{eff}} \frac{\partial Y_j}{\partial x_i}) + S_{Y_j}$$

Here $\Gamma_{Yj,\mathrm{eff}}$ is the effective mass diffusion coefficient flux of species Y_j and S_{Yj} is the rate of production by chemical reaction of the species inside the control volume. Since the obstacles inside the control volume are solid and therefore impervious to mass transfer, no additional source term appears in this equation. The only influence of obstacles, other than reduced volume and areas, is enhanced mixing through the surfaces of the control volume and distortion of the flame inside the volume. This distortion is modelled by multiplying the reaction rate by a combustion enhancement factor E_T.

Turbulence: In the above equations (1) and (3) the effective turbulent viscosity μ_{eff} and the effective turbulent transport coefficient $\Gamma_{\Phi,\mathrm{eff}}$ must be modelled. For this we use the k-ε turbulence model which determines the distribution of the kinetic energy of turbulence, k, and its rate of dissipation, ε. The viscosity and transport coefficients are modelled according to:

$$\mu_{\mathrm{eff}} = \mu_l + C_\mu \rho \frac{k^2}{\varepsilon} \; ; \quad \Gamma_{\Phi,\mathrm{eff}} = \mathrm{musub}\frac{\mathrm{eff}}{\sigma_\Phi}$$

An effective Prandtl/Schmidt number σ_Φ has been introduced. C_μ is a constant taken to be 0.09 (Launder and Spalding[25]). The Schmidt numbers σ_k and σ_ε are given the values 1.0 and 1.3, respectively, whereas the other Schmidt/Prandtl numbers are put equal to 0.7. The transport equations that determine the distribution of k and ε has additional source terms as follows. The generation rate of turbulence G, consists of two parts. One part G_s, is related to the stresses on the surfaces of the obstructed control volume and another part, G_R is related to the internal frictional resistances caused by the solid obstructions inside the volume. The additional source term in the k-equation reads:

$$R_k = G = G_s + G_R \ ; \quad G_s = \beta_v \sigma_{ij} \frac{\partial U_j}{\partial x_i} \ ;$$

$$G_R = \sum_i C_s \mu_{eff} U_i^2 \bullet A_w^2 + \sum_{k,\,i} C_{T_{k,i}} \bullet \left| R_{k,U_i} \right| \left| U_i \right|$$

The first part G_s is the same production rate term which is found for non-obstructed flows (when $\beta_v = 1.0$). The second part G_R is similar to that proposed by Sha and Launder[16]. G_R are expressed in two ways: one part which uses the turbulent viscosity and another part that uses the drag force. C_s and C_T are constants. A_w is the wetted area of the obstacles inside the control volume per unit volume. This assumption indicates therefore that $C_{T,k,i}$ gives the fraction of the pressure drop in the x_i-direction across obstruction no. k that goes into generation of turbulent kinetic energy. In densely packed areas the ε-equation is not solved. Instead, a fixed length scale is prescribed and the dissipation is related to the length scale and the turbulent kinetic energy according to:

$$l = C_l \bullet D_h \ ; \quad \varepsilon = C_\mu^{3/4} \frac{k^{3/2}}{l}$$

The typical obstacle dimension, D_h, has been taken: 1) as the inverse of the specific area ($1.0/A_w$), 2) as the obstacle diameter (D) or 3) as the pitch between obstructions (P). The constants has the following values: C_T = 0.025 to 1.0; C_s = 2.0; and C_l = 0.1 to 1.0.

Combustion modelling using the "eddy dissipation model": The combustion is treated as a single step irreversible chemical reaction with finite reaction rate between fuel and oxygen. For a homogeneous premixed system the mixture fraction will be constant in the domain of interest and consequently only the Y_{fu} equation needs to be solved. The rate of combustion may be modelled according to the 'eddy-dissipation' concept by Magnussen and Hjertager[26] with the ignition/extinction modification introduced by Hjertager[2] and the quasi-laminar combustion modification introduced by Bakke and Hjertager[18]. If the local turbulent Reynolds number, based on the turbulent velocity and length scale, is less than a critical value, the rate of combustion is calculated according to:

$$S_{fu} = \beta_v \bullet E_L A_{lam} \frac{u^o_{lam}}{\delta_{lam}} \rho Y_{lim}$$

Here Y_{lim} is the smallest of three mass fractions, namely fuel, Y_{fu}, oxygen Y_{O2}/s, or mass fraction of fuel already burnt, $Y_{fu,b}$. u^o_{lam} and δ_{lam} are the laminar burning velocity and thickness of the laminar flame. E_L is the enhancement factor related to the cellular wrinkling of the laminar flame and this factor is proportional to the radius of flame propagation up to a maximum radius of 0.5 m. The enhancement factor is 1.0 for a radius of 0 m and is 2.5 for radii larger than 0.5 m. The constant A_{lam} is calculated according to the proposal from van den Berg[33]. This involves integration over the combustion products and insuring that the flame speed is according to the specified one (EXSIM-94)[5]. If the local turbulent Reynolds number is larger than the critical value the rate of combustion is calculated according to the eddy dissipation approach modified by the extinction/ignition criteria. Two time scales are defined, namely the turbulent eddy mixing time scale, τ_e, and the chemical time scale, τ_{ch} :

$$\tau_e = \frac{k}{\varepsilon} \; ; \quad \tau_{ch} = A_{ch} e^{\left(\frac{E}{RT}\right)} \bullet (\rho Y_{fu})^a \bullet (\rho Y_{O_2})^b$$

Also, an ignition/extinction criterion is defined when the two time scales are in a certain ratio $(\tau_{ch}/\tau_e)^* = D_{ie}$. The rate of combustion is thus calculated as:

$$S_{fu} = 0 \text{ when } \frac{\tau_{ch}}{\tau_e} > D_{ie} \; ;$$

$$S_{fu} = \beta_v E_T A \bullet \frac{\varepsilon}{k} \rho Y_{lim} \text{ when } \frac{\tau_{ch}}{\tau_e} \leq D_{ie}$$

A and D_{ie} are two constants with values 20 and 1000, respectively. E_T is a combustion enhancement factor to take account of the break-up and acceleration of the flame when multiple obstacles are inside the control volume.

<u>**Combustion modelling using "turbulent burning velocity relations":**</u> Catlin et al[10] introduce a combustion model that uses turbulent burning velocity relations, u_t and specification of a thickness of the turbulent flame, δ_t. These relations are put in a form so that the reaction rate of the progress variable c, including quench may be deduced (Catlin and Lindstedt[34]). The rate of reaction and turbulent diffusion coefficient for the progress variable equation read:

$$S_c = \bar{\rho}\left(\frac{u_t \Lambda_2}{\delta_t \Lambda_1}\right) c^4 (1-c) \left(\frac{\rho_u}{\rho_b}\right)^2 \; ; \quad \Gamma_c = \left(\frac{\mu_1}{\sigma_c}\right) + \bar{\rho}\left(\frac{u_t \delta_t}{\Lambda_1 \Lambda_2}\right)$$

Here Λ_1 and Λ_2 are the burning velocity and flame thickness eigenvalues. Both values are evaluated from a number of one-dimensional numerical calculations of planar flame propagation in a flow field with constant levels of turbulence (velocity and length scale). These calculations demonstrated that unique expansion-ratio-independet values of Λ_1 and Λ_2 of 0.346 and 3.575, respectively. In these analyses the flame thickness was taken equal to a turbulent length scale and the turbulent burning velocity was taken as:

$$\delta_t = l = c_\mu^{3/4}\frac{k^{\frac{3}{2}}}{\varepsilon}\ ;\quad \frac{u_t}{u^o_{lam}} = 0.875K^{0.392}\frac{u'}{u^0_{lam}}\ ;$$

$$K = 0.157\left(\frac{u'}{u^o_{lam}}\right)^2 Re_l^{-1/2}\ ;\ u' = \left(\frac{2}{3}k\right)^{1/2}\ ;$$

$$Re_l = \frac{\rho u' l}{\mu_{lam}};\quad u^o_{lam} = u^o_{lam}(p_o,\ T_o)\bullet\left(\frac{T_u}{T_o}\right)^2\left(\frac{p}{p_o}\right)^{1/2}$$

The unstrained laminar burning velocity $u^o_{lam}(p_o,T_o)$ was taken equal to 0.41m/s for stoiciometric methane-air mixtures. The turbulent burning velocity in Eq. 10 was taken from Bray[32]. For values of the Karlowitz number, $K<0.2$ and $Re_l<3000$ a burning velocity relation derived by Gulder was used[35].

<u>Combustion modelling using "the laminar flamelet model":</u> Bray[32] has proposed a combustion model that also solves for the progress variable c. The turbulent flame is assumed to consist of thin reacting interfaces between reactants and products and that these interfaces have the structure of strained laminar flames. For such a description to be valid the laminar flame thickness must be less than the Kolmogorov length scale of turbulence. The relation for the reaction rate in the c equation reads:

$$S_c = \rho_r\, u^o_{lam}\, I_o\, \Sigma$$

Here Σ is the specific flamelet area, u^o_{lam} is the unstrained laminar burning velocity and $(u^o_{lam}\ I_o)$ is the mean laminar burning velocity averaged over the flamelet surface. The factor I_o depends on the curvature and strain rates over the flamelet surface. The influence of strain rate is accomodated through a presumed pdf of strain rates obtained from an assumed lognormal distribution of dissipation rates. It is possible to use a "library" of strained flame calculations including detailed description of reaction mechanisms and molecular transport to calculate I_o. It is possible to set up transport equation for the specific flamelet area Σ. We will here only show an algebraic expression which is based upon square wave spatial distribution. I_o and Σ may thus be

$$I_o = I_o(\bar{\varepsilon},\sigma)\ ;\ \ \Sigma = \frac{g\bar{c}(1\bar{c})}{\bar{\sigma}_y L_y}\ ;\ \ L_y = C_L l\left(\frac{u^o_{lam}}{u'}\right)^n$$

expressed as:
Here ε is the mean dissipation, σ is the standard deviation of the lognormal distribution, g is a number of order unity, L_y is the integral length scale of the square wave and σ_y is the mean direction of a direction cosine defining the flamelet orientation realtive to the c=constant contour; C_L and n are constants for a given flame. The resultant rate of reaction of c may be written as:

$$S_c = 12.15\rho_r I_0(K)\frac{(1+\tau)c(1c)}{(1+\tau c)^2}\frac{\varepsilon}{k} \quad ; \quad I_0(K) = \frac{0.117}{1+\tau}K^{0.784}$$

Here the constants have been given the values: n=1, C_l=1, g=1.5, σ_y=0.5, σ =2.0. τ is the heat release parameter defined through $T_p=T_r(1+\tau)$. T is temperature, p is product and r is reactants. The Karlowitz number, K and the expression for I_o are deduced from Eq. 10. It should be noted that the above expression is the same as the eddy-break-up expression modified by the factor $I_0(K)$ where chemical influences comes through this factor.

3. Idealized geometries (MERGE/EMERGE)

Background Several European research institutions had acknowledged that a fundamental approach for assessing and improving the capabilities of the CFD codes were badly needed. Therefore the socalled MERGE and EMERGE programmes were launched. MERGE is an acronym for Modeling and Experimental Research into Gas Explosions and EMERGE stands for Extended MERGE. The following 9 research groups participated: TNO - Prins Maurits Lab. (TNO), The Netherlands; British Gas (BG), UK; Shell Research Ltd., TRC (SRL), UK; Imperial College (ICSMT), UK; Christian Michelsen Research a.s. (CMR), Norway; Telemark Institute of Technology and Telemark Technological R&D Centre (HiT-TF/Tel-Tek), Norway; Battelle, Germany; INERIS, France; Fraunhofer ICT, Germany (only EMERGE). The work was sponsored by the CEC and MERGE lasted for two years from mid 1991 to mid 1993 (Mercx[36]) and EMERGE was carried out in 1994 and 1995. The objective of the CFD modelling part of the MERGE project was to provide simulation data for each of the experimental tests and hence be able to assess and improve the prediction quality. The CFD part of the MERGE work was divided into three phases, namely:

Phase 1. Evaluation of the various submodels of the CFD models.
Phase 2. Verification of the integrated CFD models against small- and medium scale experimental data produced in the project.
Phase 3. Prediction of the large-scale experimental data, prior to the experimental period.

Validation (Phase 2) This phase was directed towards validation of the porosity/distributed resistance (PDR) models and combustion models against explosion data in small scale (TNO) and medium scale (BG) experiments. The drag and combustion models used are taken from Popat et al.[11] and Hjertager et al.[37].

"Blind" prediction of large scale experiments (Phase 3) After the validation work (Phase 2) was completed the groups that had the CFD capability available (CMR, TNO, Tel-Tek, and BG) were asked to submit calculations of two of the large scale test geometries. The groups agreed on dates and submitted the results to the coordinator, TNO, in good time before the tests were performed (TNO submitted their predictions to BG).

After the tests were performed the measured pressure pulses were released to the modellers. In Figure 1, predictions of all the codes (EXSIM, FLACS, REAGAS and COBRA) have been compared with experiments. It can be observed that all predictions except some of the REAGAS predictions all fall within the band of a factor of two.

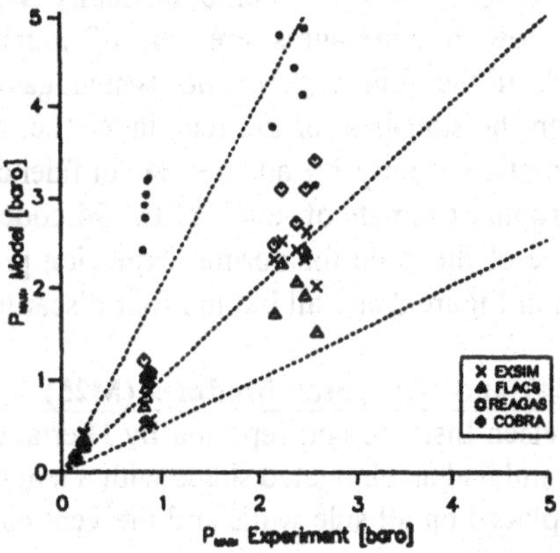

Figure 1. Summary of the comparison of the large-scale peak explosion pressure predictions and measurements of Type E and C* of the MERGE geometries[11]. EXSIM-92 code (HiT-TF/Tel-Tek), FLACS code (CMR), REAGAS code (TNO), COBRA code (BG).

4. Simulation of offshore modules

General Gas explosions occuring on offshore structures may have severe consequences as demonstrated by the Piper Alpha accident (Petrie[38]). The reasons for the severe consequences in offshore situations are mainly due to the densely packed areas that may be found on platforms. In order to enable proper design and operation of such installations, means of quantifying and subsequently reducing the hazard is a key point in the safety procedure. The oil industry and research institutions have for a long time recognised this. Large resources are invested in building up the knowledge base and tools are developed for predicting gas explosions.

Simulation cases The EXSIM-94 code have been validated against experimental data from several experimental programmes performed by well known research groups (Sæter[39] and Sæter et al.[5]). The data sets chosen include the large scale SOLVEX test series performed by Shell Research and reported by Bimson et al.[40]; the 35 m^3 module tests performed by DNV and partly reported by Cates and Samuels[39]; the 1:5 scale Compressor and Separator module tests performed by Chr. Michelsen Institute (now Christian Michelsen Research a.s.) and reported by Hjertager et al.[42] and the 1:12 scale Troll Process module tests performed by Shell Research and reported by Samuels[43]. A total of 40 cases have been used in the validation study. A summary of the results will be described in the following.

<u>**Quantification of Uncertainty between Experiments and Simulations**</u> In Figure 2 all the 40 simulation cases are compared with corresponding experimental data. The figure shows the maximum predicted overpressure as function of maximum observed overpressure. A completely ideal simulation model would place all points along the diagonal. If we use all the 40 cases we find that the EXSIM-94 code show an average underprediction of 15% and the predictions indicate that there is a 95% confidence that the mean maximum overpressure will lie within ±146% of the predicted value. It can be seen from Figure 2 that EXSIM has some difficulty with cases at very low overpressures, but very low overpressures are not of much interest in practical application of the model. If the four experiments which gave overpressures below 20mbar are excluded, then the statistics for the remainder (i.e. retain 90% of the data) show an average overprediction of only 3% and a 95% confidence interval of ±46% . It is believed that this uncertainty estimate of the EXSIM-94 code gives the most correct picture of the performance of the code for normal explosion pressure prediction. This confidence limit is drawn in Figure 2 and all but the four discarded cases fall within this proposed intervall.

<u>**CMR Compressor (M24) and Separator Modules (M25)**</u> These experiments were performed by Chr. Michelsen Institute and reported by Hjertager et al.[42]. The two test modules were both 50m^3 and had an elongated shape with a length to heigth ratio of 3.2. The vent areas could be placed on all side walls and the vent parameter $A_v/V^{2/3}$ varied from about 0.45 to about 4.8. The fuel-air clouds

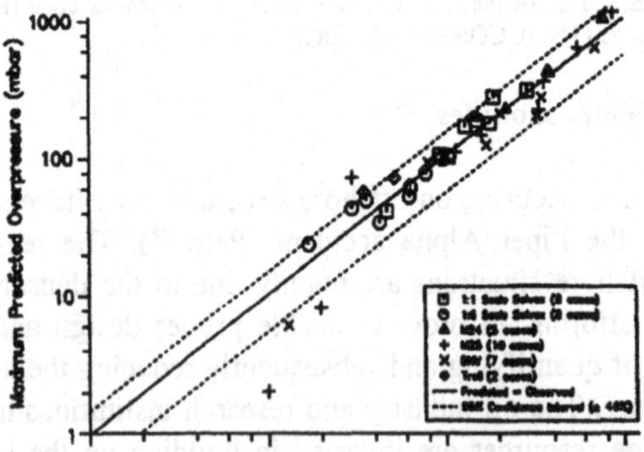

Figure 2: Maximum predicted overpressure versus Maximum observed Overpressure[5].

used were homogeneous and covered the open space inside the module. Both methane-air and propane-air clouds that were stoichiometric (equivalence ratio equal to 1.0) were used. The ignition point was varied from central ignition on either the lower or upper decks or at the end. The internal layout were replica of two different typical offshore modules, namely one Compressor (M24) and one Separator (M25) layout. In order to check the influence of grid resolution on the predicted pressures 11 different grid resolutions across the module width are tested. These range from 5 cells to 25 cells across the module width. Sæter et al[5] found that EXSIM-94 predictions are fairly grid independent. Based on this all cases were calculated using 15 cells across the module.

Influence of variable venting Predictions were done to see the variation of peak explosion pressures inside the modules for centrally ignited clouds as function of the vent parameter, $A_V/V^{2/3}$. It was noted that the predicted general trends are in good accordance with the measurements. The computer model is able to predict the following characteristic features found in the experiments: 1) the variation of peak pressures with the vent parameter and 2) the influence of internal process equipment on the violence of the explosion. Although the general trends are predicted well, it is also noted that there are discrepancies between experiments and simulations. This is especially seen for the cases with vent parameter larger than about 2.5.

Influence of scaling Figure 3 shows comparison between predicted and measured peak pressures for methane/air clouds as function of scale. The extrapolation to full scale is also included. It is noted that the predicted peak pressure increases with scale. In 1:5 scale the peak pressures were about 350 mbar, whereas the pressures for 1:1 scale is about 600 mbars.

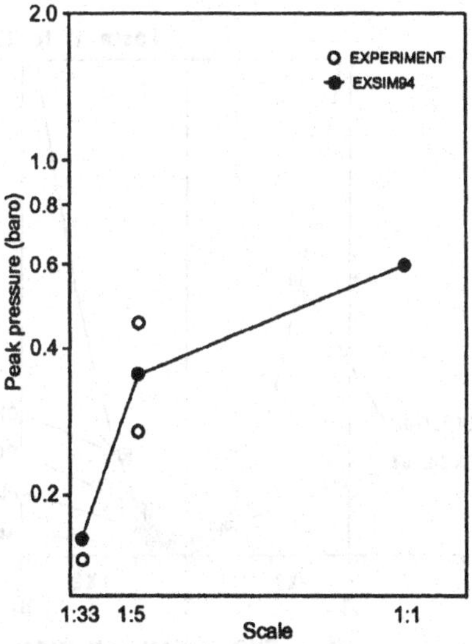

Figure 3 Peak pressure versus scale for methane-air explosions. Comparisons between measurements and simulations.

Summary of medium scale experiments The performance of the EXSIM-94 code has been quantified by predicting 40 experimental offshore cases. Good grid independence of the predicted pressure pulses have been demonstrated for two different offshore geometries. The statistical tool given by Sæter et al[5] has been applied to the 40 simulation cases performed using the EXSIM-94 code. It was found that the EXSIM code had some problems with cases at very low pressures. When the four cases with pressures lower that 20mbar were excluded in the data set, the uncertainty of the EXSIM-94 code are determined as follows: 1) the code shows on average an overprediction of 3% and 2) there is a 95% confidence that the mean maximum overpressure lie within ±46% of the predicted value.

Full-scale experiments In the last couple of years the oil industry has invested in large-scale facilities for obtaining full-scale gas explosions data relevant for offshore situations. The research programme and results are reported by Selby and Burgan[44]. A series of experiments were designed to be a scaled up versions of the compressor module tests done by Hjertager al.[42]. The interior of the modules had low density obstructions and high density obstructions. Modelers were invited to submit blind predictions in two Stages. The results in Stage I indicated that most of the models had problems with the high density cases. In Stage II the modellers updated their models and again did blind predictions of three new experiments. Figure 4 gives an overview of the statistical results for the model performance for the three-dimensional CFD models AutoReaGas, EXSIM-94 and FLACS, together with some simplified models (EXPLODE, CHAOS, COMEX+NVBANG) for the three final experiments. It is seen that the CFD models are closest to the point (1.0, 1.0), which is the point for a perfect model.

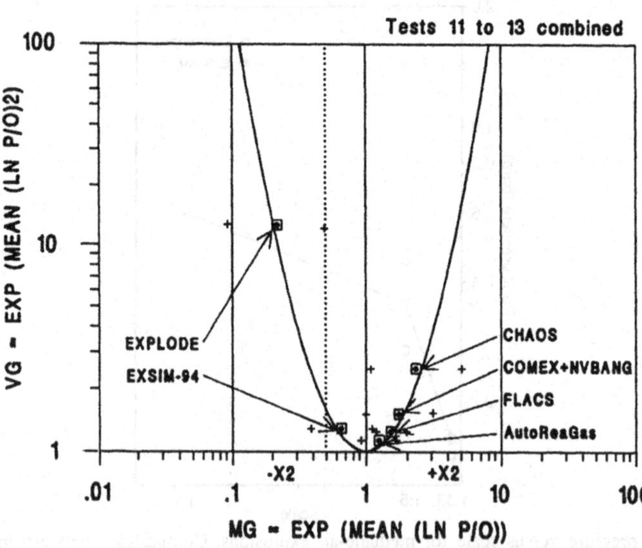

Figure 4 Geometric mean (MG) and Geometric variance (VG) for model performance for maximum overpressure prediction. Tests 11, 12, 13 (from Selby and Burgan[44])

5. Concluding remarks

Computer models capable of analysing the processes which occur in turbulent gas explosions inside complex congested geometries were presented. Several computations were reported which compare the computer models against several sets of experimental data relevant for offshore situations. The agreement between predictions and measurements is in general good. Although the status of many 3D codes makes it possible to carry out detailed scenario calculations, further work is needed: 1) to develop and verify the porosity/distributed resistance model for explosion propagation in high density obstacle fields; 2) to improve the turbulent combustion model especially for subgrid application and 3) to develop a model for deflagration to detonation transition. More experimental data are needed to enable verification of the model in high-density geometries using homogeneous filling of the whole or part of the obstructed volume as well as non-homogeneous fuel/air clouds and to validate model predictions at large and full scale. This is particularly needed for onshore process plant geometries.

6. Acknowledgement

The work on gas explosions at AAUE/Tel-Tek is financially supported by Shell Research Ltd and was supportet by the Commision of the European Communities (CEC) STEP and ENVIRONMENT programmes.

7. References

1. Hjertager, B.H.: Simulation of transient compressible turbulent reactive flows, Comb. Sc. and Techn., 41, pp. 159-170, 1982.
2. Hjertager, B.H.: Numerical simulation of flame and pressure development in gas explosions, SM study No. 16, University of Waterloo Press, Ontario, Canada, pp 407-426, 1982.
3. Hjertager, B.H.: Computer modelling of turbulent gas explosions in complex 2D and 3D geometries, J. Hazardous Materials, Vol. 34, pp 173-197, 1993.
4. Hjertager, B.H., Solberg, T., Nymoen, K.O.: Computer modelling of gas explosion propagation in offshore modules, J. Loss Prev. Process Ind., Vol 5, No 3, pp 165-174, 1992.
5. Sæter, O., Solberg, T., Hjertager, B.H.: Validation of the EXSIM-94 gas explosion simulator, 4th Int. Conf. on Offshore Structures - Hazards, Safety and Engineering Working in the New Era, London, 12-13 Dec., 1995.
6. van den Berg, A.C, van Wingerden, C.J.M. and The, H.G.: Vapor cloud explosion blast modeling, in International conference and workshop on modeling and mitigating the consequences of accidental releases of hazardous material, arranged by CCPS/AIChE, New Orleans, USA, May 20-24, pp. 543-562, 1991.
7. van Wingerden, C.J.M., Storvik, I, Arntzen, B., Teigland, R., Bakke, J.R., Sand , I.Ø. and Sørheim, H.-R.: FLACS-93, a new explosion simulator, 2nd Int. Conf. on Offshore Structural Design against Extreme loads, pp. 5.2.1-5.2.14, 1993
8. van Wingerden, C.J.M., Hansen, O.R., Storvik, I.: On the validation of a numerical tool used for eplosion and dispersion predictions in the offshore industry, BHR Conference Series Publ., Vol 15, pp 201-219, 1995.
9. van Wingerden, C.J.M., Hansen, O.R, Teigland, R.: Prediction of the strength of blast waves in the surroundings of vented offshore modules, 4th Int. Conf. on Offshore Structures -Hazards, Safety and Engineering Working in the New Era, London, England, 12-13 December, 1995.

10. Catlin, C.A., Fairweather, M, Ibrahim, S.S.: Predictions of turbulent premixed flame propagation in explosion tubes, Comb. and Flame, Vol. 102, pp 115-128, 1995.
11. Popat, N.R., Catlin, C.A., Arntzen, B.J., Hjertager, B.H., Solberg, T., Sæter, O., Lindstedt, R.P., van den Berg, A.C.: Investigations to improve and assess the accuracy of computational fluid dynamic (CFD) based explosion models, J. Haz. Mat., Vol. 45, pp 1-25, 1996.
12. Guidelines for model developers, Model evaluation group set up by the CEC, DGXII, May 1994.
13. Model evaluation protocol, issued by the Model evaluation group set up by the CEC, DGXII, May 1994.
14. Gas explosion model evaluation protocol, presented at Seminar on Evaluation of Models for Gas Explosions, Imperial College, London, April 18-19, 1996.
15. Patankar, S.V, Spalding, D.B.: A calculation procedure for the transient and steady-state behavior of shell-and-tube heat exchangers, in Heat Exchangers: Design and Theory Sourcebook, edited by N.H. Afgan and E.V. Schlünder, McGraw-Hill, pp 155-176, 1974.
16. Sha, W.T., Yang, C.I., Kao, T.T, Cho, S.M.; Multi-dimensional numerical modeling of heat exchangers, Journal of Heat Transfer, 104, pp 417-425, 1982.
17. Sha, W.T., Launder, B.E.: A model for turbulent momentum and heat transport in large rod bundles, ANL-77-73, 1979.
18. Bakke, J.R, Hjertager, B.H.: Quasi-laminar/turbulent combustion modelling, real cloud generation and boundary conditions in the FLACS-ICE code, CMI No. 865402-2, 1986. also in Bakke's dr.scient. thesis "Numerical simulation of gas explosions in two-dimensional geometries", University of Bergen, Bergen, 1986.
19. Bakke, J.R., Hjertager, B.H.: The effect of explosion venting in obstructed channels, in Modeling and simulation in engineering, Elsevier Science Publication, pp. 237-241, 1986.
20. Bakke, J.R., Hjertager, B.H.: The effect of explosion venting in empty volumes, Int. Journal of numerical methods in engineering, 24, pp 129-140, 1987.
21. Kjäldman, L., Huhtanen, R.: Numerical simulation of vapour cloud and dust explosions, in Numerical Simulation of Fluid Flow and Heat/Mass Transfer Processes, Vol. 18, Lecture Notes in Eng., pp. 148-158, 1986.
22. Marx, K.D., Lee, J.H.S, Cummings, J.C.: Modeling of flame acceleration in tubes with obstacles, Proc. of 11th IMACS World Congress on Simulation and Scientific Computation, Vol. 5, pp. 13-16, 1985.
23. Martin, D.: Some calculations using the two-dimensional turbulent combustion code FLARE, SRD Report R373, UK Atomic Energy Authority, 1986.
24. van den Berg, A.C.: REAGAS - a code for numerical simulation of 2-D reactive gas dynamics in gas explosions, PML-TNO Report PML 1989-IN48, 1989.
25. Launder, B.E., Spalding, D.B.: The numerical computation of turbulent flows, Computer methods in applied mechanics and engineering, 3, pp 269-289, 1974.
26. Magnussen, B.F., Hjertager, B.H.: On the mathematical modeling of turbulent combustion with special emphasis on soot formation and combustion, 16th Symp.(Int) on combustion, pp. 719-729, 1976.
27. Patankar, S.V., Spalding, D.B: A calculation procedure for heat, mass and momentum transfer in three-dimensional parabolic flows, Int. J. Heat and Mass Transfer, 15, pp 1787-1806, 1972.
28. Spalding, D.B.: A general purpose computer program for multi-dimensional one- and two-phase flow, Mathematics and Computers in Simulation, IMACS, XXII, pp. 267-276, 1981.
29. Boris, J.P., Book, D.L.: Flux-corrected transport I, J. Comp. Phys., 11, pp. 38, 1973.
30. Gudonov, S.K., Math. Sb. Vol 47, pp. 271-290, 1959.
31. Connel, I.J., Watterson, J.K., Savill, A.M., Dawes, A.M., Bray, K.N.C: An unstructured adaptive mesh CFD-approach to predicting premixed methane-air explosions, proc. of 2nd Int. Specialist Meeting on Fuel-Air Explosions, Christian Michelsen Research a.s., Bergen, June 26-28, 1996.
32. Bray, K.N.C.: Studies of the turbulent burning velocity, Proc. R., Soc. London, Vol. 431, pp. 315-355, 1990.
33. van den Berg, A.C.: Private communication, 1992.
34. Catlin, C.A., Lindstedt, R.P.: Comb. and Flame, Vol 85, pp 427-439, 1991.
35. Gulder, O.L.: Turbulent premixed flame propagation models for different combustion regimes, 23rd Symp (Int.) on Combustion, Combustion Institute, Pittsburg, pp 743-750, 1990.
36. Mercx, P: Final report on the MERGE project, 1994.
37. Hjertager, B.H., Solberg, T., Sæter, O.: CEC MERGE project: Numerical Studies - Final Report, Tel-Tek Rapport nr.: 905004-3, december 1993.
38. Petrie, J.R.: Piper Alpha technical investigation, Interim report, Dept of Energy, Sept. 1988.

39. Sæter, O.: Dr.Ing. thesis, Telemark Institute of Technology, to appear, 1998.
40. Bimson, S.J., Bull, D.C., Cresswell, T.M., Marks, P.R., Masters, A.P., Prothero, A., Puttock, J.S., Rowson, J.J. and Samuels, B.: An experimental study of the physics of gaseous deflagration in a very large vented enclusure, Paper at the 14th Int. Coll. on the Dyn. of Expl. and Reactive Syst. , Coimbra, Portugal, 1993.
41. Cates, A.T., and Samuels, B.: A simple assessment methodology for vented explosions, J. Loss Prev. Process Ind., Vol. 4, pp. 187-196, 1991.
42. Hjertager, B.H., Fuhre, K., Bjørkhaug, M.: Gas explosion experiments in 1:33 and 1:5 scale offshore separator and compressor modules using stoichiometric homogeneous fuel/air clouds, Journal of loss prevention in the process industries, 1, pp 197-205, 1988.
43. Samuels, B.: Explosion hazard assesment of offshore modules using 1:12th scale models, in Major hazards onshore and offshore, Trans IChemE, Vol 71, part B, pp. 21-28, 1993.
44. Selby, C.A., Burgan, B.A.: Joint Industry Project on Blast and Fire Engineering for Topside Structures, Phase 2, Final Summary Report. SCI-P 253, The Steel Constructions Institute, 1998.

DETONATION HAZARDS OF GASEOUS MIXTURES

À.À.VASIL'EV
Lavrentyev Institute of Hydrodynamics SB RAS,
Novosibirsk State University
630090 Novosibirsk, Russia
gasdet@hydro.nsc.ru

The critical initiation energy E. for detonation wave (DW) is used as the basic parameter in estimate of detonation hazards. At experimental definition of Å. the blast explosion approximation is proposed to use. The similar technique is identically suitable both to the various initiators with of large distinctions in the spatial - temporary energy-release characteristics (flame igniters, electrical or laser spark, exploding wire, explosive, high-speed bullet...), and to various mixtures. The priority at theoretical Å. definition is given up to «multi-point» initiation model based on DW transverse waves collisions as the micro-initiators of a detonation («hot» points). For various fuel-oxygen and fuel-air mixtures the dependence of critical initiation energy Å. on percentage of fuel in a mixture is determined, that allows to find out comparative detonation hazards of various combustible systems at plane, cylindrical and spherical initiation.

1. Introduction.

A great number of gas-dynamics, chemical and physical parameters of chemically active systems are important in hazard problem. The main purpose of this paper is demonstration of correct experimental and calculation methods for estimation of explosive and detonation hazards.

The consequences of large-scale accidents at emergency explosions of gaseous fuels directly depend on characteristic modes of a mixture burning: from low-speed (around m/s) laminar or turbulent flame up to high-speed (km/s) detonation. Of course, there are many other important factors such as geometric form of fuel-air charge, mixture homogeneity, different obstacles, etc. The main hazard parameters can be calculated using classical detonation theory.

It is well known that conservation laws of mass, impulse and energy

$$\rho_0 u_0 = \rho u$$
$$P_0 + \rho_0 u_0^2 = P + \rho u^2$$
$$H_0 + u_0^2 / 2 = H + u^2 / 2 + Q$$

V.E. Zarko et al. (eds.), Prevention of Hazardous Fires and Explosions, 93–108.

have two types of stationary (self-sustaining) solutions, namely, subsonic deflagration (point F in Fig.1) and supersonic detonation (point D) (ρ - density, u - flow velocity, P - pressure, H - thermodynamic part of enthalpy, Q - chemical energy-release of a mixture, the index 0 relates to initial condition).

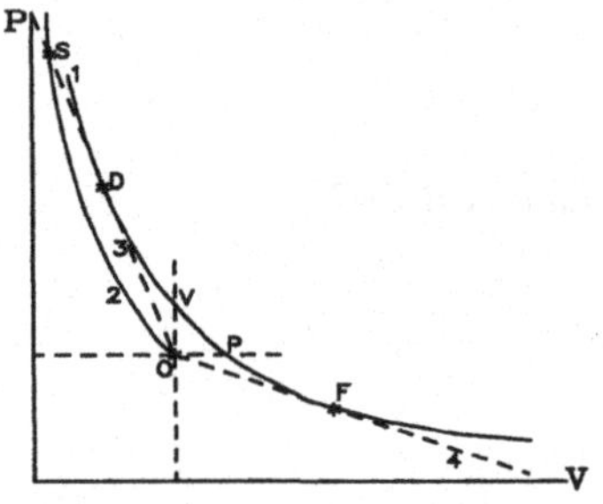

Fig. 1

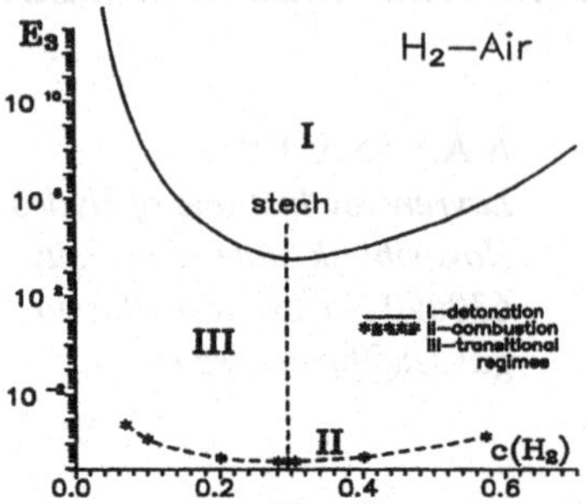

Fig. 2

The final states of reacting system (with chemical energy-release Q) are represented by adiabatic curve 1. The F and D are the points of a contact of straight lines 3 and 4 with adiabatic curve 1 (the lines start from the point O corresponding to initial condition. On adiabatic curve 1 there are two characteristic points, which are important at estimations of fire- and explosion hazard: Ð - the combustion at constant pressure, V - instant explosion in a volume. The distinction in values of basic gas-dynamics parameters of reaction products in points F,P,V and D determines the character of accident consequences with gaseous fuels. At estimations of danger of explosions it is necessary to remember that the detonation wave (DW) except a point D for detonation products is characterised by a point S - condition at DW front prior to the beginning of chemical reaction (shock adiabatic curve 2 for DW chemical spike).

It is necessary to remind that the real DW in gaseous mixtures essentially differs from that predicted by the plane classical model. Real DW represents multifront pulsing system which consists of shock waves, rarefaction waves, contact discontinuous, local zones of chemical reaction, etc.(for example, see [1-2]). Despite heterogeneity and non-stationarity the certain ordered structure with characteristic linear scale - so-called size of an elementary cell *a* - is characteristic of DW.

According to modern classification the excitation of fast chemical reaction can be carried out by three basic ways: 1) - the weak initiation when the initiator only ignites a mixture and low-speed combustion (deflagration) occurs; 2) - the intermediate regimes when mixture is ignited on initial stage only and then due to natural or artificial acceleration of a flame front the subsequent transition from deflagration to detonation (DDT) proceeds; 3) the strong initiation when self-sustaining DW is formed in immediate proximity from a initiation point. In the first case the critical flame initiation energy E_{flame} is used as the main hazard parameter traditionally (Fig.2 - dashed line II with the minimal energy of ignition $Å_{flame} = 1.7\ 10^{-5}$ J (for a hydrogen-air mixture at spherical initiation [3]). For the second case the quantitative criterion of DDT conditions still has not been formulated because of multi-parameter nature of DDT phenomenon. The initiation of detonation wave is multi-parameter process too, but for «ideal» (see below) initiators only one parameter - the critical initiation energy $Å_*$ for detonation may be considered as a measure of explosion hazard: the less $Å_*$, the more hazardous a combustible mixture (see Fig.2 - solid line with the minimal energy of DW initiation $Å_* = 3.35\ 10^3$ J [4]). The «ideality» criteria is a necessary condition for correct experimental comparison between power capabilities of the various initiators by virtue of its distinction in spatial - temporary characteristics of energy-release.

Among basic tasks of detonation hazard evaluation it is necessary to mention the following:

1) problem of energy and power equivalence of various initiators (explosive charges, high-velocity bullet, electrical or laser spark, etc.);
2) correct definition of $Å_*$ with account of temporary $Å_t$ and spatial $Å_r$ characteristics of energy-supply $Å = Å(t,r)$;
3) the $Å_*$ optimisation ways from the explosive- and ecological safety points of view;
4) classification of various mixtures by a degree of detonation hazards (the degree of mixture activity in dependence of $Å_*$ or a).

2. Experimental measurement of critical initiation energy.

The range of critical energies for various fuel - oxygen (FOE) and fuel - air (FAE) explosive mixtures is very large. Therefore the only individual type of initiator cannot be used the DW process triggering for all cases. As a rule, each initiator has its specific features of energy-release. Consequently creation of simple and universal technique for determining a source effective energy (which must be suitable for various initiators as well as for various FOE and FAE) becomes a prime task.

At initiation the explosive mixture absorbs during some finite time interval t_0 in finite space domain V_0 some quantity of energy $Å_v$ released by the initiator (which is only η-portion of the $Å_{00}$ energy originally stored in the initiator):

$$E_v = \int_0^{t_0}\int_0^{V_0} \varepsilon(t,V)\cdot dt\cdot dV = \eta E_{00},$$

where $\varepsilon(V, t)$ is a function describing spatial - temporary law of energy-release.

From the formal mathematical point of view this integrate equality represents typical functional of two variable in a variation task about optimisation of energy $Å_v$ at movable limits of integration. At mutual influence of the temporal and spatial energy-release factors the requirement of minimisation of initiation energy up to value $Å_{min}$ is reduced to minimisation of power density up to $\varepsilon_{min} = \varepsilon(r_*, t_*)$. At the account only of temporal factor (V_0 = const) it is required simultaneously with $Å_{min}$ to optimise the initiator power $\varepsilon_{min} = \varepsilon(t_*)$. Last conclusion is confirmed experimentally in [5] at initiation of a mixture $Ñ_2Í_2 + 2.5Î_2$ at variation only of the electrical discharge duration (at the fixed pressure). It was revealed that if time of discharge $t_0 \le t_*$ the required for initiation energy $Å_t = \text{const} \cong Å_{min}$ while at $t_0 > t_*$ («stretching» discharge) $Å_t$ exceeds $Å_{min}$ and grows with increase t_0 (Fig. 3). The value $Å_{min}$ is accepted as the critical energy $Å_*$ of a detonation initiation. The temporal parameter t_* is individual for each mixture and depends on initial pressure P_0. The condition $t_0 \le t_*$ represents for the given mixture a criterion for instantaneous initiator. In this case excitation of multifront DW is determined only by energy.

The influence of spatial components of energy-release Å$_r$ is reduced to optimisation of density of supplied energy $\varepsilon_{min} = \varepsilon(r_*)$ simultaneously with Å$_{min}$. In other words, the optimality of each mixture is characterised by some temporary t_* and spatial r_* scales of initiation.

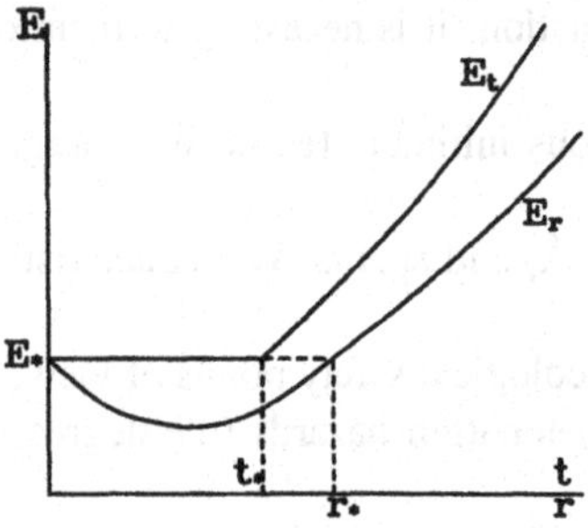

Fig. 3

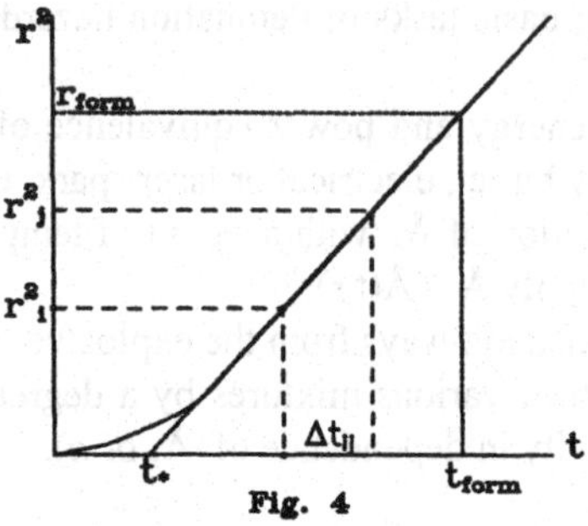

Fig. 4

Thus, correct experimental determining Å* for each individual initiator requires the control of spatial - temporary characteristics of energy-release. It appreciably complicates measurements, however without information about Å$_t$ and Å$_r$ the error in Å* can achieve several orders of magnitude. Unfortunately, understanding of this is not typical for the combustion community. For example, till now the energy of self-ignition is defined as a difference between initial and final energy of the condenser battery - Å$_{flame}$ = 0.5Ñ (U_0^2 - U_f^2) - without the account of losses in *RLC*-elements of electrical circuit.

In order to elaborate an universal approach in [6] the experimental method to determine the energy absorbed by mixture was developed and used in practice. The method is based on processing the trajectory r(t) of an explosive wave excited by initiator, from the point of view of strong explosion model (at an initial stage of a movement) [7]:

$$r(t) = \left(E_0 / \alpha_\nu \rho_0\right)^{1/(\nu+2)} t^{2/(\nu+2)}$$

Here E_0 - energy of explosion, ρ_0 - initial density, ν=1,2,3 for plane, cylindrical and spherical symmetry, α_ν - parameter dependent from ν and an adiabatic index γ. The effective initiator energy E_0 is given by

$$E_0 = \alpha_\nu \rho_0 \left[\left(r_i^{(\nu+2)/2} - r_j^{(\nu+2)/2}\right) / \Delta t_{ij}\right]^2,$$

where r_i and r_j are the co-ordinates of a shock wave in the moments divided by time interval Δt_{ij} (Fig.4). If for an explosive mixture 2Í$_2$ + Î$_2$ a critical initiation regime (simultaneous influence of initiator and internal chemical energy-release) is observed at some pressure Ð*, the effective energy only of the initiator at given Ð* is defined on a trajectory r(t) in an inert mixture 2Í$_2$ + N_2. It ensures similarity of densities ρ_0 and adiabatic indexes γ_0, that is condition of similarity of mixture parameters and their profiles in the DW formation area. The $\AA_0$ value, determined in such inert mixture, represents the critical initiation energy of a multifront detonation $\AA_*$.

The approach described allows to determine adequately the effective energy of the various initiators and to establish the power equivalents between them. The validity of the approach has been checked up for such initiators as the electrical discharge, exploding wire, explosive charges. It was found that under identical conditions (mixture composition and initial pressure, initiation geometry) the results of experiments correlate satisfactory with the authentic data of other researchers [4].

3. Peculiarities of near-critical modes of DW initiation.

In Fig.5 two typical smoke prints for cylindrical DW initiation are presented. They demonstrate a threshold character of critical energy $\AA_*$: with initiator energy $\AA<\AA_*$ the detonation burning ceases, and with $\AA\geq\AA_*$ a self-sustaining DW is formed. The similar conclusion follows from examination of a cellular picture of prints: cell structure is typical only for detonation process while an ordinary flame has not such structure.

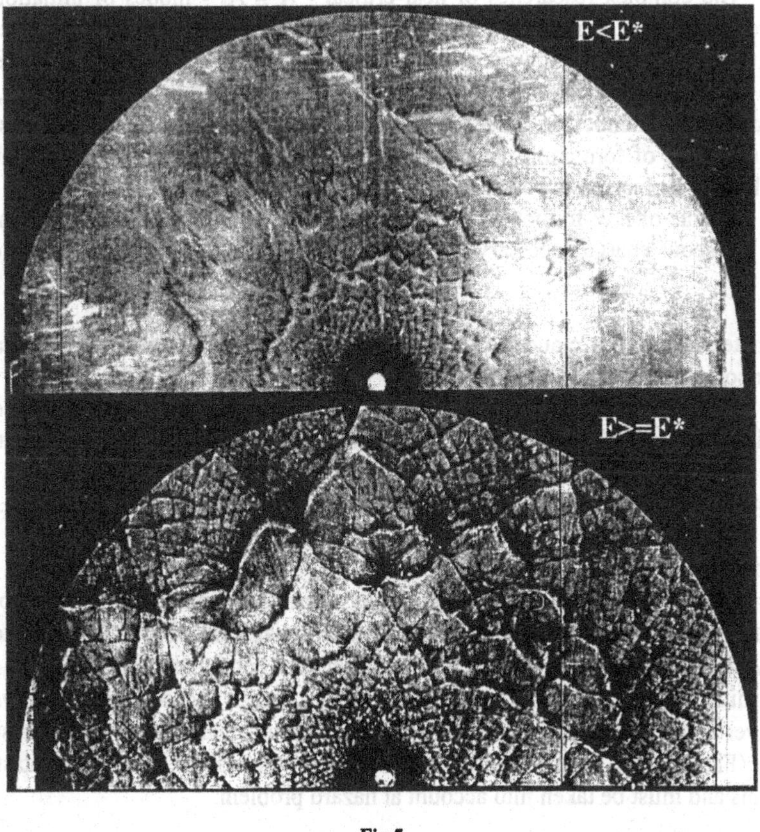

Fig.5

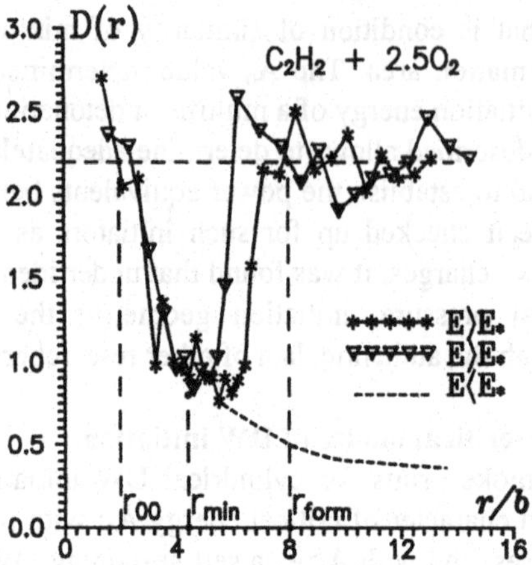

Fig. 6

Fig. 6 Velocity dependence D

The complex researches of near-critical - Å ≅ Å• - modes of initiation have revealed some peculiarities in DW behaviour. Examining of initiating wave from radius) one may recognise a wave over-pressure at an initial stage - $D > D_0$, decrease of a wave velocity below detonation one, $D < D_0$ up to some minimal value D_{min}, sharp duplication of a number of transverse waves (TW) at the DW front after passing a minimum, stage of formation the self-sustained DW. At E<E• DW decays and only low-velocity combustion is observed.

Radius of DW formation at critical initiation regime characterises the minimal size of a gaseous charge necessary for correct determination of critical initiation energy Å•:

$$r_{form} \approx 2r_{min},$$

at diameter of gaseous cloud $d_0 < 2r_{form}$ the received E-values are underestimated in comparison with Å•.

Experimentally and with the help of calculation [8] the strong influence of an explosive charge shell on gaseous mixture initiation is found, which manifest itself in change of energy redistribution from detonation products to a shock wave in a nearest zone of charge (most critical for initiation).

4. Detonation in mixture jet.

The critical charge diameter d• is defined as minimal diameter of free (without any boundary wall) cylindrical charge for stationary propagation of DW [9]. At d<d• DW destroys and only unsteady regimes of high-velocity combustion are observed at any power initiator. So, if a diameter d of free gaseous charge exceeds the critical value d• (for investigating mixture), the DW propagates along charge axis in self-sustaining regime (Fig.7) [10]. Such gaseous charge-jets may appear in numerous catastrophic situations and must be taken into account at hazard problem.

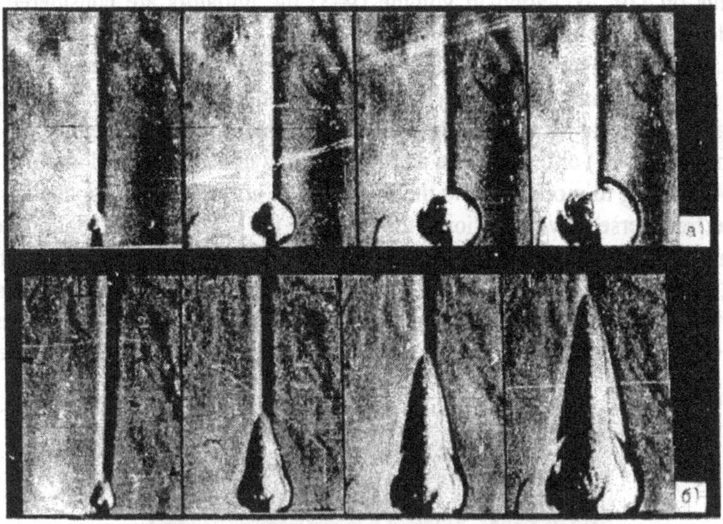

Fig.7

Approximate relation

$$d_* \approx 2.5d_{**}$$

may be proposed for estimation [10], where d_{**} is the critical diameter of diffraction re-initiation (see next paragraph).

5. Spatial initiation.

The influence of spatial distribution of energy Å_r on the value of Å_* (DW excitation by non-ideal sources) till now is theoretically investigated rather avariciously [11-12]. Only experimental effect of initiation energy reduction at replacement of the pointed electrodes of a digit interval on flange (disk) received practical application in the method for determining flame self- ignition energy [3,13].

The technique of diffraction re-initiation of a multifront detonation (DRMD) was chosen as basic one for an experimental research of a Å_r role (for example, [14-15]). The essence by this rather simple technique (widely used in laboratory researches) consists in the following: preliminarily generated in rectilinear tube of constant section the quasi-plane DW is "passing" into researched volume of an explosive mixture and is transformed to a spherical detonation, or degenerates in an attenuating blast wave. Quantitatively the diffraction re-initiation criterion (at the fixed pressure of a mixture) is formulated through a critical diameter d_{**}: at diameter of initiating tube $d < d_{**}$ attenuation is observed while at $d > d_{**}$ DW re-initiation one observes. In this case d_{**} represents itself as an energy equivalent of spherical initiation Å_{3*}.

With reference to laboratory researches only DRMD-technique allows rather simply and over a wide range to vary the geometrical sizes and form of an energy-release zone (membrane with through passage apertures of the various form) [16]. In experiments on DW excitation in volume (v=3) the initiators are considered, whose form modelled: a) a charge as a thin disk of radius R; b) a circular charge with radiuses R_1 and R_2; c) multi-point scheme as the circuit representing n separate charges of radius r, located on the certain way from each other (for example, on a circle of radius R); d) a single plane charge as a rectangular L*xl*; e) system of linear charges L*xl*, located parallel (on distance z) or under some angle (including under the closed circuit); etc..

According to modern views the main role in initiation and propagation of DW belongs to transverse wave collisions, for example [17]. The DW excitation occurs at the expense of total energy TW collisions, concentrated on a surface *S* of an initiating wave, i.e. $Å_{3*} \sim S$. The *S* constancy at a variation of the initiator form provides spatial redistribution of inputted energy. The ratio of critical pressures $Ð_{*i}$ or areas (at identical $Ð_{*i}$) may be used as parameter of efficiency *g*: the less $Ð_{*i}$, the higher *g* (or greater *g* corresponds smaller *S*).

Fig.8

The results of investigations unequivocally testify that the influence of the spatial factor $Å_r$ is inadequate to temporary factor $Å_t$. Instead of constancy level $Å_t$ = const = $Å_*$ in area $t_0 \leq t_*$ (see page 3) the dependence $Å_r$ has the reliably fixed U-figurative form with the minimal value $Å_{min} < Å_*$ at some optimum ratio of the geometrical sizes of the initiator and some chemical-physics characteristics of a mixture (Fig. 3). The greatest efficiency *g* of spatial distributing initiators (in comparison with the concentrated charge) is typical for the circular initiators, linear as the closed triangle and multi-pointed... At an optimum ratio *g* is increased up to the order.

The photograph 8 illustrates the multi-point initiation of cylindrical DW at essential smaller pressure of a mixture as compared with «by the homogeneous initiator» ($Å_* \sim 1/P_0$). Collective interactions of shock waves from the local micro-initiators play the important role in the similar re-initiation schemes. The mathematical modelling of two-dimensional dynamics of circular initiation [18] has revealed an effect of energy cumulation on a charge axis and formation of the original new initiator which «works» with some delay relatively primary explosion. Under certain conditions such two-stage initiation scheme appears more effective, than single-cycle explosion of a uniform charge as a disk. The calculation results qualitatively prove to be true experimental.

6. Transition of deflagration to detonation and its optimisation.

The more essential effect of spatial energy-release influence is established for cases of weak and intermediate initiation: mixture firing (deflagration) with the subsequent transition of deflagration to detonation. The standard criterion of explosion hazard at weak initiation till now is not formulated, that is caused by stochastic character of DDT and by multi-parametric physical pattern of the DDT phenomenon.

Quasi-plane flames were investigated ([17]) in tubes of constant cross-section with diameter d = 20÷250 mm for acetylene, hydrogen, ethylene, propane, methane, acetone steam, petrol... in mixtures with oxygen or air and at various schemes of artificial turbulization of a flame front: depending on type of turbulence elements, blocking degree of tube cross-section, laws of turbulence elements distribution in tube section and its axis. The analysis of the research results has allowed to develop highly effective multi-section DDT-accelerator for quasi-plane flames and to optimise its design in such extent that on laboratory scales it became possible to perform transition of deflagration to self-sustained DW for such hard-initiated mixture like methane-air. The DDT-accelerator allows to avoid using powerful explosive charges as the initiator and to transfer research on testing FAE and their classification in laboratory instead of proving ground conditions.

Intensification of burning in expanding (ν=2,3) flames was carried out not only at the expense of artificial turbulence of a flame at its propagation through turbulent arrangements, but also at the expense of spatial redistribution of ignition sources. The results of research have shown that DDT for an expanding flame can be successfully realised via multi-point initiation schemes even in case when single (total) ignitor provides only propagation of a flame front without appreciable acceleration. The spatial variation of the ignition centres allows under certain conditions to reduce the DDT-transition length considerably (by the orders of magnitude). The elementary schemes of such ignition and transition of deflagration to detonation have been investigated, their comparative efficiency is given in [17].

7. Initiation by high-speed bullet.

Using the aerodynamics hypothesis, known as «plane sections» (for example, [19-20]), in [17] the criterion of detonation excitation by a high-velocity bullet (HVB) is offered: the work of aerodynamic resistance forces on length unit of a HVB trajectory in explosive mixture should exceed the minimal energy of cylindrical initiation of a multifront detonation -

$$c_x \cdot \rho_0 w^2 \cdot \pi d^2 / 8 \geq \beta E_{2*} \equiv \beta \cdot A_2 \rho_0 D_0^2 b^2$$

Here c_x - factor of aerodynamic resistance HVB, ρ_0 - density of a mixture, w - relative velocity of HVB and mixture, d - midsection diameter (maximal section of a body in a plane, perpendicular direction of flight), *b* - longitudinal cell size of a multifront DW, β

- effectiveness ratio of spatial excitation, $A_2 = f(\gamma, E_{act}/Q, \varepsilon, \alpha_2)$ - coefficient from «multi-point» initiation model (see below). After transformations the following expression is obtained

$$d/b \geq \sqrt{8A_2\beta/(\pi c_x)} \cdot D_0/w$$

The above expression gives interrelation between HVB aerodynamic characteristics and physical-chemical parameters of an explosive mixture. For particular mixture and fixed pressure the values D_0, A_2 and b are constant, therefore, if HVB during flight does not change the form and orientation (c_x = const), the criterion of detonation initiation has a simple form

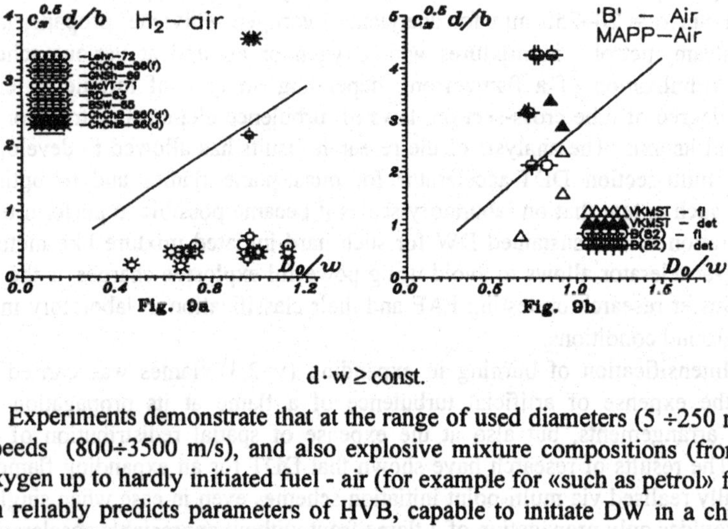

Fig. 9a Fig. 9b

$$d \cdot w \geq \text{const}.$$

Experiments demonstrate that at the range of used diameters (5 ÷250 mm) and HVB speeds (800÷3500 m/s), and also explosive mixture compositions (from active fuel - oxygen up to hardly initiated fuel - air (for example for «such as petrol» fuel)) the criterion reliably predicts parameters of HVB, capable to initiate DW in a chemically active mixtures [17,21-23] (Fig.9 for FAE, for example).

8. Detonation cell models.

At present some cell models are known and can be used for estimation of detonation cell sizes and construction of hazard raw of combustible mixtures. The first type - numerical modelling (for example - computer code program of Naval Research Laboratory (USA) for two-dimensional channel with constant cross-section, [24]). Second type - "cell" models for ideal DW. For example, in [25] the induction zone length $l_{10} = (D_0 - u_{10})\tau_{10}$ is calculated for DW. After that cell size a is calculated by $a = k\, l_{10} = 29\, l_{10}$ (τ - induction period, index 10 corresponds to induction zone parameters for ideal Chapman-Jouguet DW). Constancy of a/l_{10} for different mixtures and pressures is the main assumption of [25], and it may be subjected to critics. In [26] for rough tube the minimal diameter d_{limit} is calculated for DW propagation and a is defined as $a = d_{limit}/\pi$. Third type - cell model [27] for real multifront DW, where longitudinal cell size b is calculated from the integral relationship

$$b^{-1} = \int_{x}^{1+y} \frac{d\zeta}{D(\zeta)\cdot\tau_{st}(\zeta)},$$

where ζ=r/b, x=r*/b... (see [27]). It must be mentioned that the last relation may be used in reverse problem for determination of τ_{st} and average kinetic constants for induction zone.

9. Multi-points model of detonation initiation.

The numerical calculations within the framework of one-dimensional DW model (for example, [11]), describe in detail a qualitative picture of one-dimensional initiation: attenuation of an initiating wave at initiator energy Å of smaller than critical value Å* and DW formation at Å ≥ Å*. In case of stoichiometric hydrogen-air mixture within the framework of «detailed» kinetics model the quantitative agreement between calculating Å* and value, fixed in experiment, is obtained in [28]. Other mixtures need performing special calculations.

Up today about twenty various approximate models of one-dimensional detonation initiation are known, which all were analysed in [17,4]. Such models allow to estimate value Å* to some accuracy.

In multifront DW in each fixed moment a period (or duration) of the induction zone strongly differ (up to two orders of magnitude) for various elements of DW front. At these conditions using the same ignition delay for all front elements (as in one-dimensional models) can produce wrong initiation condition, because the ignition event is caused not by the average, but local temperature in the most «hot» point. Such points in real DW are the domains of collision of transverse waves. Taking into account similar non-one-dimensional collisions of shock-wave configurations of real detonation front it is possible to lower appreciably a level of critical energy (in comparison with one-dimensional models). Such model of multi-point initiation - MPI - was offered first in [29] and then was further developed in [30-31]. In the most completed version MPI-model is presented in [17]. According to last version (for ν= 1,2,3)

$$E_1 = \frac{16\varepsilon^2\alpha\sqrt{\pi}}{\gamma_0 - 1}\cdot\frac{E_{act}}{Q}\cdot\rho_0 D_0^2 b = A_1\cdot\rho_0 D_0^2 b,$$

$$E_2 = \frac{16\varepsilon^2\alpha\sqrt{\pi}}{\gamma_0 - 1}\cdot\frac{E_{act}}{Q}\cdot\rho_0 D_0^2 b^2 = A_2\cdot\rho_0 D_0^2 b^2,$$

$$E_3 = \frac{512\varepsilon^2\alpha\cdot \mathrm{tg}\varphi}{(\gamma_0 - 1)^2}\cdot\left(\frac{E_{act}}{Q}\right)^2\cdot\rho_0 D_0^2 b^3 = A_3\cdot\rho_0 D_0^2 b^3.$$

where $Å_{act}$ - effective activation energy of an induction period (within the framework of the average description based on the Àrrhenius equation), Q- chemical energy-release in DW, tg $\varphi = a/b$, α - parameter of strong explosion model [7].

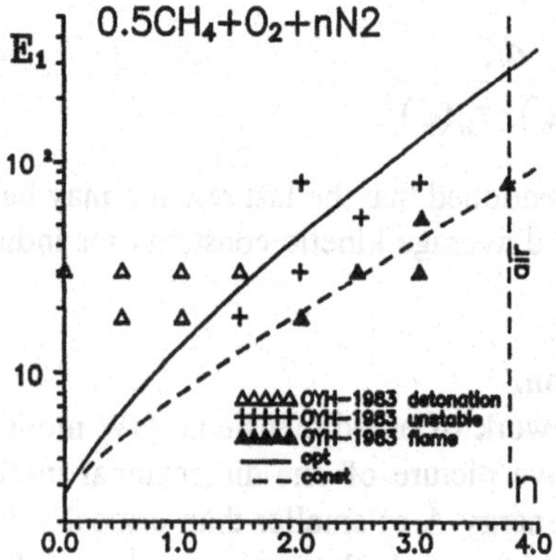

Fig. 10a

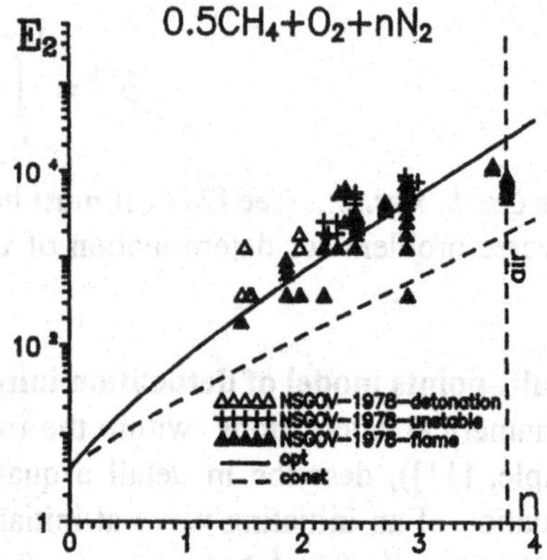

Fig. 10b

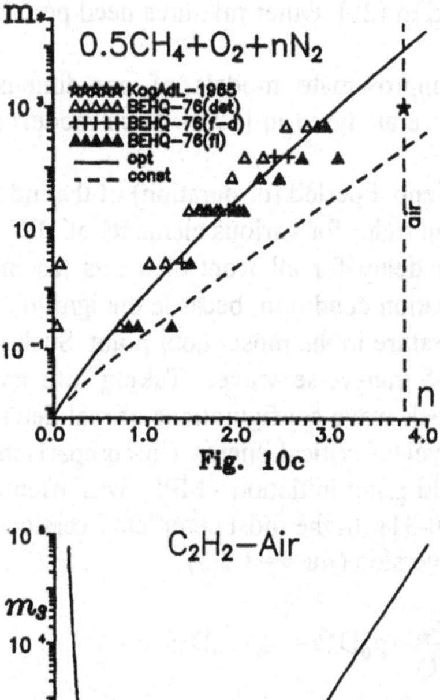

Fig. 10c

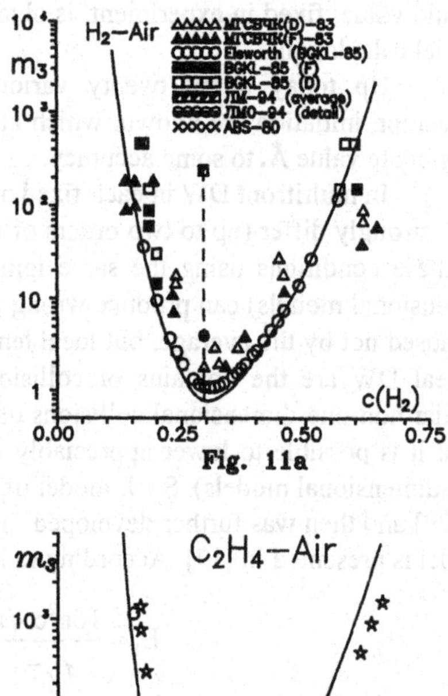

Fig. 11a

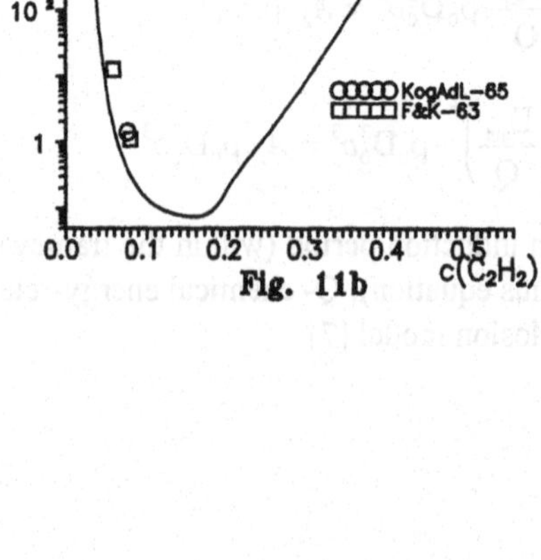

Fig. 11b

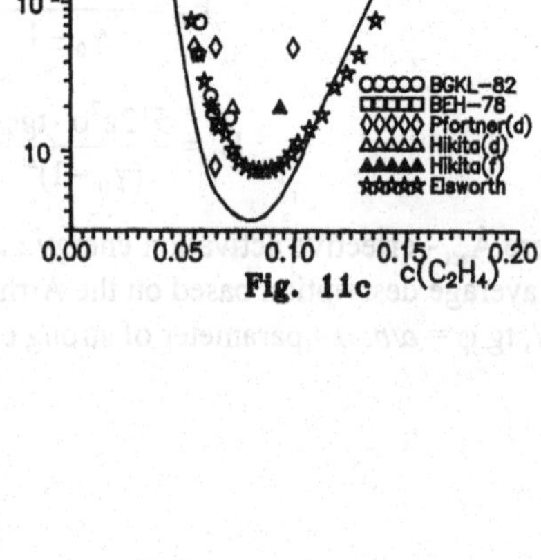

Fig. 11c

From other approximate models it is possible to recommend for estimations only few [17,4]. In [32-33] the following relationships are offered

$$R_{cr} \cong 4(\nu - 1)k^2(\sigma_{10} + 1/\sigma_{10} - 2)\lambda_{10} \cdot E_{act} / RT_{10},$$

$$E_{*\nu} = \alpha_\nu P_0 (8R_{cr} / \nu)^\nu,$$

where k = I/U - relation of thermodynamic enthalpy I and internal energy U, $\lambda_{10} = (D_0 - u_0)\tau_{10}$ - size of induction zone behind a shock wave propagating with speed of an ideal detonation D_0, σ - degree of compression (density) ratio behind such wave.

Last version [34] defines the critical energy of spherical initiation

$$E_c = 2197\pi\gamma_0 P_0 J M_{CJ}^2 a^3 / 16,$$

where Ð$_0$ - initial pressure, Ì$_{CJ}$ - Mach number of ideal DW (velocity D_0), J≅Q/(νD$_0$).

In [25] at first stage an induction zone length l_{10} is calculated with the help of model of detailed kinetics. The critical energy of spherical initiation was considered proportional l_{10}:

$$E_{*3} = B\lambda_{10}^3,$$

where coefficient B was determined from experimental data of Å$_{*3}$ for some particular mixture composition and then was considered constant for any other mixtures with given fuel.

The formulas of other models give a much more divergence with experimental data and therefore they are not presented here.

10. Detonation limits.

The spinning detonation with one transverse wave is the limiting regime of stationary DW propagation in circular tube, thus [35]

$$d_s \cong \pi a$$

For rectangular channel lxδ limiting regime of DW propagation is determined by relationship

$$l_s = (k+1)a/2\pi,$$

where k=l/δ [35]. For galloping detonation (quasi-stationary regime with power longitudinal pulsation of DW-front) and low-velocity detonation (supersonic regimes with velocities up to average velocity for galloping DW) the limiting criteria may be formulated in the near future.

11. DW diffraction

The critical channel size (ν=2) and critical tube diameter (ν=3) for diffraction re-initiation of multifront DW may be defined by the approximate formulas [17]:

$$l_{**}/a = 2/(\gamma - 1)\, E_{act}/Q, \qquad d_{**} = \sqrt{\pi} \cdot l_{**}/4.$$

The relations $l_{**}/a \approx$ const =10 and $d_{**}/a \approx$ const =13 from [2,15,25] are useful for approximate estimation.

12. Gas-dynamic detonation parameters.

For hazard evaluation task the main detonation parameters - DW velocity, pressure and temperature of products,... - are calculated by traditional methods (for ideal gas model with chemical equilibrium of products).

13. Calculating results. Comparison with experiment.

The most interesting and important parameters of a multifront detonation were determined practically using computer code "SAFETY" [36]: DW velocity, pressure, temperature, density, mass and sound speed of products, their equilibrium composition, parameters of an induction zone, cell size, critical energy of DW initiation for plane, cylindrical and spherical cases, critical parameters of diffraction re-initiation, HVB parameters, geometrical limits for DW propagation in channels of arbitrary cross-section, etc...

Due to restricted volume of the paper it is impossible to demonstrate all calculating and experimental results, so only critical initiation energies data are presented in this report. In Fig. 10, the critical initiation energy for plane E_1 and cylindrical E_2 cases and critical explosive charge weight m_{3*}, necessary for initiation of a spherical detonation, are presented for stoichiometric fuel-oxygen mixtures diluted with nitrogen [17]. The critical explosive charge weights m_{3*}, necessary for initiation of a spherical detonation in FAE, are presented in Fig. 11 at various fuel concentration [4].

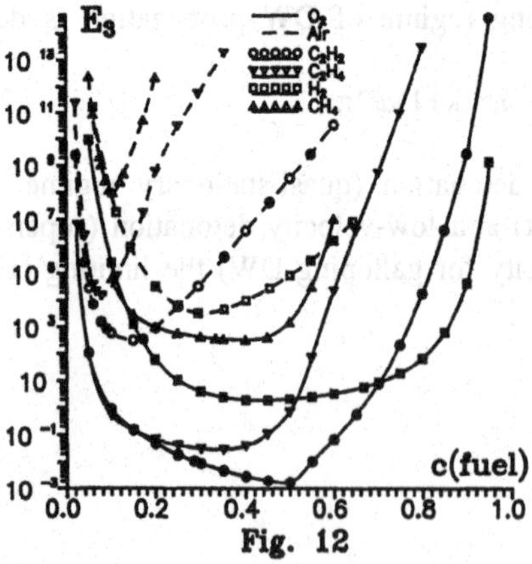

Fig. 12

Figure 12 gives a hazard efficiency graph for some typical fuels which illustrates summary data about Å$_{3*}$ (dependence for Å$_{2*}$ and Å$_{1*}$ are qualitatively similar Å$_{3*}$). For stoichiomet-ric FAE the sequence of detonation activity of various fuels (hazard raw) looks as follows: acetylene, hydrogen, ethylene oxide, ethylene, ÌÀÐÐ, ethane, propane, butane, methane. Some calculating results are presented in [4,37], other data will be published in the nearest time.

The comparison of calculating and experimental data shows their good agreement. It means that the «multi-point» initiation model for today most adequately corresponds to correct experimental data. Similar conclusion was made independently in [34]). Following on «affinity» are one-dimensional approximate models of Zhdan-Mitrofanov [32-33] and Lee («surface») [34].

14. Conclusions.

The analysis of various models and their comparison with experiment for such different in structure and chemical activity fuels like hydrogen, methane, ethylene and acetylene in mixes with oxygen and air in a wide range of various parameters (pressure, concentration of fuel, adding degree of inert gases) have shown that the available experimental data are described most adequately by the model of «multi-point» initiation.

For various fuel-oxygen and fuel-air mixtures the dependences of critical initiation energy of detonation Å$_{*}$ on fuel content in mixture are constructed that allow to establish quantitatively a degree of their detonation hazard.

15. References.

1. Shelkin K.I., Troshin Ja.K. (1963) *Gasdynamics of combustion* (in Russian), USSR Academy Sci, Ìoscow.
2. Vojtzehovsky B.V., Mitrofanov V.V., Topchian M.E. . (1963) *Structure of detonation front in gases*(in Russian), Siberian Branch USSR Academy Sci, Novosibirsk.
3. *Fire-explosion hazard of substances both materials and means of their suppression*(1990) (in Russian), reference book in 2 volumes, ed. by A.N.Baratov and À.Ja. Êîrolchenko Chemistry, Moscow.
4. Vasil'ev A.A. (1997) Gaseous fuels and detonation hazards, (N.Eisenreih eds.), *Combustion and Detonation* Proc. 28-th Fraunhofer ICT-Conference, Germany.
5. Bach G.G., Knystautas R., Lee J.H. (1970) *Initiation criteria for diverging gaseous detonation*, 13-th Symp. (International) on Combust. pp.1097-1110.
6. Vasil'ev À.À. (1983) *Research of critical initiation of a gas detonation*(in Russian), Phys. Comb. Expl. **19**, ' 1. pp.121-131.
7. Korobejnikov V.P. (1973) *Tasks of the theory of point explosion in gases* (in Russian), Science, Moscow.
8. Vasil'ev A.À., Zhdan S.À. (1981) *Parameters of a shock wave at explosion of a cylindrical charge in air* (in Russian), Phys. Comb. Expl . **17**, ' 6. pp.99-105.
9. Hariton Ju.B. (1947) *About detonability of explosives* (in Russian), in "Voprosi teorii vzrivchatih veshestv". Moscow-Leningrad. Izd. AN USSR. -Vip.1. pp.7-28.
10. Vasil'ev A.A., Zak D.V. (1986) *Detonation in gaseous jet* (in Russian), Phys.Comb.Expl. **22**,4. pp.82-88
11. Levin V.À., Ìàrkov V.V. (1975) *Occurrence of a detonation at concentrated energy input*(in Russian), Phys. Comb. Expl.. **11**, ' 4. pp.623-633.
12. Uljanitsky V.Ju. (1980) *The closed model of direct initiation of a gas detonation in view of instability. II. Undot initiation* (in Russian), Phys. Comb. Expl. **16**, ' 4. pp.79-89.
13. Matsui H., Lee J.H. (1976) *Influence of electrode geometry and spacing on the critical energy for direct initiation of spherical gaseous detonations*, Combust. Flame. **27**. pp.217-220.

14. Zeldovitch Ja.B., Êîgarko S.Ì., Simonov N.N. (1956) *An experimental research of a spherical gas detonation* (in Russian), J. Tech. Phys. **26**, ¹ 8. pp.1744-1768.
15. Ìitrofanov V.V., Soloukhin R.I. (1964) *About diffraction of amultifront detonation wave* (in Russian), Repp. USSR Academy Sci.. **159**, ¹ 5. pp.1003-1006.
16. Vasil'ev À.À. (1989) *Spatial excitation of amultifront detonation* (in Russian),Phys. Comb. Expl.. **25**, ¹ 1. pp.113-119.
17. Vasil'ev À.À (1995). *Near-critical modes of a gas detonation* (in Russian), Novosibirsk.
18. Vasil'ev À.À., Demchenko V.V. (1993) *Geometry of the initiator and explosion safety of combustible mixtures* (in Russian), Chem. Phys. **12**, ¹ 5. pp.709-711.
19. Iljushin À.À. (1956) *The law of plane sections in aerodynamics of large supersonic speeds*(in Russian), Appl. Math. Mech. **20**. pp.6-12.
20. Chernij G.G. (1959) *Current of gas with large supersonic speed* (in Russian) Ìoscow.
21. Vasil'ev À.À., Kulakov B.I., Ìitrofanov V.V., Silvestrov V.V., Òitov V.Ì. (1994) *Initiation of explosive gaseous mixtures by a high-speed body* (in Russian), Izvestija RAS. **338**, ¹ 2. pp.188-190.
22. Vasiljev A.A. (1994) *Initiation of gaseous detonation by a high speed body,* Shock Waves**3**, ¹ 4. pp.321-326.
23. Vasil'ev A.A. (1997) *Modeling of detonation combustion of gas mixtures using a high-velocity projectile* (in Russian), Phys. Comb. Expl. **33**, ¹ 5. pp.85-102.
24. Kailasanath K., Oran E.S., Boris J.PP., Young T.R. (1985) *Determination of detonation cell size and the role of transverse waves in two-dimensional detonations,* Comb. Flame. **61**. pp.199-209.
25. Westbrook C.K., Urtiew PP.A. (1982) *Chemical kinetic prediction of critical parameters in gaseous detonations,* 19-th Sympp. (International) on Combust. pp.615-623.
26. Gelfand B.E., Frolov S.M., Nettleton M.A. (1991) *Gaseous detonations - a selective review,* Prog. Energy Combust. Science. **17**. pp.327-371.
27. Vasiljev A.A., Nikolaev Ju. A. (1978) *Closed theoretical model of a detonation cell,*Acta Astr. **5**. pp.983-996.
28. Levin V.A., Markov V.V., Osinkin S.F. (1995) *Initiation of detonation in hydrogen-air mixture by spherical TNT charge* (in Russian), Phys. Comb. Expl **31**, 2. pp.91-95.
29. Vasil'ev À.À. (1978) *An estimation of initiation energy of a cylindrical detonation* (in Russian),Phys. Comb. Expl. **14**, ¹ 3. pp.154-155.
30. Vasil'ev À.À., Nikolaev Ju.À., Uljanitsky V.Ju (1979). *Critical energy of initiation of a multifront detonation* (in Russian), Phys. Comb. Expl. **15**, ¹ 6. pp.94-104.
31. Vasil'ev À.À., Grigorjev V.V. (1980) *Critical conditions of distribution of a gas detonation in sharp-divergented channels* (in Russian), Phys. Comb. Expl.. **16**, ¹ 5. pp.117-125.
32. Ìitrofanov V.V. (1983) *Some critical phenomena in detonations connected to losses of a pulse*(in Russian) , Phys. Comb. Expl.. **19**, ¹ 4. pp.169-174.
33. Zhdan S.À., Ìitrofanov V.V. (1985) *Simply model for account of energy of initiation of a heterogeneous and gas detonation* (in Russian), Phys. Comb. Expl.. **21**, ¹ 6. pp.98-103.
34. Benedick W.B., Guirao C.M., Knystautas R., Lee J.H. (1986) Critical charge for direct initiation of detonation in gaseous fuel-air mixtures, in Bowen,Leyer and Soloukhin (eds) *"Dynamics of Explosion",* v.106 of "Progress in Astronautics and Aeronautics", N.Y., pp.181-202.
35. Vasiljev A.A. (1991) The limits of stationary propagation of gaseous detonation , inBorisov A. (eds), *Dynamic structure of detonation in gaseous and dispersed media,* v.5 of "Fluid Mechanics and its applications", Kluwer Academic Publishers, Dordrecht-Boston-London, pp.27-49.
36. Vasil'ev A.A., Valishev A.I., Vasil'ev V.A., Panfilova L.V., Topchian M.E. (1997) *Parameters of detonation waves at higher pressure and temperature* (in Russian), Chemical Physics. **16**,11. pp.114-118.
37. Vasil'ev A.A. (1997) The experimental methods and calculating models for definition of the critical initiation energy of multifront detonation wave, *Proc. 16th ICDERS.* University of Mining and Metallurgy, AGH, Cracow, Poland. pp.152-155.

EVALUATION OF HAZARD OF SPRAY DETONATION

V.V. MITROFANOV, S.A. ZHDAN
Lavrentyev Institute of Hydrodynamics SD RAS,
Novosibirsk, 630090, Russia

1. Introduction

Phenomenon of propagation of self-sustaining detonation-like waves in suspensions of condensed fuel or oxidiser in atmosphere of another component is called heterogeneous detonation. In contrast to gaseous detonation being investigated since the end of XIX century, the first experimental evidences of detonation wave (DW) existence in a shock tube in suspension of kerosene droplets, diethylciclohexane and diesel fuel in gaseous oxygen were obtained early in the 1960-s [1,2].

Throughout the time an extensive cycle of experimental studies of DW propagation in shock waves was performed. Determined there, are basic detonation peculiarities in a system of fuel droplets and gaseous oxidiser (detonation velocity dependence on droplet size, mixture content, inert gas addition) and structure of stationary DW. Showed there, is the difference of character of change in a reaction zone of pressure and mixture content for heterogeneous and gaseous detonation. It has been analysed dynamics of deformation and drop break-up behind a shock wave (SW) as well as times of break-up, supersonic flowing around droplets and ignition in the reaction zone of steady DW.

Analysis of experimental and theoretical study of spray detonation in shock tubes is presented in surveys [3-6]. Formulated there, is the necessity to investigate the non-steady regimes of spray detonation appearance at different ways of initiation and to perform analysis of cylindrical and spherical detonation characteristics. It implies determination of the detonation velocity dependence on drop sizes, wave structure and initiation conditions.

One of the most important problems of heterogeneous (gas-droplets) detonation is initiating the detonation process by intensive SW formed as a result of local energy release during a short period of time. Along with experimental studying of the heterogeneous detonation initiating, the methods of mathematical simulating the above problem on the basis of physico-mathematical models can be successfully used, which describe dynamics of processes that proceed at initiating the reacting gas-droplets medium.

V.E. Zarko et al. (eds.), Prevention of Hazardous Fires and Explosions, 109–121.

2. Problem formulation

Let us consider quiescent flammable gas-suspension consisting of carrying gas phase (GP), which is oxidiser under pressure p_0 and with density ρ_0, and disperse condensed phase (c-phase), which is uniformly distributed system of poly-disperse liquid fuel droplets. At a time moment t = 0 in a point, along the line or plane of symmetry an explosion occurs, which means that the finite energy E_0 is released instantly. As a result of an explosion an intense SW is formed, which propagates in space. It is necessary to determine a motion of a gas-droplet reacting media behind the SW front and a law of its motion in further moments of time (t > 0). As a result of c-phase inertia a relaxation zone is formed behind the SW front, where under action of high-velocity gas flow droplets of c-phase start to accelerate, deform and break-up. Evaporated fuel mixes with gaseous oxidiser and chemically reacts with heat being released. At certain conditions transition to the self-sustaining spray detonation is possible.

Mathematical models of non-steady spray detonation are formulated on the basis of equations of heterogeneous medium mechanics [7] for gas-liquid mixtures at the following assumptions: distances at which the averaged flow characteristics change considerably are much higher then distances between droplets; viscosity effects are important only in the processes of gas and condensed phases interaction; eigen pressure and compressibility of drops and collisions between them can be neglected; for poly-disperse fuel sprays the distribution of droplets by sizes is described roughly by the finite number k of groups, in each of them droplets of equal sizes are consisted.

Considering that in spray of fuel droplets in a gaseous oxidiser for the oxidiser-fuel ratio, near to a stoichiometric one, a volume concentration of the c-phase is low ($\alpha_d \approx 10^{-3} \div 10^{-4}$), the system of differential equations for multi-velocity two-phase media is divided into two subsystems conjugated via right parts. The first sub-system describes the behaviour of gas-phase

$$\rho_{1,t} + r^{-\nu}(\rho_1 u_1 r^{\nu})_r = \sum_{i=2}^{k+1} \rho_i M_i ,$$

$$(\rho_1 u_1)_t + r^{-\nu}(\rho_1 u_1^2 r^{\nu})_r + p_r = \sum_{i=2}^{k+1} \rho_i (M_i u_i - F_i), \tag{1}$$

$$[\rho_1(e_1 + u_1^2/2)]_t + r^{-\nu}[\rho_1 u_1 (e_1 + u_1^2/2 + p/\rho_1) r^{\nu}]_r =$$

$$= \sum_{i=2}^{k+1} \rho_i [M_i (e_i + u_i^2/2) - u_i F_i - Q_i],$$

The second system describes the behaviour of *i*-th group of droplets

$$\rho_{i,t} + r^{-\nu}(\rho_i u_i r^{\nu})_r = -\rho_i M_i , \quad \rho_i = \alpha_i \rho^0_2 ,$$

$$(\rho_i u_i)_t + r^{-\nu}(\rho_i u_i^2 r^{\nu})_r = -\rho_i (M_i u_i - F_i), \quad i = 2,\ldots, k+1, \tag{2}$$

$$(\rho_i e_i)_t + r^{-\nu}[\rho_i u_i e_i r^{\nu}]_r = -\rho_i (M_i e_i - Q_i).$$

Here ρ_1 , u_1 , p, e_1 are the density, mass velocity, pressure, internal energy of gas phase ; ρ_i , α_i, u_i , e_i are the average density, volume concentration, mass velocity, internal energy of i-th group of c-phase; and M_i , F_i , Q_i are the intensities of the mass, force and thermal interactions of a mass unit of i-th group correspondingly. Indexes $\nu = 0,1,2$ stand for planar, cylindrical and spherical cases of symmetry correspondingly.

Let us supplement (1), (2) with equations of phases state

$$p = \rho_1 RT_1 , \quad \rho^0_2 = \text{const}, \qquad e_1 = (\tilde{n}_1 - R)T_1 + Z\,q, \; e_i = c_2 T_i + q, \quad (3)$$

where R is the universal gas constant; $\tilde{n}_1$,T_1 are the specific heat and temperature of gas phase; Z is the mass concentration of fuel vapour; c_2 ,T_i are the specific heat and temperature of i-fraction of droplets; q is the thermal effect of chemical reactions per unit mass of fuel, respectively.

Let us apply Eqs. (1-3) to detonation processes in gas-drop reacting mixtures. Note that DW in such mixtures consists of shock wave propagating in gas phase and complex relaxation zone in a fluid behind a wave. The next processes occur in this zone: acceleration of droplets, their deformation and possible break-up, phase transitions, convection and diffusion mixing of the components, chemical reactions. Part of above processes, due to variety of local conditions, can be described only approximately in the framework of definite approximate assumptions and models.

Due to the fact that when a drop is passed over by a high-velocity gas flow behind SW the thermal boundary layer is considerably thinner than the viscous one which is broken off at a drop periphery [8], a heat exchange between gas and drop can be neglected ($Q_i = 0$). Then it follows from the last equation of system (2), that along the motion trajectory of i-fraction of c-phase an internal energy of drops and their temperature do not change. The heat release in every elementary volume after ignition is a continuous function of time and it is limited by the rate of liquid phase transition into gaseous one.

In Ref. [10] at the additional simplifying assumptions: the mass interaction goes according to mechanism [8] of the boundary layer break off from drop surfaces by high-velocity gas flow; evaporation of micro drops, mixing in gas phase and chemical transformation proceed instantly with constant heat effect of chemical reactions q, - closing of the system of equations (1), (2) is performed. The problem of spot explosion in the flammable poly-disperse gas-droplet mixture with droplet sizes lower then 100 μm was numerically solved. Dynamics of transition to the mode of spherical heterogeneous detonation is determined by calculation.

Carried out in [11,12], is the generalisation of unsteady model by considering additionally deformation and break-up droplets [9] in a flow behind the SW front for mono- and poly-disperse spray. This allows to extend the range of applicability of the model until droplet size ≈1 mm and to determine DW reaction zone lengths. The latter turned out close to experimental ones. It was revealed interesting peculiarity of a flow hydrodynamics: non-monotonous behaviour of unsteady DW velocity. Decreased at the initial stage of the process, DW "slips" through value of the steady heterogeneous detonation velocity D_0 for mixture, attains minimum D_{min} at distance $\lambda_f = r_f / r^0 \approx (\nu+1)/8$, and then slowly approaches D_0 from below. Here $r^0 = (E_0 / \sigma p_0)^{1/(\nu+1)}$ is the dynamic radius, E_0 is the explosion energy, p_0 is the initial pressure in a mixture. With decreasing the spot explosion energy or with increasing droplet size in fuel spray, value of velocity minimum D_{min} decreases monotonously. Similar problem without considering break-up and drop deformation was solved later in Refs.[13, 14].

It was proposed in mathematical models [10 - 14], that behind the front of unsteady SW during all the process chemical ignition delays and combustion time in gaseous phase are small as compared with time of mechanical droplet rupture, and attention was mostly paid to an influence of mechanical relaxation and phase transitions to DW evolution. It is correct if velocity of unsteady DW $D(t) \geq D_0$. However, as it was shown in Ref. [11] on the basis of analysis of non-monotonous behaviour of unsteady DW velocity with values of D_{min} considerably less then D_0, taking into account of ignition delay in a reaction zone is necessary. Indeed, decreasing D_{min} with decreasing explosion energy or with increasing fuel drop sizes in spray causes an exponential growth of chemical ignition delays which can become comparable with time of mechanical droplet rupture and ever surpass it.

The model of unsteady heterogeneous detonation with chemical ignition delays (t_{ig}) was formulated in [15, 16]. It was supposed that behind the SW front during the induction period the energy does not release. System of equations (1) was added with equations describing chemical ignition delay and change of the fuel vapour mass concentration for this period:

$$Y_t + u_1 Y_r = - 1/ t_{ig}, \quad t_{ig} = K \ \rho_1^{\,s} p^{-l} \exp(\varepsilon / RT_1), \qquad (4)$$

$$Z_t + u_1 Z_r = (1 - Z) \sum_{i=2}^{k+1} \rho_i M_i / \rho_1 + w_1 , \qquad (5)$$

where K, s, l are the constants; ε is the activation energy; Y is the induction period portion; Z is the mass concentration of fuel vapour; Y=1 and Z=0 for the SW front. The end of induction period is characterised by the moment of time when magnitude Y according to Eq. (4) equals to zero. If Y > 0, then $w_1 = 0$, otherwise $w_1 = - K_1 \rho_1 Z \exp(- \varepsilon_1 / RT_1)$.

It was obtained from analysis of numerical solutions of the gas-droplet

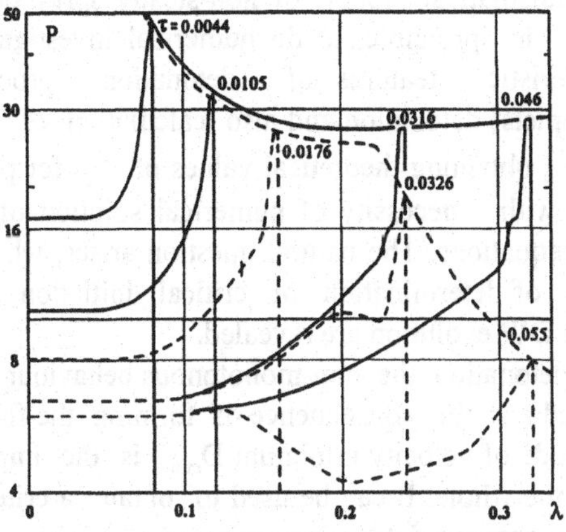

Fig. 1. Pressure dynamics of cylindrical DW in dependence on dimensionless radius $\lambda = r/r_0$.

detonation initiation problem [15, 16], that there exists a critical energy E_c. When $E_0 \approx E_c$ DW evolution considerably changes. If explosion energy $E_0 > E_c$, then DW dynamics (continuous curves in Fig.1) is similar to the one obtained in [10-14]. Wave velocity in some time after beginning of the process reaches minimum $D_{min} < D_0$, and then starts to increase, approaching value of steady detonation D_0 velocity "from below". If $E_0 < E_c$, then a point of detonation "break off" exists (see Fig. 1), that is such value of shock front radius r^*_f, that at $r_f > r^*_f$ wave damping happens (dash curves). Value of r^*_f decreases with decreasing E_0. Magnitude of E_c depends on ignition delay and fuel droplet size. It was discovered [15-18], that a problem of "breaking off" the heterogeneous detonation is solved at a dimensionless distance

$$\lambda^*_f = r^*_f / r^0 \approx (\nu+1)/8 .$$

Studied numerically [15-19], are common and characteristic features of unsteady detonation process, as well as initiation energy of spherical, cylindrical and plane gas-droplets detonation in sprays of hydrocarbon fuels with sources of different volumetric density of energy: spot explosion [15,16], explosive charge [17, 18], detonation of reacting gas mixture [19]. Generalisation of the initiation model for non-spot explosion sources allows performing the comparison of numerical calculation results with respect to critical energies and experimental data [20] in initiating the cylindrical spray detonation in mono-disperse kerosene-oxygen mixture by an explosive charge, which were performed in sector shock tube at fixed concentrations (the stoichiometric coefficient $\phi = 0,33$) and fuel drop size ($d_0 = 400$ μm). Obtained by calculations [18], are values 0,57 ÌJ/m < E_c < 0,75 ÌJ/m. The experiment gives $E_c \approx$ 0,34 ÌJ/ m. Correlation of calculated results for the one-dimensional model of heterogeneous detonation initiation and the experiment with accuracy up to coefficient 2 can be considered good, taking into account that kinetic data in fuel ignition have considerable errors.

The above mathematical models of non-steady spray detonation realised as computer codes allow, in principle, to do numerical investigation and analysis of common and characteristic features of detonation process evolution for heterogeneous (gas-droplets) detonation and also calculation of its critical initiation energies (E_c). However, obtaining theoretical values of E_c for particular gas-droplet mixture is connected with necessity of numerical solution of system of the non-stationary differential equations. The natural question arises, whether it is possible to simplify the procedure of determination of critical initiation energies, if definite features of the transient DW evolution are revealed.

At initiating detonation the non-monotonous behaviour of wave velocity in a gas-droplet reacting media is the consequence of forming the finite thickness zone of energy release. Magnitude of velocity minimum D_{min} is the important characteristic of transient process as the whole. It can be used to obtain a criterion of transition to the self-sustaining heterogeneous detonation regime.

The problem about an intense explosion in mono-disperse fuel droplet - gaseous oxidiser mixture contains two independent parameters with length dimension: $r_0 = (E_0/p_0)^{1/(\nu+1)}$ is the dynamic radius and d_0 is the droplet diameter, the velocity magnitude D_{min} depends on. Characteristic time for droplet break down [5] is $t_0 = d_0 (\rho^0_2/\rho_1)^{0,5}/u_1$. In DW moving with velocity D_0 it correlates with characteristic length

$$\lambda_0 = t_0 D_0 = d_0 (\rho^0_2/\rho_{10})^{0,5}/(\sigma_0^{0,5} - 1/\sigma_0^{0,5}),$$

where $\sigma_0 = \rho_1/\rho_{10}$ and $u_1 = D_0 (1 - 1/\sigma_0)$ are the compression degree and gas velocity behind SW, correspondingly.

To obtain quantitative dependencies of velocity minimum D_{min} of non-steady DW on the determining parameters r_0, d_0 and geometry of problem (ν) in heterogeneous medium, calculations were performed in spherical, cylindrical and plane symmetry. Analysis of calculated data on velocity minimum allows to state, that D_{min}/D_0 depends only on dimensionless complex $x = \lambda_0/r_0$ and does not depend on a problem geometry. The obtained values of D_{min} for mono-disperse sprays of hydrocarbon fuels can be presented as follows

$$D_{min}/D_0 = 1 - 69\,x. \qquad (6)$$

For poly-disperse fuel sprays consisting of k groups of droplets, the k+1 independent parameters (d_{i0} (i =2 ,...,ê+1); r_0) exist with length dimension, and it is not clear a priori, if correlation (6) is correct in this case. By comparison results of D_{min} calculations for mono- and poly-disperse sprays [21], it is possible to show that every poly-disperse spray with the Nukiyama-Tanasawa distribution function [22] can be correlated with an average diameter of drops d_{av} of mono-disperse spray, for which velocity minimum of non-steady DW D_{min} is the same as for the initial poly-disperse fuel droplet spray. An immediate test of calculated results for D_{min} showed that $d_{ave} = d_{3,2}$, where

$$d_{3,2} = \sum_{i=2}^{k+1} n_{i0}\, d_{i0}^3 / \sum_{i=2}^{k+1} n_{i0}\, d_{i0}^2 = \alpha_d / \sum_{i=2}^{k+1} \alpha_{i0}/d_{i0}.$$

3. Generalising correlation

Analysis of numerous variants of numerical solution of the problem of initiating for mono-disperse and poly-disperse sprays of liquid fuels in air or oxygen with droplets of diameter d_0=(50-700) μm showed that in all cases the critical "dynamic radius" of the direct detonation initiation (by intense shock wave) almost does not depend on space dimension:

$$r^o(\nu) = (E_{\tilde{n}\nu}/p_o)^{1/(\nu+1)} \cong \text{const} = r_c\ . \qquad (7)$$

Here $E_{c\nu}$ is the critical initiation energy, p_o is the initial pressure; ν = 0,1,2 for planar, cylindrical and spherical case of symmetry, correspondingly. The known for authors experimental data in initiating sprays of liquid hydrocarbons in air by point (concentrated) and cord charges of condensed explosive quite correlate well with Eq.(7). Similar dependence was noted by J.H. Lee at processing experiments in gas detonation initiation [23].

Constancy of r_c for different ν is not strictly proved theoretical conclusion, this is only generalisation of available results of experimental and numerical simulation convenient for approximate calculations. This is important because which is when knowing initiation energy in a case of spatial symmetry, it is possible to obtain r_c and then an initiation energy in other cases of symmetry as

$$E_{\tilde{n}\nu} \cong p_o r_c^{(\nu+1)}\ . \qquad (8)$$

Another useful correlation generalising numerical solutions is the ratio of critical radiuses of detonation wave $r_{c\nu}$ and r_c (the first corresponds to the minimum of DW velocity). This correlation is as follows:

$$r_{c\nu}/\ r_c \cong (\nu+1)/8\ . \qquad (9)$$

Using Eqs. (8,9), one may calculate initiation energies per unit area of critical radius wave front for different ν:

$$E_{\tilde{n}0}/2 = 0{,}5\, p_o r_{\tilde{n}},\ E_{C1}/(2\,\pi r_{c1}) = 0,\,64\, p_o r_{\tilde{n}},\ E_{c2}/(4\pi(r_{c2})^2) = 0,\,57\, p_o r_{\tilde{n}}.$$

It is seen that these values turn to be closed to each other in all the three cases of symmetry. That is, correlation (8,9) accord with the law of the surface energy density constancy in critical states of wave proposed by J.Lee et al. [24] to calculate the gaseous detonation initiation.

Critical parameters can be obtained from correlation (6), if it is modified taking into account chemical ignition delays of evaporated fuel $t_{ig} = \hat{E}\ (\rho_{10}\ /\rho)^m\ (T_{10}\ /T)^n \exp(\varepsilon/RT)$. Decreasing a wave velocity lower then $D_{\hat{\imath}}$ in the process of initiating by an explosion source is caused mostly by energy consumption for creating zone of compressed by shock but still not consumed substance behind a shock wave (the "chemical peak zone"). At instant evaporation and consumption of fuel broken off from droplets, this zone is of $2\lambda_0$ and only slightly depends on wave velocity in a real velocity range (0,7 ÷1,0) $D_{\hat{\imath}}$. Additional chemical ignition delays increase [25] the thickness of this zone by value $x_{ig} = x_0\, y^M \exp[\beta(y^\alpha - 1)]$, where $y = D_0/\ D_{min}$, $\beta = \varepsilon/RT_*$, T_* is the temperature in gas compressed by shock for an ideal wave moving with velocity D_0, Ì = α m - 1 + γ (n + 1), α = dlnT/dlnD, γ = dlnσ /dlnD. Correction, we obtain the equation for y from correlation (6):

$$(y-1)/y = \frac{34{,}5}{r_0}\{2\lambda_0 + x_0 y^M \exp[\beta(y^\alpha - 1)]\}, \qquad (10)$$

This has solution, if parameter r_0 is not less than the definite critical value r_c which is derived from the joint solution of Eq. (10) and equation (10) differentiated by y. Correlation for critical velocity y_c is the following

$$\grave{A} = 0{,}5[(y-1)(M+\alpha\beta y^\alpha)-1]\, y^M \exp[\beta(y^\alpha-1)], \quad A = \lambda_0/x_0. \qquad (11)$$

After solving (11), we obtain r_c from equation

$$r_c = 34{,}5\, x_0\{2A + y_c^M \exp[\beta(y_c^\alpha - 1)]\}/(1-1/y_c). \qquad (12)$$

For A = 0 correlation (11, 12) determine critical parameters y_g, r_g for a gaseous fuel-oxidiser mixture initiation. If in the mixture fuel is already in the gaseous phase, we obtain gaseous detonation. It is convenient to rewrite Eq. (12) as follows

$$\frac{r_c}{r_g} = \frac{1-1/y_g}{1-1/y_c}\{\frac{2A + y_c^M \exp[\beta(y_c^\alpha - 1)]}{y_g^M \exp[\beta(y_g^\alpha - 1)]}\}. \qquad (13)$$

In the common case it is necessary to solve Eq. (11) by iterations and substitute the obtained results into Eq. (12). Therefore, knowing correlation for chemical ignition delays, one may determine detonation initiation parameters in heterogeneous (gas-droplets) or gas mixture from by formulae (11-13).

Let us analyse the solution behaviour of Eq. (13), designating as $\eta = r_c/r_g - 1$. According to Eqs. (11), (13) η = F(α, β, M, A) is the four parameter function. For most of real detonating mixtures α, β, M change in ranges $\alpha \in$ [1.4÷1.6], $\beta \in$ [5÷10], M $\in$ [-0.5÷ 0.5]. The results of solution of Eq. (13), in which α, β, M varied in the above ranges, are presented in Fig. 2, as dependencies η(À) (width of hatched

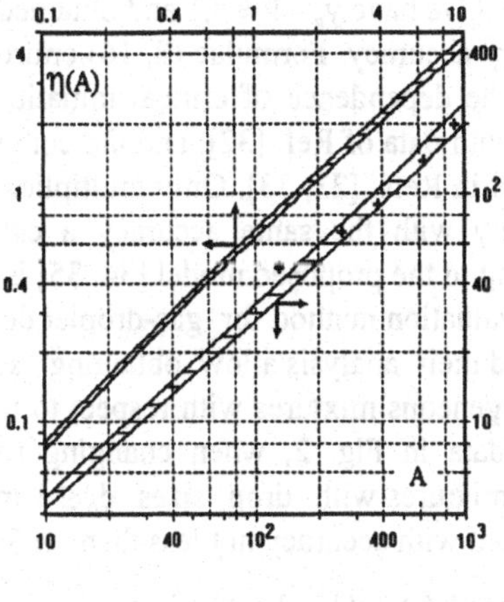

Fig. 2. Dependence $\eta(\grave{A}) = r_c/r_g - 1$ on parameter $A = \lambda_0/x_0$.
(+ data [17] for the same kinetic constants in induction period).

strip demonstrates the degree of change from α , β , M). Shown on the same figure (by sign +), are the results of detailed numerical solution of the initiation problem [17] in oxygen suspensions. It is seen that function η increases monotonously with A and depends weakly (in the limits ± 15%) on α, β, M . Moreover, for fuel-air mixtures with fixed A ($A \in [0 \div 10]$) range of change η is less then 6%. That is, parameter $\grave{A} = \lambda_0 / x_0$ is the basic dimensionless parameter that makes influence on the critical radius value. This fact is principally important for developing approach for determining critical initiation energies of heterogeneous detonation. This allows for every concrete gas-droplet mixture, using graphical dependence η (A) (Fig. 2), to obtain energy values of heterogeneous detonation initiation instead of solution of transcendent equation system (11 -13).

Unfortunately, at present a knowledge about chemical ignition delay dependencies and accuracy for wide range of gas reacting media is not sufficient. That is why, it is useful to define a size of elementary cell on the gaseous detonation front [26-30] as:

$$b = C\beta\, \sigma_0\, x_0 \,. \qquad (14)$$

This is expression generalised on the basis of data [28-31] for cell length at the self-sustaining average detonation velocity $D = D_0$. According to data [28-31], coefficient C = 1,4 ±0,6. Moreover, for air mixtures C is lower than for oxygen ones, let us take for them value C = 1.1 and 1.7 correspondingly.

Transition to pure gas mixtures without droplets in equations (11,12) is realised at A=0. Expanding the obtained expressions into series by degrees (y-1) and reserving only quadratic terms, we obtain approximate expressions for y_g , r_g

$$y_g = 1 + 2\xi^{-1}/(1 + (1 + 4\alpha^2\beta/\xi^2)^{0,5}), \qquad (15)$$

$$r_g = 92\, x_0/(y_g - 1) \approx (105\alpha/C\sigma_0)\, b\,, \qquad (16)$$

where $\xi = \alpha\,\beta + M$. At $\beta \geq 7$ we have $y_g - 1 \approx 0{,}1$ and obtained approximate solution for y_g and r_g is of the necessary accuracy. Formulae (8, 16) entirely determine an initiation energy of gas mixtures. The dependence of critical initiation parameters on width of induction zone correlates with data of Ref. [32], and the common form of correlation is analogous to ones obtained in Refs. [23, 33]. Only multipliers in front of $b^{(\nu+1)}$ or $x_0^{(\nu+1)}$ rather vary. Approximately with the same accuracy a calculation correlation with experimental data according to the proposed model [34, 35] is observed.

The proposed evaluation method for gas-droplet detonation initiation energy, carried out calculations and their analysis allow obtaining solution of the problem of an activity order of heterogeneous mixtures with respect to the explosion (detonation) transition. According to data in Fig. 2, when changing parameter $\grave{A} \in (0 \div 10)$ (it corresponds to fuel-air mixtures with drop sizes $d_0 \leq 1$ mm), dependence $\eta(\grave{A}) = r_c/r_g - 1$ can be approximated with accuracy not less then $\pm 5\%$ by the formula

$$\eta(\grave{A}) = 1{,}6\grave{A}/\ln(1 + 1{,}6A) - 1\,. \qquad (17)$$

According to Eq. (17) with increasing parameter $\grave{A} = \lambda_0/x_0$, function $\eta(\grave{A})$ increases lower then the linear one, therefore an activity order in heterogeneous (gas-drop) media, which is constructed for equal values of drop diameters d_0 (more exactly, for equal values of λ_0) coincides with an activity order of corresponding gas (vapour) reacting mixtures. That is, the statement can be made:

"If an activity order for gas (vapour) mixtures is as follows

$$r_{g,1} \rightarrow r_{g,2} \rightarrow r_{g,3} \rightarrow \cdots,$$

then for every λ_0 an activity order for gas-droplet mixtures with respect to explosive (detonation) transformation in air is of the same type:

$$r_{c,1} \rightarrow r_{c,2} \rightarrow r_{c,3} \rightarrow \cdots."$$

Let us consider examples of calculations of critical initiation energy for gas-droplet detonation in fuel- air mixtures.

Example 1. Knowing critical radius of gas detonation r_g and activation energy ε, it is possible to determine critical initiation radius of heterogeneous detonation. Calculation will be made for stoichiometric ($\phi = 1$) gasoline-air mixture.

According to experimental data obtained by Alekseev et al. [36,37], evaporated gasoline-air mixture ($\phi = 1$) is initiated by 0.3 kg of TNT, that converting to critical radius $r_g = (E_0/p_0)^{1/3}$ gives $r_g = 2{,}3$ m. An activation energy of gasoline [38] $\varepsilon \approx 35$ kcal/mole, therefore for D_0=1800 m/sec $\beta = \varepsilon/R\dot{O}_* = 11{,}4$; $\alpha = 1{,}446$, $\sigma_0 = 6{,}6$. Parameter $\grave{A} = \lambda_0/x_0$ can be expressed using formula (16) via r_g and β: $\grave{A} = 150\,\beta\lambda_0/r_g$. For gasoline of density $\rho^0_2 = 700$ kg/m^3 from correlation for λ_0 we obtain $\lambda_0 = 11{,}85\,d_0$(m). Then from Eq. (17) we determine critical initiation radii of heterogeneous detonation $r_{\text{ñ}}$ in dependence on initial diameter d_0 of gasoline drops in fuel spraying. Values of $r_{\text{ñ}}$ (m) for a number of values d_0 (μm) in the stoichiometric gasoline-air mixture are given below

d_0, μm	0	50	100	200	400	800	1000
À	0	0,44	0,88	1,76	3,52	7,04	8,8
$r_{\text{ñ}}$, m	2,3	3,04	3,69	4,84	6,85	10,33	11,94

Example 2. Knowing critical radius of heterogeneous detonation $r_{h,1}$ for given value of drop diameter d_{01} and activation energy ε, it is possible to determine critical initiation radii of gas r_g and heterogeneous r_c detonation. Consider this for stoichiometric ($\phi = 1$) kerosene-air gas-drop mixture.

Density of liquid kerosene $\rho^0_2 = 830$ kg/m³, its initial volumetric concentration $\alpha^0_2 = 10^{-4}$, velocity of ideal CJ detonation $D_0 = 1790$ m/sec. According to the experimental data [36] the gas-drop ($d_{01} = 100$ mkm) kerosene-air mixture ($\phi = 1$) is initiated by mass of TNT $\approx 4{,}5$ kg, that being converted to critical radius $r_{c,1} = (E_0/p_0)^{1/3}$ gives $r_{c,1} = 5{,}7$ m. Activation energy for kerosene [39] $\varepsilon \approx 40$ kcal/mole, therefore for $D_0 = 1790$ m/sec : $\beta = \varepsilon/R\dot{O}_* = 13{,}1$, $\alpha = 1{,}446$, $\sigma_0 = 6{,}6$. For density $\rho^0_2 = 830$ kg/m³ $\lambda_{01} = 12{,}14 \cdot d_{01} = 1{,}21 \cdot 10^{-3}$ m. Substituting values of β, λ_{01}, $r_{c,1}$ in Eq. (17), we obtain the correlation for critical radius of gas detonation r_g

$$r_{c,1} = 240\beta\lambda_{01} / ln(1 + 240\beta\lambda_{01}/r_g),$$

from which we obtain $r_g = 4.0$ m. Then doing as in example 1, we determine critical initiation radii of heterogeneous detonation r_c for any value of initial drop diameter of kerosene drops d_0.

d_0, μm	0	50	100	200	400	800	1000
Á	0	0,298	0,597	1,193	2,386	4,77	5,96
r_h, m	4,0	4,89	5,70	7,15	9,71	14,16	16,21

4. Conclusion

Thus, generalisation of dependencies for characteristic parameters in detonation initiation problem, previously obtained in calculations for heterogeneous medium, allows formulating simple approximate method and obtaining analytical expressions for estimating initiation energies for gas and for gas-droplet mixtures taking into account chemical ignition delays. The formulae obtained for gas mixtures turned out similar to the most precise ones known in literature. For gas-drop systems with longer time of droplet break-up as compared with chemical induction time in the CJ wave, a calculation according to the above method correlates with results of detailed numerical solution of initiation problem [17] in oxygen suspensions. In the intermediate case of comparable times of break up and chemical induction, this method gives sufficiently reliable interpolation of initiation energies on the fine-disperse sprays

5. References

1. Webber W.T. Spray combustion in the presence of a travelling wave. - In: 8thSymp. (International) on Combustion. Pasadena, Calif., 1960. Baltimore, 1962, p.1129-1140.
2. Cramer F.B. The onset of detonation in a droplet combustion field. - In: 9thSymp. (International) on Combustion, New York - London: Acad. press, 1963, p.482-487.

3. Dabora E.K., Weinberger L.P. Present status of detonations in two-phase systems/ Acta Astr., 1974, V 1, N 3/4, p.361-372.
4. Borisov A.A., Gelfand B.E. Paper survey on detonation in two-phase systems. Arch. Termodyn. i Spalan., 1976, Vol. 7, N 2, б.273 - 287.
5. Gelfand B.E. Contemporary state and investigation problems of detonation in the system of liquid drop-gas. /Chemical physics of combustion processes and explosion. Detonation.Chernogolovka, 1977, p. 28-39 (in Russian).
6. Nettleton M.A. Shock-wave chemistry in dusty gases and fogs: A review. - Combust. and Flame, 1977, vol.28, N 1, p.3-16.
7. Nigmatulin R.I. Multy-phase dynamics. - M., Nauka, 1987.- Parts 1,2 (in Russian).
8. Ranger A.A. and Nicholls J.A. Aerodynamic shattering of liquid drops/ AIAA J., 1969, V 7, N 2, pp.285-290.
9. Borisov A.A., Gelfand B.E. et al. The reaction zone of two-phase detonations. Astr. Acta - 1970.- V. 15, N 5/6.- P. 411-419.
10. Zhdan S.A. Calculation of spherical heterogenious detonation. / Fizika Goreniya Vzryva, 1976, No. 4, pp. 586-593 (in Russian).
11. Zhdan S.A. Calculation of heterogenious detonation considering deformation and fuel drop decomposition. / Fizika Goreniya Vzryva, 1977, No. 2, pp. 258-263 (in Russian).
12. Zhdan S.A. Spot explosion in combustible two-phase medium. /Continuus dynamics. Novosibirsk, 1977, issue 32, p. 36-46 (in Russian).
13. Eidelman Sh., Burcat A. The evolution of a detonation wave in a cloud of fuel droplets/ Part 1, AIAA J, 1980, Vol. 18, N 9, pб. 91-99.
14. Burcat A., Eidelman Sh. The evolution of a detonation wave in a cloud of fuel droplets/ Part 2, AIAA J, 1980, Vol. 18, N 10, pб. 104-108.
15. Zhdan S.A., Mitrofanov V.V. Calculation of critical initiation energy of heterogeneous detonation. In: Detonation.Critical phenomena. Physical-chemical transformations in shock waves. Chernogolovka, 1978, p. 50-54 (in Russian).
16. Mitrofanov V.V., Pinaev A.V. and Zhdan S.A. Calculations of detonation waves in gas-droplet systems/ Acta Astron., 1979, Vol. 6, N 3/4, pp. 281-296.
17. Zhdan S.A. Calculation of heterogeneous detonation initiation by condensed explosion charge/ Fizika Goreniya Vzryva, 1981, V. 17, No. 6, pp. 105-111 (in Russian).
18. Zhdan S.A. Detonation waves in gas-droplet systems upon explosion of a cylindrical charge./ Abstracts of V111 ICGERS, Minsk, 1981, p. 50.
19. Voronin D.V., Zhdan S.A. Calculation of heterogeneous detonation initiation in a tube by hydrogen-oxyden mixture explosion. / Fizika Goreniya Vzryva, 1984, V. 20, No 4, p.112-116.
20. Nicholls J.A., Bar-Or R., Gabrijel Z. et al. Recent experiments on heterogeneous detonation waves. - AIAA Paper, 1979, No. 288, 4 p.
21. Zhdan S.A., Mitrofanov V.V. Simple model for calculating initiation energy of heterogeneous and gas detonation./ Fizika Goreniya Vzryva, 1985, v. 21, No. 6, pp. 98-103.
22. Williams F.A. Combustion Theory. Palo Alto-London, 1964.
23. Lee J.H. Initiation of gaseous detonation. - In: Ann. Rev. of Phys. Chem., 1977, v.28, p. 75-104.
24. Lee J.H. , Knystautas R. and Guirao C.M. The link between cell , critical tube diameter, initiation energy and detonability limits. Proceedings of the Intern. Conf. of Fuel-Air Explosions, Univ. of Waterloo Press, Montreal, Canada, 1982, pp. 157-188.
25. Mitrofanov V.V. Sone critical phenomena in detonation, connected with pulse wastes. /Fizika Goreniya Vzryva, 1983, v. 19, No 4, p. 169-174 (in Russian).
26. Voitsekhovskii B.V., Mitrofanov V.V., Topchiyan M.E. The Structure of Detonation Front in Gases. Izdat. Sibir. Otdel. Akad. Nauk SSSR, Novosibirsk, 1963, 168 p. (in Russian).
27. Mitrofanov V.V., Subbotin V.A. On mechanism of detonation combustion in gases. In: "Gorenie i Vzryv". M., Nauka, 1977, p.447-453 (in Russian).
28. Vasil'ev A.A., Nikolajev Yu.A., Ulyanitsky V.Yu. Calculation of cell parameters of multy-front gas detonation. / Fizika Goreniya Vzryva , 1977, v. 13, No 3, p. 404-408 (in Russian).
29. Vasil'ev A.A., Nikolajev Yu.A., Ulyanitsky V.Yu. Critical initiation energy of multy-front detonation. - / Fizika Goreniya Vzryva, 1979, V. 15, No 6, p.94-104 (in Russian).

30. Manzhaley V.I., Subbotin V.A., Stcherbakov V.A. Stability boundaries and connection of cell size of gas detonation with kinetic constants of explosive gas mixtures. In: Chemical physics of combustion and explosion. Detonation. Chernogolovka, 1977, p.45-48 (in Russian).
31. Manzhaley V.I. Unstability of detonation front in gases. Thesis for Ph.D. degree. Novosibirsk, 1981, 147 p (in Russian).
32. Zel'dovitch Ya.B., Kogarko S.M., Simonov N.I. Experimental study of spherical gaseous detonation. Journal of technical physics, 1956, v. 26, No. 8, pp. 1744-1768 (in Russian).
33. Ulyanitskiy V.Yu. On a rolc of "flash" and co-stroke of transversal waves in forming multy-front structure of detonation waves in gases. FGV, 1981, v. 17, No. 2, pp. 127-133 (in Russian).
34. Bull D.C., Elworth J.B., Hooper G. Initiation of spherical detonations in hydrocarbon/air mixtures. Acta Astron., 1978, Vol. 5, No. 11-12, pp. 997-1008.
35. Vasil'ev A.A. Investigation of gas detonation critical initiation. /Fizika Goreniya Vzryva, 1983, v. 19, No 1, p. 121-131 (in Russian).
36. Alekseev V.I., Dorofeev S.B. et al. Experimental study of detonation initiation in motor fuel sprayed in air. Preprint IAE-4872/13, Moscow, Atominform, USSR, 1989.
37. Alekseev V.I., Dorofeev S.B. et al. Experimental study of large - scale unconfined fuel spray detonations. - In: Dynamic Aspects of Explosion Phenomena, edited by A.L.Kuhl, et al., vol. 154, Progress in Astron. and Aeron., AIAA, Washington, 1993, p.95-104.
38. Stcherbakov V.A. Experimental study of sell size connection with kinetic parameters of gas explosive mixtures. Graduation work, NSU, Novosibirsk, 1977, 38 p (in Russian).
39. Stchetinkov Ye.S. Gas combustion physics. M., Physmathgiz, 1966 (in Russian).

DISPERSION-INITIATION AND DETONATION OF LIQUID AND DUST AEROSOLS-EXPERIENCES DERIVED FROM MILITARY FUEL-AIR EXPLOSIVES

M. SAMIRANT
French-german research institute of Saint-Louis 5, rue de General Cassagnou, BP 34, F-68301 Saint-Louis Cedex, France

1.Introduction

Fuel-air explosives consist of a gaseous combustible or a fine distribution of solid particles or liquid droplets intimately mixed with air. A thermal or a chemical energy source may lead this medium either to a combustion eventually to a deflagration or a detonation. For different explosive mixture the type of reaction after the initiation depends on: the chemical composition, on the initiation conditions, on the eventual flow in the fluid, on a more or less heavy confinement and on the physical state of dispersion of the fuel. The behaviour of purely gaseous media have been intensively studied and modelized (1-5). On the other hand there has been a lot of research activities on the combustion of two-phase mixtures containing solid particles or liquid drops which have found widespread industrial application (1, 6-8).

In this paper I will present the experimental methods used at the French-German Institute in Saint-Louis (ISL) to characterise flammable and detonable aerosols. These experimental values are used as input data and calibration quantities for computer models of the behaviour of these systems.

We develop techniques for characterising the dispersion of an aerosol under controlled or uncontrolled conditions. In the resulting reactive medium we try to define the reached physical state after dispersion : density, diameter, distribution and velocity of particles or droplets, fuel evaporation richness of the gaseous phase and of the two phases mixture. These are the key parameters for predicting violent reactions subsequent to a controlled or uncontrolled initiation (9).

For our specific purpose a liquid fuel was dispensed by a reduce amount of high explosive resulting in drops of large diameter (100µn-1000µn) with initial velocities of about 100-400 ms^{-1}. These initial conditions correspond for our actual purpose to the explosion of a fuel tank due to overpressure or instant boiling (BLEVE) or to the rupture of a pipe with a high velocity flow. The resulting aerosols are very instable. Due to gravity there is a rapid sedimentation and the droplets may agglomerate or evaporate. In order to determine the composition of this aerosol at a given time, we need short time measurement with high repetition rates to follow the evolution of the medium with a typical time interval of less than 10^{-3}s. If we need to monitor the initial phase of the

V.E. Zarko et al. (eds.), Prevention of Hazardous Fires and Explosions, 123–134.

rupture of the liquid and the dispersion of the droplets we need measurements with a typical time interval of 10^{-6}s.

The same time scales are found to monitor the reactions following an initiation which leads either to a combustion or to a detonation. The combustion is a slow process, which propagates with velocities of about 10 m s^{-1} without resulting in a noticeable pressure increase in an unconfined medium. If there is an important energy flux and particularly in confined medium a deflagration may be induced. Its propagates by heat transfer and diffusion. This slow rate mechanism limits the propagation velocity below the velocity of sound in the gas. Fortunately most of the casual uncontrolled aerosol explosions are deflagrations resulting in overpressure of - 100Kpa-800 Kpa.

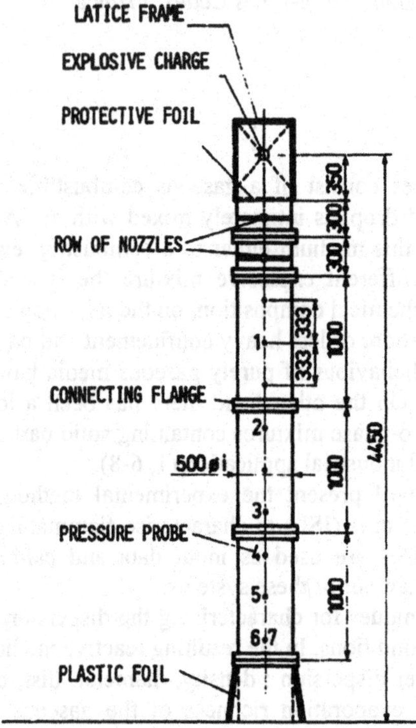

Fig. 1 Vertical Shock tube (mm)

2.Investigations in a vertical shock tube

2.1 SHOCK TUBE

For experimental spray detonation studies under defined conditions a vertical shock tube as shown in fig. 1 was built.

The tube is approximately 4 m long, has an inner diameter of 50 cm and is open at both ends (only closed by thin plastic foils to avoid convection). The fuel spray is generated

by means of 12 fuel injectors located ringwise at the top of the tube. In falling down inside the tube, the droplets agglomerate. Only at the bottom part of the tube the spray has a volume fraction of about 10^{-4}. A strong blast wave from a 1 kg explosive charge suspended 35 cm above the upper tube end initiates the detonation.
Seven pressure gages are mounted in the tube wall. They are mainly used for observing the shock transition. In addition, an optical probe for Doppler measurements is placed in the vicinity of the tube axis closc to the lower tube end. It allows different physical quantities to be recorded in the reaction zone.

2.2 SPRAY CALIBRATION

A photographic technique was used for determining the number density as well as the diameter distribution of the droplets in a test volume selected near the tube axis in is lower part. By means of an optical system the spray was stroboscopically illuminated by an intense parallel laser light bundle with a rectangular cross section of 5 mm x 40 mm. Photographs taken from a direction perpendicular to the light bundle exhibit traces of each particle in a well defined volume. Since the droplets are ideal spheres, there exists a rather accurate relation between their diameters and their sinking velocities. By analysing 12 photographic records in measuring the distances separating the „dots“ for each trace and in separately counting the traces for each species, the distribution of the liquid fuel as a function of the droplet diameters was determined (Fig. 2). The total volume fraction of the liquid was found to be about 1.1×10^{-4}. This value, however, can only be regarded as an average one. The individual records show that at a scale of about 25 cm^3 - which is the size of the photographed volume - a considerable deviation up to about 50% is possible.

From the liquid volume fraction, the stoechiometric equivalence ratio s_ℓ of the liquid fuel component can be calculated. s_ℓ Is dependent on both the fuel and spray temperature. It was stoechiometric or slightly understoechiometric for all the fuels investigated.

In order to achieve a definite oechiometric equivalence ratio of the vaporised fuel component s_v, the vapor was always saturated by intermittently spraying several times prior to the experiment until an equilibrium temperature of the spray was reached. From this temperature, s_v was calculated.

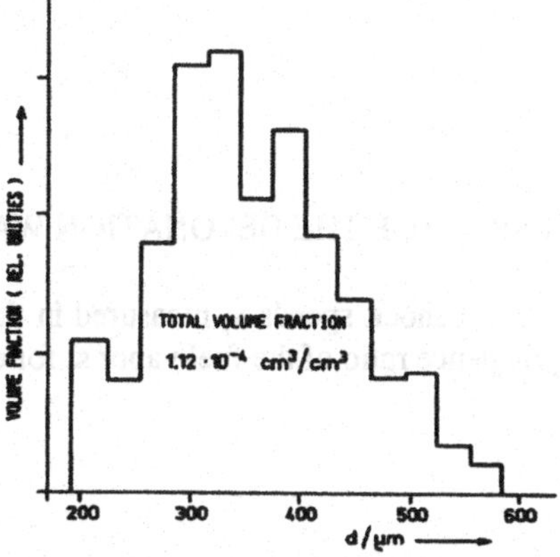

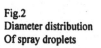

Fig.2
Diameter distribution
Of spray droplets

2.3 LASER DOPPLER SYSTEMS

For non-intrusive measurements in the reaction zone of the detonation waves, an optical probe had to be built working in the environment of high pressures and temperatures (see Fig. 3). A light fiber (monofiber, stepped index, 200 µm in diameter) transmits continuous monochromatic laser light from an Argon-ion laser (in monomode operation) into the probe. The light bundle is focussed at the measuring point. Light scattered from particles passing through this point is collected and transmitted by a second light fiber into a particular highly resolving spectrometer [15,16]. The Doppler shift of the scattered light is proportional to the particle velocity component along the axis of symmetry of the probe. The spectrometer generates an electrical signal following proportionally and in real time (with a time resolution better than 1 µs) the frequency variations of the Doppler shifted light. The optical components are shielded from the flow by solid rods and bear sapphire windows heated to about 200°C in order to avoid vapor condensation.

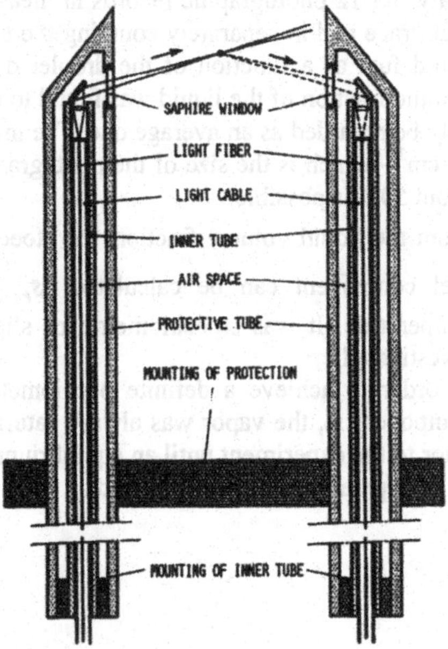

Fig. 3
Optical probe for
Laser Doppler
mesurements

2.4 SPEED OF THE DETONATION WAVES

In Fig. 4 shock speeds x_s measured in the lower part of the tube are plotted versus the equivalence ratio of the fuel vapor s_v for each point.

From all the experiments performed only those were retained for which a Chapman-Jouguet detonation characterised by a constant $\dot{x}_s$ was reached, that is, all events showing a decaying shock wave were excluded.

For all the liquids investigated (methanol, ethanol, propanol-1, hexanol and decane), the reaction energies of the stoechiometric mixtures are nearly the same (627 cal/g for methanol and between 647 cal/g for the other mixtures). For this reason a plot in a common diagram is justified.

The dashed line connects the points of the experiments in the absence of droplets: In this case there is only an amplification for $s_v > 0.3$. For $s_v < 0.3$, amplification is exclusively due to the liquid fuel component.

Of greatest interest are here the three shots in decane and hexanol sprays. They clearly demonstrate that a real two-phase Chapman-Jouguet detonation does exist in sprays with mean droplet diameters of about 350 µm (and diameter distribution given in Fig. 4) having vapor concentrations as low as $s_v = 0.06$.

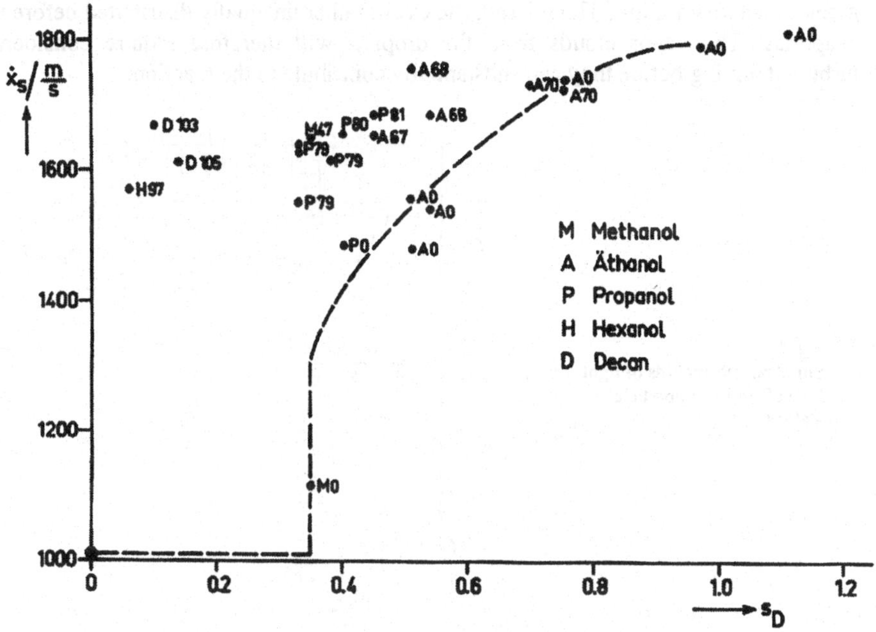

Fig. 4 Constant shock speeds in the lower part of the tube

2.5 INVESTIGATION OF THE REACTION ZONE

Prior to reacting with the oxygen, the fuel in liquid form has to be mixed with the air and distributed down to the molecular scale. The first step is the atomisation of the droplets by stripping of the boundary layer in the flow behind the shock wave. Thus, from each droplet a cloud of microparticles results. After rapid vaporisation of the small liquid particles in the high temperature gas, the droplets lead to regions of high fuel vapor concentration in the flow behind the shock. A quick distribution of the gaseous fuel is only possible by turbulent mixing because the diffusion is much too slow to be effective enough.

The optical investigations of the reaction zone give some insight into the different processes mentioned above. An example of simultaneous recordings of the scattered light level and the particle velocity is given in Fig. 5. The period of light scattered (from the microparticles) is always limited to about 10 µs. The records taken show strong variations and are not reproducible. Some of them contain a large peak as exemplified in Fig. 5. The particle velocity determined from the Doppler shift of the scattered light rises within 2 µs and shows a spike which is probably due to the primary reaction of the relatively high fuel vapor component of the two-phase mixture. Whereas the spike does not always appear, a rise time as short as 2 µs is typical for all velocity records. From these measurements, the following informations can be drawn: atomisation and evaporation of the liquid fuel occur very rapidly so that 10 µs after the shock has passed, the liquid is completely transformed into vapor. The diameter of the microparticles having a velocity relaxation time, which does not exceed 2 µs, cannot be greater than about 2 µm. The microparticles are rather unequally distributed before they evaporate. The vapor clouds from the droplets will therefore require considerable turbulent mixing before they can substantially contribute to the reaction.

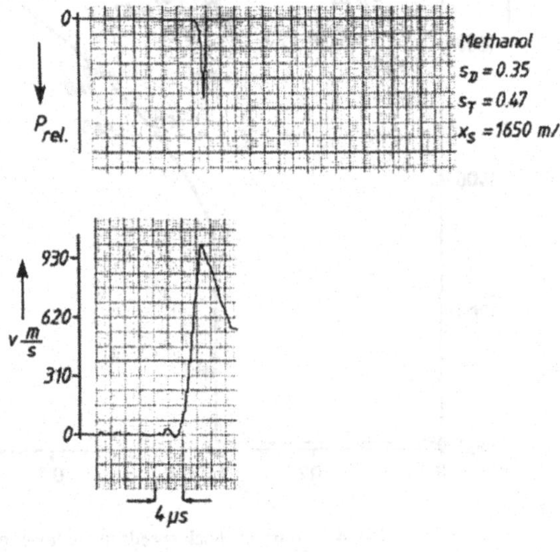

Fig. 5
Simultaneous records of light Level P and mictoparticle Velocity v

3. **Open range studies**

3.1 EXPLOSIVELY DISPERSED CANISTER

As presented in the introduction our liquid fuel was dispersed with high explosive. The techniques used to caricaturise this dispersion can be applied to every point like source of dispersion of a liquid fuel in air.

Fig. 6 gives a scheme of the canister used for the explosive dispersion of the liquid and its dimensions for two volumes 1.4 litters and 5 litters. The canisters are made of tin with a 0.25-mm wall thickness. The explosive is a low density cartridge (0.9 g cm^{-3}) with a low detonation velocity made up of a mixture of PETN-polyurethane 50/50.

Fig. 6
Canister for liquid dispersion

	A	B	C
1.4 l	110mm	140mm	160mm
5 l	180mm	190mm	210mm

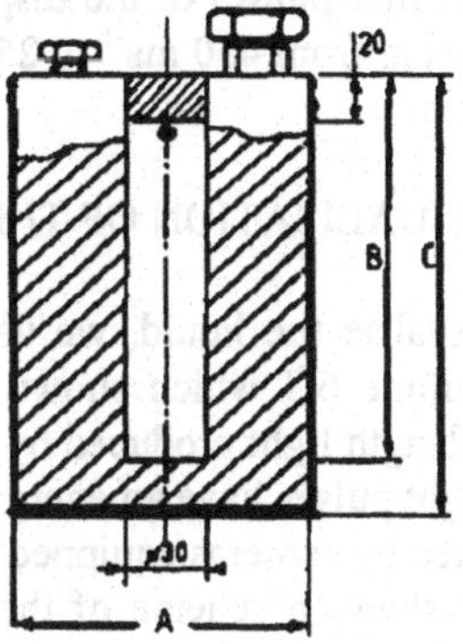

3.2 VISUALISATION BY FLASH-X-RAY PHOTOGRAPHY

The explosively dispersed liquid is made absorbing for the X-rays by dissolution of iodine (5 mass per cent). At different times after the detonation of the explosive X-ray pictures are taken and thus allow to follow the motion of the liquid masses (see Fig. 7). These photographs are treated by an image processing system. Figure 8 shows a picture of methanol dispersion 1.2 ms after initiation, the centre being on the left side of the picture at a distance of 390 mm.

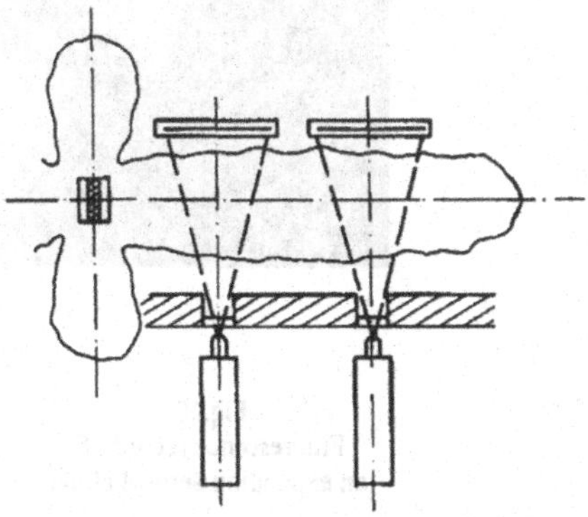

Fig. 7
X-Rays flash radiography set up for an aerosol

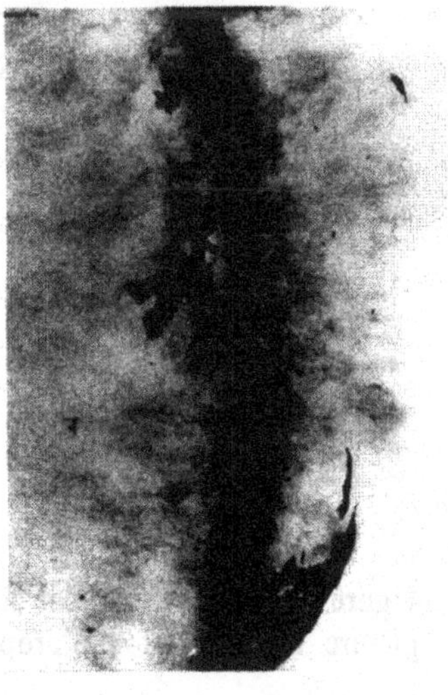

Fig.8
X-Rays flash radiography of the expending fuel torus

If we analyse this picture, we come to the result that after 1.2 ms the liquid shell is quasi unbroken. Its apparent thickness of 3 mm corresponds to that of a torus with a main radius of 390 mm and a height of 200 mm having the same volume of 1.4 litters as our dispersion canister (initial height of the canister 150 mm). After 1.6 ms the liquid shell is broken and the mean size of the drops is 5 mm.

For the first phases of the dispersion, we can thus define a velocity of the liquid front decreasing from 450 ms^{-1} to 250 ms^{-1}, a breaking time of 1.3 ms and an initial drop size of 5 mm.

3.3 VISUALISATION OF THE AEROSOL CLOUD BY LASER FLUORESCENCE

To visualise the liquid, we use a method by fluorescence. The liquid is coloured by Rhodamine 6G which shows an orange fluorescence when illuminated by a shorter wave-length light produced by a dye laser (Coumarine 504) radiating in the green range. The light pulses have an energy of 1.1J and a duration of 0.7 μs. The fluorescent light is recorded by cameras equipped with filters to eliminate the direct laser radiation.

Fig. 9 shows a scheme of the method used for visualising a sheet in the cloud. This method has the advantage that it only shows the liquid fuel and not the gas and the condensed water.

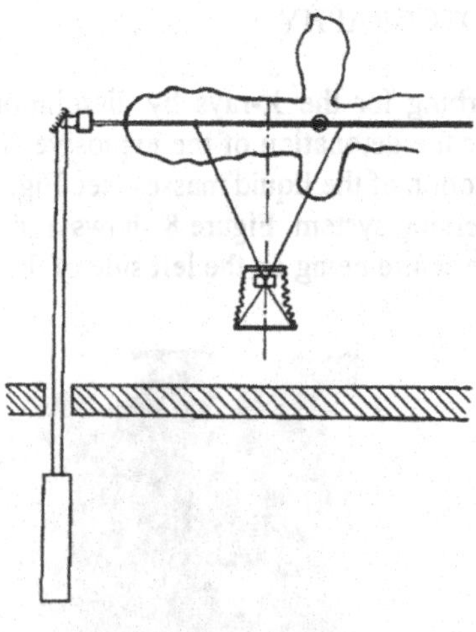

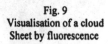

Fig. 9
Visualisation of a cloud
Sheet by fluorescence

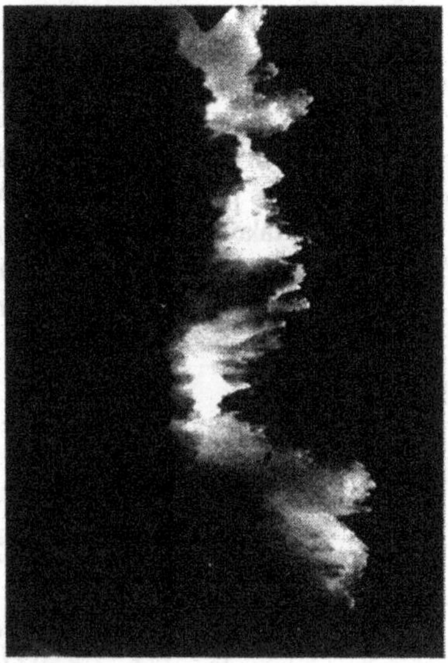

Fig.10
Fluorescence record of
an expanding aerosol cloud

Figure 10 shows a typical sight of the expanding cloud. We can determine on this picture the size of the droplets and calibrate the results of the dispersion codes.

However, the velocity measurement of the liquid masses needs a picture sequence taken during the same experiment, as we will see in the next paragraph.

3.4 CINEMATOGRAPHIC RECORDING OF THE DISPERSION

To follow continuously the development of an aerosol cloud, we have realised a cinematographer recording of the phenomena. The set-up is the same as in Fig. 9, only the used optical equipment is different. The photographic cameras are replaced by a rotating-drum camera moving the film at constant velocity. Illumination is made by a repetitively pulsed copper-vapor laser giving pulses of 20 ns duration, 2.5 mJ energy and with a repetition rate between 1000 Hz and 10000 Hz. This laser was developed at ISL[17]. The recorded films allow the measurement of the velocity of the liquid masses.
We used these films as a base to realise a motion picture. Each exposure has been rephotographed on a 16-mm film and synchronized with the film perforations in order to allow a continuous movement. For this synchronization, we used a C.N.C. machine supporting both film and camera, which allowed to recenter each picture.
This continuous projection of the dispersion shows some interesting behaviours. Particularly at the beginning of the dispersion, the liquid front has a wave-like aspect due to the presence of twelve splitters, which formed the canister. After some milliseconds of propagation, these curved fronts interfere and the interference points send out rapid Mach waves which in turn interfere and so on, thus creating an intense mixing of the air-fuel aerosol. These local phenomena are important for the formation of all combustible or detonable mixtures. They can only with difficulty be interpreted from fixed views.

3.5 DIMENSIONS SHAPES AND VELOCITIES OF DROPLETS

3.5.1. *Diffraction methods*

The set-up in Fig. 11 uses a cylindrical laser beam traversing a small aerosol volume, the geometrical position of which is chosen prior to dispersion. The light is collected by an objective, which forms the figure of the diffraction on an array of 512 photodiodes. This array is scanned and the signal is digitalized and shaped. Measurement frequency is 2500 analysis per second. The aerosol test volume is delimited by windows, which are swept by laminar air flux preventing any particle depositions.
There is a large variation of intensity between the centre and edge of the diffraction image. To reduce the dynamic necessary for the measurement system, we used a progressive filter with a large variation of the absorption made of a chrome thin film with a variable thickness vacuum deposited on a glass plate. This filter allows improving the measurements at small diffraction angles, which correspond to large drop diameters. Data exploitation is fully automatic by a microcomputer.
For example, the dispersion of Xylene leads after 9 ms to droplets of diameters between 10 μm and 80 μm (bimodal distribution). After 400 μs, the distribution is trimodal 10-70 and 200 [μm] (agglomeration). After 800 μs, the sizes are less than 70 μm (evaporation).

Fig. 11
Measurement of droplets diameters by diffraction
1:Laser 2:Telescope 3:Measuring zone 4 Objective
5:Filter 6 CCD

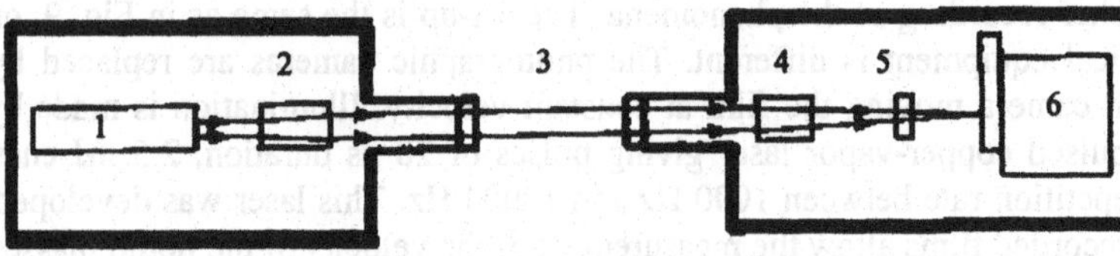

3.5.2. *Holographic method*
Recording of the hologram
A narrow light-beam is produced by a pulsed laser and enlarged up to the dimensions of the investigated field (up to 100 cm^2 area). The beam passes through the aerosol cloud and illuminates the holographic plate. One part of the incident light is diffused by droplets and the hologram is formed by its recombination with the non-diffused part.
This method yields very good pictures if the drop density is low, if their size is over 10 µm and if the ambient medium has no index gradient. If the ambient medium is strongly perturbed (combustion), we use a different set-up in which the reference-beam path is outside the object field and joins the diffused beam on the holographic plate.

Reconstruction of the images
After development, the hologram is placed in the beam of a continuous laser. Behind the plate a real image of the holographic field is formed. This image is enlarged and observed on a TV monitor. In contrast to other particle size measurement set-ups holography yields the shape of the droplets. Their sizes are measured by comparison to a reference grid, which is holographed under the same conditions. If two exposures are recorded on the same hologram, one gets two images of each particle and the distance between these two images yields the velocity of the droplets. We can also observe variations in the size (evaporation)or in the shape between the two expositions. The data gained are: size, shape, three-dimensional position of each droplet and its velocity. The measurable maximum velocity depends on the acceptable blur on the picture. With a ruby- or YAG-laser (pulse duration 20 ns) this velocity limit is always more than 100 ms^{-1}.

3.6 LASER-DOPPLER VELOCIMETER METHOD

The laser-Doppler system already used in the vertical shock tube could also be applied to investigate the flow phenomena during the cloud formation after the pyrotechnical

dispersion of the liquid fuel. The main component scattering the laser light seems to be a micromist of secondary particles from a condensation of water vapor in air rather than the fuel aerosol.
During the time interval between 10 ms and 30 ms after ignition of the explosive charge for the fuel dispersion an average radial velocity of approximately 25 m/s was found in most parts of the cloud being superimposed by large velocity fluctuations in all directions.

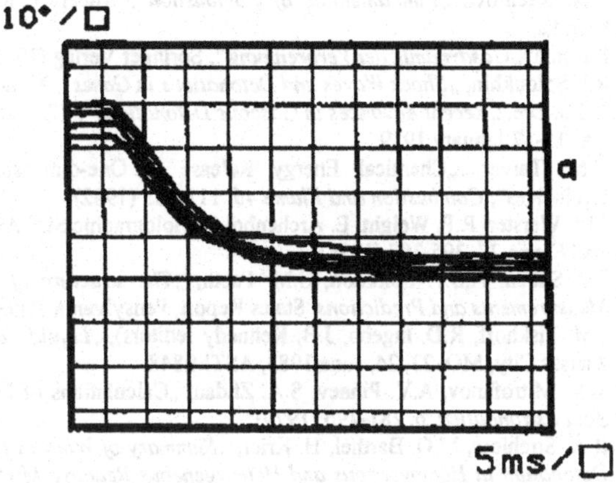

Fig. 12
Thermal signal of
4 grouped sensors

3.7 THERMAL MEASUREMENTS

We have developed thermocouples with a low thermal inertia (10^4°Cs^{-1} in air flow at 100 m s^{-1}). They are mounted on slender gages or on the surface of a half-sphere inside the boundary layer.
To study local temperature variations and the effect of droplet impacts on the gages, we realise arrays with four couples in line with a 2.5 mm spacing.
We equally develop mixed gages with a miniaturised light barrier the beam of which coincides with the tip of the thermocouple and detecting droplets flying across.
The use of gages with multiple measurement points allows to verify that the temperature is homogeneous in a volume of about one cubic centimetre and that the drop impacts occur only rarely. The reproducibility of the different thermocouples (rise time, calibration voltage, temperature) has also been confirmed by these measurements.

4. Conclusion

We have presented a variety of ISL specific methods for investigating combustible or detonable aerosols[18], nevertheless, we also use conventional techniques. High-speed cinematography (up to several 10^6 pictures s^{-1}) is used for studying the deflagrations and the detonations coupled with pressure gage arrays. One type is a slender gage working in one direction and another type is a plane gage being able to detect any grazing shock wave. Such a gage array placed at different distances from the reaction point (between 2

and 20 times the radius of the cloud) allows a precise measurement of the energy released by the reactions.
All these together allow a detailed study of the explosive aerosols and of their behaviour from the droplets' size scale (sprayers, diesel-injectors, ...) to the scale of large clouds.

5. References

1. R.A. Strehlow, „*Fundamentals of Combustion*", Robert E. Krieger Publishing Company, Malabar Florida.
2. Bartlnia, „*Gasdynamik der Verbrennung*", Springer Verlag (1975)
3. R.I. Soloukhin, „*Shock Waves and Detonations in Gases*", Mono Book Corporation, Baltimore (1966)
4. J.H.S. Lee, „*Recent Advances in Gaseous Detonations*", *17th Aerospace Sciences Meeting*, Neworleans, LA, 15-17 January 1979.
5. C.M. Tarver, „Chemical Energy Release in One-dimensional Detonation Waves in Gaseous Explosives", *Combustion and Flame 46*, 111-133 (1982).
6. J.M. Werster, R.P. Weight, E. Archenhold, „Holographic Size Analysis of Burning Sprays",*Combustion and Flame 27*, 395-397 (1976).
7. J.S. Shuen, A.S.P. Solomon, G.M. Faeth, „*The structure of Evaporating and Combusting Sprays. Measurements and Predictions*, States Report, Pensylvania States Union, March 1982.
8. J.M. Tiskhoff, R.D. Ingebo, J.B. Kennedy (editors), „*Liquid Particle Size Measurement Techniques*", Kansas City, MO, 23, 24, June 1983, ASTM 848.
9. V.V. Mitrofanov, A.V. Pinaev, S.A. Zhdan, „Calculations of Detonation Waves in Droplet Systems", *Acta Astronautica, 6*, 281-296 (1979).
10. R.A. Strehlow, M.O. Barthel, H. Krier, „*Summary of Work in Initiation Combustion and Transition to Detonation in Homogeneous and Heterogeneous Reactive Mixtures*", Interim Report, June 1978, May 1979, AFOSR 77-3336.
11. D.A. Lundgren, M. Lippmann, F.S. Harris, Jr. W.E. Clark, W.M. Marlow, M.D. Durham, „*Aerosols Measurements*" University Press of Florida, Gainesville, 1979.
12. I.O. Moen, A. Sulmistras, B.M. Mjertager, J.L. Bakke, „Turbulent Flame Propagation and Transition to Detonation in Large Fuel-Air Cloud", *21st International Symposium on Combustion*, München, West Germany, 3-8 August 1986.
13. I.O. Moen, D. Bjerkevedt, A. Jenssen, P.A. Thibault, „Transition to Detonation in a Large Fuel-Air Cloud, „*Combustion and Flame 61*, 285-291 (1985).
14. D.C. Bull, M.A. McLeod, G.A. Mizner, „Detonation of Unconfined Fuel Aerosols", *Progress in Astronautics and Aeronautics* Vol. 75, 48-60 (1981).
15. G. Smeets, „Michelson Spectrometer for Instantaneous Doppler Velocity Measurements",*J. Phys. E: Sci. Instrum.* 14, 838-845 (1981).
16. G. Smeets, G. Mathieu, „*Optische Doppler-Messungen mit dem Michelson-Spektrometer*", ISL-Report R 123/83 (1983).
17. Hirth, „*Les lasers à vapeurs métalliques et leurs applications*", ISL-Report CO 218/84 (1984)
18. Samirant, G. Smeets, C. Baras, M. Royer, LR. Oudin *Dynamic Measurements in Combustible and Detonable Aerosols Propellant Explosives Pyrotechnics*, Vol. 14, pp. 47-56, 1989.

HYDROGEN FIRE AND EXPLOSION SAFETY
OF ATOMIC POWER PLANTS

Yu. N. SHEBEKO
All Russian Scientific Research Institute for Fire Protection
Balashiha-3, Moscow Region, 143900, Russia

1. Introduction

Formation and accumulation of hydrogen in rooms and equipment of atomic power plants with PWR and BWR is a process, which takes place both in normal and accidental regimes of operation. If no protective measures were not undertaken, an accident with combustion of hydrogen contained mixtures is possible. One of such accidents took place in Three Mile Island (TMI-2) atomic power plant in 1979. After this accident an interest to the problem of hydrogen safety of atomic power plants has elevated significantly. Many investigations have been executed in this research area (see, for example, [1-5]). The experimental and theoretical investigations were carried out also in All Russian Scientific Research Institute for Fire Protection. The main results of these investigations are presented in this paper.

2. The Peculiarities of Hydrogen Propagation in a System of Interconnected Rooms

One of the important aspects of hazard of rooms with presence of hydrogen is distribution of H_2 concentration. This distribution depends on peculiarities both of hydrogen propagation between connected rooms and in each room. The hydrogen concentration distribution in a single room has been studied rather well (see, for example, [6]), but a process of hydrogen propagation in a system of rooms connected with each other has been investigated much more poor. The results of experimental investigation of this process are presented in this section.

The main part of the set-up is a chamber in the form of vertical cylinder of volume of 20 m^3 (height 5.48 and diameter 2.2 m). Some partitions are placed into the chamber, which divide the chamber on nine model rooms. The model rooms have the following volumes: V_1=1.2 m^3, V_2=1.4 m^3, V_3=1.16 m^3, V_4=0.99 m^3, V_5=12.6 m^3, V_6=2.14 m^3, V_7=5 · 10^{-2} m^3, V_8=0.23 m^3, V_9=0.23 m^3. These model rooms are connected with each other by perforations, which have areas F indicated in Table 1.

V.E. Zarko et al. (eds.), Prevention of Hazardous Fires and Explosions, 135–149.

The experimental set-up has the following systems;
- system for supply of hydrogen into the chamber;
- system for removal of hydrogen from the camber;
- system for measuring the hydrogen concentrations.

Hydrogen supply was each from laboratory tanks of volume of 40 dm^3 and initial pressure up to 15 MPa through special pressure reductors and valves with measurement of gas flow rate. Hydrogen was introduced into the chamber through a metal tube, placed in the lower part, and then was supplied to a required model room by a special flexible tube. Metal and flexible tubes have internal diameter of 10 mm.

Hydrogen removal from the chamber was made by evacuation of the volume of the chamber to a residual pressure not higher than 0.1 kPa.

The system for measurements of hydrogen concentrations includes 20 catalytical detectors, source of direct electrical current and multi-channel oscilloscope. The error in determination of H_2 concentrations did not exceed 0.2 % (vol.).

TABLE 1. Perforation areas between model rooms (m^2)

$V_i \backslash V_j$	V_1	V_2	V_3	V_4	V_5	V_6	V_7	V_8	V_9
V_1	0	$3 \cdot 10^{-2}$	$3.45 \cdot 10^{-1}$	0	$8.1 \cdot 10^{-2}$	0	0	$2.8 \cdot 10^{-1}$	0
V_2	$3 \cdot 10^{-2}$	0	0	$3.6 \cdot 10^{-1}$	$15 \cdot 10^{-1}$	0	0	0	$2.8 \cdot 10^{-1}$
V_3	$3.45 \cdot 10^{-1}$	0	0	0	$4.1 \cdot 10^{-2}$	$6.8 \cdot 10^{-1}$	0	$1.4 \cdot 10^{-1}$	$1.4 \cdot 10^{-1}$
V_4	0	$3.6 \cdot 10^{-1}$	0	0	$4.3 \cdot 10^{-2}$	$5.5 \cdot 10^{-1}$	0	$1.4 \cdot 10^{-1}$	$1.4 \cdot 10^{-1}$
V_5	$1.8 \cdot 10^{-2}$	$1.5 \cdot 10^{-1}$	$4.1 \cdot 10^{-2}$	$4.1 \cdot 10^{-2}$	0	0	0	0	0
V_6	0	0	$6.8 \cdot 10^{-1}$	$5.5 \cdot 10^{-1}$	0	0	$7 \cdot 10^{-2}$	0	0
V_7	0	0	0	0	0	$7 \cdot 10^{-2}$	0	0	0
V_8	$2.8 \cdot 10^{-1}$	0	$1.4 \cdot 10^{-1}$	$1.4 \cdot 10^{-1}$	0	0	0	0	0
V_9	0	$2.8 \cdot 10^{-1}$	$1.4 \cdot 10^{-1}$	$1.4 \cdot 10^{-1}$	0	0	0	0	0

In experiments variations in hydrogen flow rate and positions of gas supply were made. The total mass of hydrogen entering experimental chamber was equal to 640±20 dm^3. Hydrogen was introduced into the model rooms V_1, V_4, V_5, V_6. In model room V_1 hydrogen was supplied horizontally at a distance of 25 cm from the floor of this room. In model room V_4 hydrogen was introduced also horizontally at distances 15 cm from the room ceiling and 10 cm from the vertical wall. In model room V_6 hydrogen was introduced upward along the axis at a distance of 25 cm from the room ceiling. In the model room V_5 hydrogen was also introduced upward at distances of 15 cm from the vertical wall and 10 cm from the room ceiling.

The following peculiarities of hydrogen spreading in the system of rooms connected with each other have been revealed in our experiments.

1. The preferred direction of hydrogen spreading is upward direction. At hydrogen bleeding in the lower part of the room system it is detected in all model rooms (including the upper model room) after a rather short time. An equalisation of hydrogen concentrations in all model rooms occurs at upward H_2 bleeding after 10-15 min from the time moment of finishing the hydrogen supply.

2. There are vertical gradients of hydrogen concentration in the model rooms. These gradients diminish in time. The highest vertical gradient was observed in the model room V_5 (this is the largest room).

3. Differences in H_2 concentrations at various points of the given model room at equal heights did not exceed 0.2 % (vol.), that is these concentrations are equal to each other in the limits of experimental error.

4. Hydrogen concentrations in volumes located at the same height are close to each other, if hydrogen is bleeding in another model volume. Difference in these concentrations does not exceed 0.5 % (vol.) during H_2 supply into chamber. After stop of hydrogen supply (time interval 5-10 min) these concentrations are equal to each other in the limits of experimental error.

5. The longest equalisation of hydrogen concentrations occurs, if H_2 is supplied in the upper level model room (volume V_5).

Data obtained in this part of the work characterise qualitatively and quantitatively peculiarities of hydrogen spreading in the system of volumes connected with each other. These data can be used for verification of computer codes describing process of hydrogen propagation.

3. Combustion Characteristics of Gaseous Mixtures at Elevated Pressures and Temperatures

Determination of combustion characteristics of gaseous mixtures containing hydrogen at elevated pressures and temperatures is very important for modeling of accidents at atomic power plants and creation of measures for fire and explosion safety ensuring. The appropriate experimental data obtained by methods described in Ref. [5] are presented in this part of the work.

Figure 1 shows flammability limits of hydrogen-oxygen mixtures with four diluents (CO_2, He, N_2, and Ar) at 0.1 MPa and two initial temperatures: 25 °C in Fig. 1a and 250 °C in Fig. 1b. Previous data [7,8] are included for comparison.

In Fig.2, flammability limits are shown for hydrogen-oxygen with three diluents (N_2, He, and steam) over a wider range of pressures and temperatures. In Fig 2b, the unusual effect of He, first observed in Fig.1, is explored in more detail. Of particular interest is the rapid change of low flammability limit, that occurs at low He concentrations, but which is not observed at higher diluent levels. The same pattern was found at atmospheric as well as at elevated pressures. This effect it is due to the influence of preferential diffusion on the combustion of very lean hydrogen mixtures (containing less than 8 % vol.of H_2 [9]).

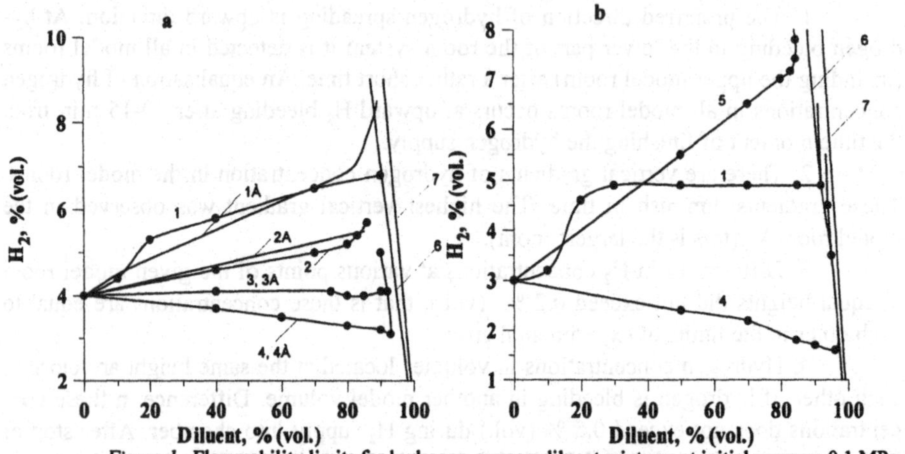

Figure 1. **Flammability limits for hydrogen-oxygen-diluent mixtures at initial pressure 0.1 MPa and temperature 20 (a) and 250 (b) °C. 1 - H_2 - He - O_2; 1a - H_2 - He - O_2 /8/; 2 - H_2 - CO_2 - O_2; 2a - H_2 - CO_2 - O_2 /8/; 3 - H_2 - N_2 - O_2; 3a - H_2 - N_2 - O_2/8/; 4 - H_2 - Ar - O_2; 4a - H_2 - Ar - O_2/8/; 5 - H_2 - H_2O - O_2; 6 - stoichiometric line; 7 - line that restricts region of possible mixtures.**

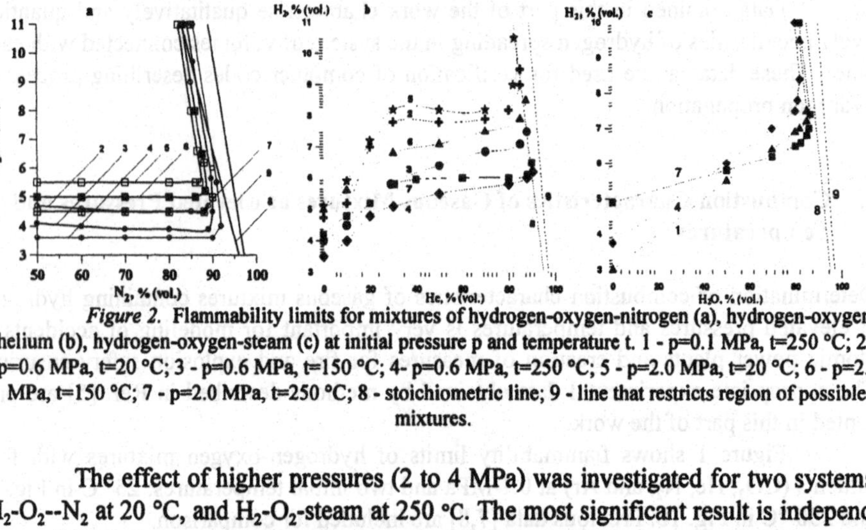

Figure 2. **Flammability limits for mixtures of hydrogen-oxygen-nitrogen (a), hydrogen-oxygen-helium (b), hydrogen-oxygen-steam (c) at initial pressure p and temperature t. 1 - p=0.1 MPa, t=250 °C; 2 - p=0.6 MPa, t=20 °C; 3 - p=0.6 MPa, t=150 °C; 4- p=0.6 MPa, t=250 °C; 5 - p=2.0 MPa, t=20 °C; 6 - p=2.0 MPa, t=150 °C; 7 - p=2.0 MPa, t=250 °C; 8 - stoichiometric line; 9 - line that restricts region of possible mixtures.**

The effect of higher pressures (2 to 4 MPa) was investigated for two systems: H_2-O_2--N_2 at 20 °C, and H_2-O_2-steam at 250 °C. The most significant result is independence of limit on pressure. Data for pressures 2, 3, and 4 MPa are close between each other, and within experimental error, data at 3 and 4 MPa coincide. This is in qualitative agreement with previous data [10, 11].

Figure 3 presents the results of determination of pressure dependence of burning velocity for stoichiometric hydrogen-air mixtures, at various temperatures and diluent (N_2) concentrations. For comparison, previous results [12] are shown for stoichiometric hydrogen-air mixtures diluted by steam. These were obtained in the same apparatus and under the same conditions of pressure and temperature. Two sets of data show qualitative and quantitative differences, as follows.

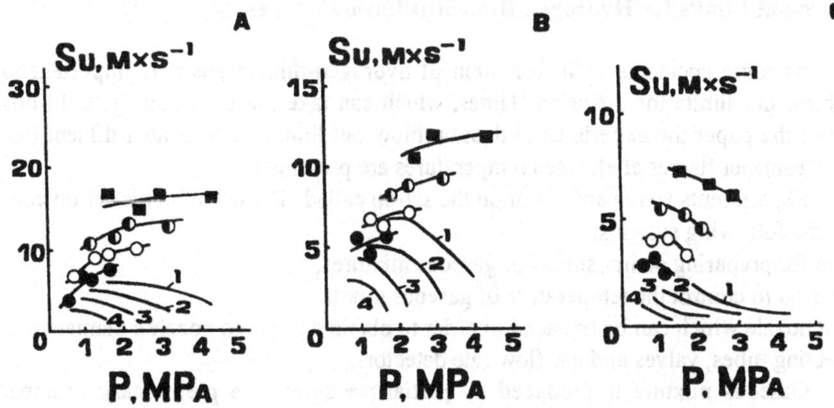

Figure 3. **Dependence of burning velocity S_u of stoichiometric hydrogen-air mixtures diluted by nitrogen and steam on pressure p at various temperatures t and diluent concentrations C=10 % (vol.) (a), 20 % (vol.) (b), 30 % (vol.) (c). Diluent-nitrogen: ●, t = 200 °C; O, t = 250 °C; ◐, t = 300 °C; ■, t = 350 °C. Diluent-steam /8/: 1 - t = 350 °C; 2 - t = 300 °C; 3 - t = 250 °C; 4 - t = 200 °C.**

First, at equal diluent concentrations, the burning velocities are higher when N_2 is the diluent. Second, when comparing two sets of data for the same diluent concentrations, there are observed clear qualitative differences in the shape of the curves. For 10% and 20% diluent concentrations, the pressure index of burning velocity is negative for steam and positive for nitrogen. This influence cannot be explained solely by differences in the molar heat capacities of the diluents. Under conditions when the formation of aerosol by condensing of steam is negligible (which we believe holds at given ranges temperature and pressure for steam levels of 1-20% (vol.)), the molar heat capacity of steam is only 1.3 times larger than that of nitrogen. But at the same time, S_u values for 20% nitrogen are considerably higher of those for 10%steam. This effect, including the change in sign of the pressure exponent for stoichiometric hydrogen-air mixtures diluted by steam, has been explained [12] in terms of the active role of water molecules in trimolecular recombination processes. According to Ref. [13], the rate constants of the trimolecular reactions

$$H + H + M \rightarrow H_2 + M,$$
$$H + OH + M \rightarrow H_2 + M,$$
$$H + O + M \rightarrow OH + M,$$
$$H + O_2 + M \rightarrow HO_2 + M,$$

which are important in the flame front, differ significantly for various chaperon molecules M. The effectiveness of the molecules H_2, O_2, and H_2O (that is, the relative rate constants of the above reactions) are in the ratio 1:0.4:6 [13]. Thus, addition of steam to hydrogen-air mixtures not only reduces the temperature, but also enhances the loss of active radicals from the flame front, in comparison to hydrogen-air mixtures, by increasing the rates of their recombination reactions. The simultaneous influence of these two mechanisms provides a better explanation of the burning velocity of combustible mixtures diluted by steam compared with nitrogen.

4. Blow-out Limits for Hydrogen-Diluent Diffusion Flames

At some accidents with formation of hydrogen-diluent jets it is important to know blow-out limits for diffusion flames, which can take place on these jets. In this section of the paper the experimental data on blow-out limits for hydrogen-diluent (nitrogen, steam) jet flames at elevated temperatures are presented.

Experiments were carried out on the set-up called "Diffusion". This set-up consists of the following systems:
- system for preparing compositions of gaseous mixtures;
- thermostat to control the temperature of gaseous mixture;
- heated nozzle which can be replaced in order to obtain a required nozzle's diameter;
- connecting tubes, valves and gas flow rate detectors.

Gaseous mixture is produced by partial pressures in a preliminary chamber with volume 20 dm^3 evacuated to a residual pressure not exceeding 0.5 kPa.

The set-up has a steam-generator with an additional heater. The generator has a volume of 25 dm^3, its maximum permissible pressure is equal to 25 MPa. A manometer separated from the volume of generator by a special separator in order to prevent steam condensation in the manometer controls pressure of steam. Similar separators were used for other manometers connected with high-temperature tubes.

A flow rate of gaseous mixture was controlled by means of pressure measurements immediately before the nozzle. The heating system of the set-up "Diffusion" has heaters of the thermostat, the steam-generator and the nozzle. Temperatures of various parts of the set-up were measured by thermocouples with recording of signals on an oscilloscope.

The nozzles have diameters 1.2, 2.1 and 3.2 mm, and it was possible to vary conditions of extinguishing diffusion jet flames by changing of nozzles. An ignition of the jet was produced by a hot wire placed near the nozzle. Extinguishing the jet was detected visually and by thermocouples.

The experimental procedure for the determination of blow-out limits was the following. The required gaseous composition was prepared by partial pressures in the preliminary chamber, which has been previously evacuated and heated to the required temperature. The thermostat, nozzle and steam generator were also heated to the required temperatures. Then the discharge of gaseous mixture through the nozzle begins, and the jet was ignited at the lowest possible velocity of gas. This velocity was increased up to the blow-out of the flame. During the whole process pressure before the nozzle was recorded continuously, and then velocity of gas was recalculated by means of the well known formulas of gas dynamics [14]. Each experiment was repeated not less than 3 times.

The relative errors of determination of all measured values did not exceed 10 %. Results of the experimental determination of blow-out velocities of jet flames of hydrogen-nitrogen mixtures as a function of temperature for various nozzles' diameters are presented in Fig. 4. An increase of the blow-out velocity Vc with an elevation of the temperature T of the gaseous mixture takes place, and this dependence is close to the linear one. A physical nature of this effect is quite clear; this is an elevation of a flame temperature with a subsequent increase of rates of chemical reactions in the flame front. An effect of a decrease of the blow-out velocity Vc with increase of concentration of a fire extinguishing agent Cn has the same physical nature.

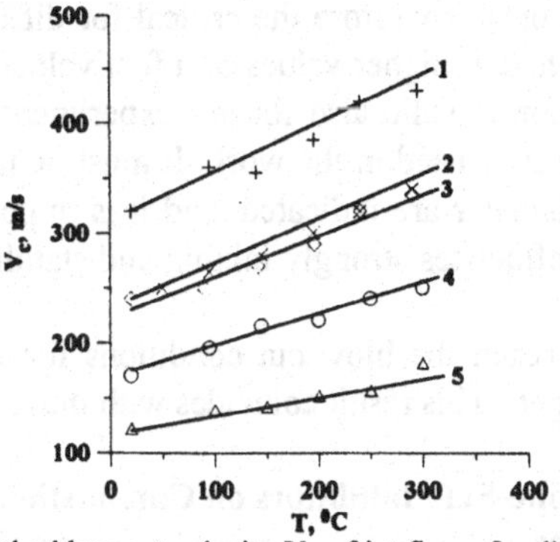

Figure 4. Dependence of the blow-out velocity V_C of jet flame for discharge of H_2-N_2 mixtures into air on temperature T for various hydrogen concentrations C_{H2} and nozzle diameter d.
1 - C_{H2}= 80 % (vol.), d = 1.2 mm; 2 - C_{H2}= 60 % (vol.), d = 1.2 mm;
3 - C_{H2}= 80 % (vol.), d = 2.1 mm; 4 - C_{H2}= 60 % (vol.), d = 2.1 mm;
5 - C_{H2}= 40 % (vol.), d = 2.1 mm;

The blow-out velocities for hydrogen-steam mixture for H_2 concentration 60 % (vol.) are presented in Table. 2.

TABLE 2. Blow-out velocities for H2/steam (60/40) mixture

Temperature, °C	Nozzle's diameter, mm	Blow-out velocity, m/s
200	1.2	260
	2.1	210
	3.2	120
250	1.2	290
	2.1	250
	3.2	230

Steam is the more effective diluent in comparison with nitrogen (compare data in Fig. 4 and in Table 2), because it has higher molar heat capacity coefficient. As in the case of hydrogen-nitrogen mixture, the blow-out velocity increases with temperature. The more rapid elevation occurs for diameter of the nozzle 3.2 mm. A cause of this effect is not clear.

It is interesting to note that the lower the diameter of the burner, the more stable a diffusion jet flame (the higher is the blow-out velocity of jet) (Fig. 4). Results qualitatively close to that obtained earlier by authors [17] when an experimental study of the stability of hydrogen-air diffusion flames was carried out. Several nozzles with different internal diameters d and lip thickness l were used (d=1.7-8.2 mm, l=0.07-0.75 mm, l/d=0.017-0.44). A variation of flow rates of hydrogen and air was made. It was found, that at some air flow rates the blow-out velocity of hydrogen increases with a decrease of the nozzle diameter d at constant l value and with an increase of the lip thickness l at constant d value. This effect was qualitatively explained by authors [17] in a following manner. With decrease of d at constant l the smaller mass flow rate of hydrogen exists at a point where stoichiometric mixtures of H_2 and air are formed (the position of a flame front). Therefore the critical for diffusion flame mass flow rate of hydrogen can be reached at higher values of a fuel velocity through the nozzle. For our opinion this explanation is valid also for our experiments because the ratios of l/d are relatively high for nozzles used in the work. It must be noted that in Ref. [14-16] only values of a burner diameter are indicated, and it is impossible for reader to determine the l/d value which influences strongly mixing and stability conditions for diffusion jet flames.

We did not reach the blow-out conditions for a jet diffusion flame at a discharge of pure hydrogen. This result coincides with those obtained in [14-17].

5. Influence of Ozone-Safe Inhibitors on Combustion Characteristics of Hydrogen-Air Mixtures

Up to now brominated halons were widely used for fire extinguishing, explosion prevention and explosion suppression. But according to the well-known Montreal convention an application of brominated halons, which can strongly destroy the ozone layer of the Earth, must be gradually finished. Therefore a problem of search of inhibitors, which can substitute brominated halons, arises. An action of halons can be explained not only as a result of dilution, but firstly as consequence of inhibition of combustion processes proceeding by chain-branching mechanism. Inhibitor molecules react efficiently with active intermediate products (atoms, radicals), which are carries of reaction chains, producing products, which are not able to regenerate carries of reaction chains.

Halons, in which number of halogen atoms is not less than number of hydrogen atoms, are as a rule non-combustible. On the other hand, combustion of gaseous mixture can at some conditions cause an oxidation of halons. If combustion of a halon is exothermic enough, associated oxidation of main combustible and the halon takes place. This effect of chemical induction has been revealed in our work.

This part of work is directed to an experimental investigation of influence of inhibitors, which are safe for the ozone layer of Earth, on combustion characteristics of H_2 - air mixtures.

Experiments have been carried on a set-up called "Variant", which is described in detail in Ref. [5]. The following inhibitors were investigated: CF_3H, C_2F_5H, C_3F_7H, $C_3F_6H_2$, CF_2ClH, C_2F_5Cl, C_2F_5I, C_4F_8 (perfluorocyclobuthane), C_4F_{10}, NAFS - III, Inh A_1.

The main results of our experiments are presented in Figs. 5, 6. In Fig. 5 flammability limits for mixtures of hydrogen-air-diluent are presented. In Fig. 6 typical dependencies of maximum explosion pressure ΔP_{max} for combustible mixtures of various compositions on a diluent concentration [Inh] are presented. For convenience, these data are presented in dimensionless form by normalisation on maximum explosion pressure ΔP_{max} of gaseous mixtures without inhibitors. The normalisation constants ΔP_{max} are the following: $\Delta P_{max\,o}$ = 290, 460, 640, 520 kPa for H_2 concentrations in air 10, 20, 30, 40, 50 % (vol.).

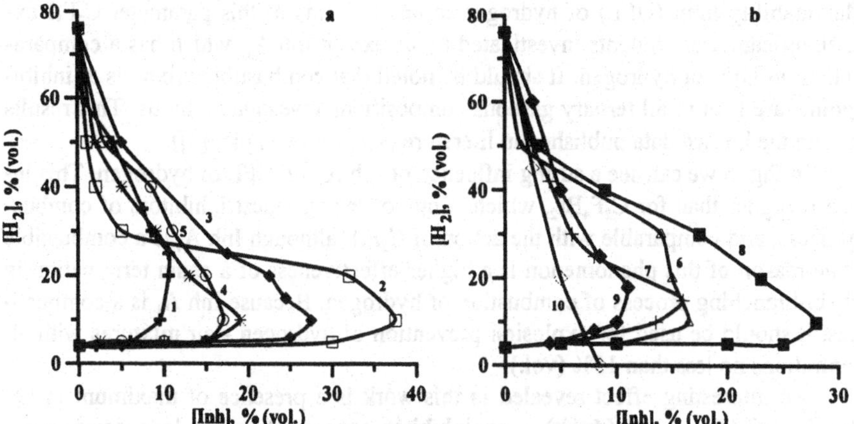

Figure 5. Flammability limits in mixtures of hydrogen-air-inhibitor.
a: 1 - C_3F_7H; 2 - C_2F_5I; 3 - CF_3H; 4 - CF_2ClH; 5 - C_4F_{10};
b: 6 - C_4F_8; 7 - C_2F_5H; 8 - C_2F_5Cl; 9 - $C_2F_4Br_2$; 10 - Inh A_1.

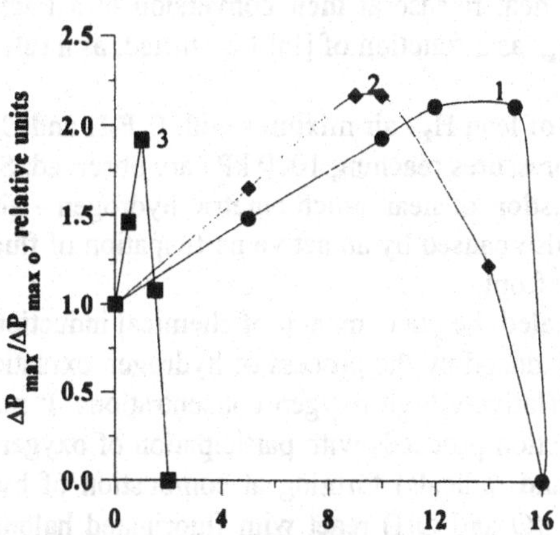

Figure 6. Typical dependencies of maximum explosion pressure of gaseous mixtures on inhibitor concentration.
1 - $[H_2]$ = 10 % (vol.), 2 - $[H_2]$ = 20 % (vol.),
3 - $[H_2]$ = 50 % (vol.), diluent - C_4F_8 (perfluorocyclobuthane).

At a transition from diluents with one carbon atom (CF_3H or CF_2ClH) to inhibitors with two C atoms the inhibition effectiveness increases, because minimum inhibition concentrations (MIC) of diluents decrease. But transition from inhibitors with two carbon atoms to diluents with three and four C atoms does not give a sufficient change in MIC. An availability of Cl atom in an inhibitor molecule causes a reduction in MIC, because energy of a chemical bond C-Cl is lower than that for chemical bond C-F.

Minimum inhibition concentrations of C_2F_5I are sufficiently higher than MIC of C_2F_5H. At the same time we should note a very strong influence of C_2F_5I on the upper flammability limit (UFL) of hydrogen in air. In terms of this parameter C_2F_5I exceeds significantly all diluents investigated by us except Inh A_1, which has a comparable action on UFL of hydrogen. It should be noted that combustible mixtures at inhibition points are lean in all ternary gaseous compositions investigated by us. This results agree with the known data published in literature (see, for example, [8]).

In Fig. 5 we can see a strong influence of Inh A_1 on UFL of hydrogen. This influence is higher than for $C_2F_4Br_2$, which is one of the strongest inhibitors of combustion process, and comparable with the action of C_2F_5I, although Inh A_1 is a combustible gas. The reason of this phenomenon is a higher effectiveness of a chain termination in the chain-branching process of combustion of hydrogen. Because Inh A_1 is a combustible gas, it should be used for explosion prevention of hydrogen - air mixtures with H_2 concentrations no less than 15% (vol.).

An interesting effect revealed in this work is a presence of maximum in dependencies of ΔP_{max} and $(dp/dt)_{max}$ on inhibitor concentration for lean combustible mixtures, despite the fact that all diluents investigated in the work are non-combustible (except Inh A_1). This effect is caused by participation of fluorinated halons in combustion processes with a heat release at their conversion in a flame front. A position of these maxima for ΔP_{max} as a function of [Inh] is shifted, as a rule, to the minimum concentration of inhibitor.

At a dilution of lean H_2 - air mixtures with C_3F_7H and C_4F_8 an anomalous high maximum explosion pressures reaching 1000 kPa are observed. Such high ΔP_{max} do not realise even at combustion of near-stoichiometric hydrogen - air mixtures in a closed vessel. This result is also caused by an active participation of fluorinated inhibitors in a heat release in a flame front.

Thus we revealed the phenomenon of chemical induction, when combustion of fluorinated halons is induced by the process of hydrogen oxidation in lean combustible mixtures containing relatively high oxygen concentrations. It means that the chemical reaction limiting induction proceeds with participation of oxygen and active intermediate products (atoms and radicals) forming at combustion of hydrogen. Two of these intermediate products (O and OH) react with fluorinated halons rather slow, because chemical bonds O-F and HO-F formed in these reaction have relatively low break energies. But reaction of separation of halogen atom from a halon molecule by atomic hydrogen is exothermic enough. Halogen radical, which is produced in this reaction, reacts with O_2 and this stage is probably limiting.

One of the possible mechanisms of a heat release in a flame front at combustion of fluorinated hydrocarbons is their conversion firstly into C_2F_4 by means of active radicals. Then tetrafluoroethylene takes part in combustion process and stipulates a heat release.

It should be noted that the phenomenon of combustion of fully fluorinated hydrocarbons induced by a hydrogen flame has been revealed in Ref. [18]. In this work combustion of H_2 - O_2 - C_2F_6 mixture was investigated. It has been found that neither C_2F_6 - O_2 mixture nor C_2F_6 - H_2 mixture are able to burn. At the same time the H_2 - O_2 - C_2F_6 mixture can generate flame at some concentrations of components, and in combustion zone some products of conversion of C_2F_6 are present (such as CF_4, COF_2 etc.) but the initial reactants are absent. This means that C_2F_6 burns in such flame.

6. Fire and Explosion Safety Ensuring by Means of Passive Catalytic Converters of Hydrogen

For the prevention of formation of hydrogen-air combustible mixtures passive catalytic H_2 - converters can be used, in which non-flame hydrogen combustion on a catalytic surface takes place. Some result of studying the process is presented in this part of the work.

Experiments have been executed on a set-up which main part is a reaction vessel with a volume of 50 dm^3 having a form of a closed vertical cylinder with a diameter of 300 mm and a height of 800 mm. A catalytic hydrogen convertor that consisted of a set of catalytic rods was placed inside the reaction vessel in its middle part. The stoichiometric hydrogen-oxygen mixture ($2H_2$+O_2) was prepared in a special high pressure mixer by partial pressures and then was introduced into the lower part of the reaction vessel. A gas flow rate was measured by a continuos pressure registration in a mixer during the whole experiment. Three hydrogen concentration detectors were placed in the lower, middle and upper parts of the reaction vessel. Signals from these detectors were recorded on a multi-channel oscillograph. The reaction vessel was connected by a tube with an intermediate cavity, which had a hydrogen concentration detector. The gaseous mixture from the reaction vessel was supplied into intermediate vessel and then was discharged into the atmosphere through the flow rate detector. Because of non-flame hydrogen and oxygen recombination on the catalytic surface and condensation of steam, formed due to the chemical reaction, the gas flow rate on the entry of the reaction is greater than that one on its exit.

The two catalytic converter structures were investigated. The structure 1 represents a combination metal tubes with internal diameter 20 and height 75 mm. The rods with a catalytic surface were placed along the axis of tubes. The catalytic surface has been made of Al_2O_3 with platinum centres and was covered by a porous teflon film permeable for gases and steam. The rods diameter and height were 5.8 and 63 mm, respectively. Each rod had an axial channel with a diameter of 1 mm. The total rods number in structure 1 was 19.

The structure 2 represents a collection of cells each in the form of equilateral triangle with a side 21 mm, placed along a perimeter with sides 70 and 140 mm. The catalytic rods were placed in the centres of these cells. There are 32 rods in this structure.

The structure 3 was quite the same as the construction 1 except the number of rods, which was equal 7.

There was one rod in each structure, for which the temperature of a catalytic surface was measured by a thermocouple. Those rods were placed in the centres of the structures 1 and 3 or near the side surface for the structure 2. Signals from thermocouples were recorded by a multi-channel oscillograph.

Some experiments with the structure 3 have been executed for the reaction vessel filled with an inert gas (nitrogen, carbon dioxide, and helium) instead of air. In these experiments only the catalytic surface temperature measurements were made.

The typical experimental data obtained for structure 1 are presented in Fig. 7. Despite the flow rate of stoichiometric hydrogen-oxygen mixture at the entry of the reaction vessel is constant, the flow at the exit of the reaction vessel after approximately 5 min from the experiment beginning drops to a zero value. This fact is due to the total hydrogen-oxygen mixture recombination and steam condensation. Simultaneously with the exit flow rate drop the hydrogen concentration in the reaction vessel decreases. The time dependencies, obtained for structures 1 and 2, are qualitatively the same.

At higher gas flow rates the catalytic surface temperature becomes high enough in order to ignite gaseous mixtures. The critical values of these flow rates were 5000 and 2000 cm^3/min for structures 1 and 2 respectively, and 260 and 62.5 cm^3/min per one rod of mentioned above structures.

Figure 8 shows the results of measurement of the catalytic surface temperature for the structure 3 at various initial gaseous mixtures. The data for air, nitrogen and carbon dioxide are qualitatively and quantitatively close to each other. Practically simultaneously with terminating the hydrogen-oxygen mixture supply the temperature reduction begins. The maximum temperature value depends on both molar heat capacity of gaseous mixture (for nitrogen this value is lower, than for carbon dioxide) and oxygen concentration in the reaction vessel volume (for air the maximum is greater than for nitrogen, because the oxidation rate of hydrogen on a catalytic surface depends on oxygen concentration).

The time dependence of a catalytic surface temperature for helium environment differs from the analogous dependencies for air, nitrogen and carbon dioxide by a more slow temperature rise rate. The temperature rise continues after the hydrogen-oxygen mixture flow stopping. This effect is caused probably by a high thermal conductivity of helium. In this case a catalytic surface heating occurs more slowly than for other gases considered. Therefore even with a hydrogen-oxygen mixture flow stopping rather large amount of hydrogen remains in the reaction vessel, and oxidation after termination of the combustible mixture supply causes the further increase of the catalytic surface temperature. The maximum value of temperature is however much lower than for other gases considered.

A comparison of the dependencies of stationary mean hydrogen concentrations in the reaction vessel C_s on a specific hydrogen-oxygen mixture flow rate per one rod W_s for the structures 1 and 2 is made. Though the structure 1 contains less catalytic rods than the structure 2 (19 instead of 32), it is more effective from the viewpoint of the lowest stationary hydrogen concentration the viewpoint of an operation possibility at more high combustible mixtures flow rates without the gas ignition. This effect is accounted by a more effective supply of a hydrogen contained gaseous mixture to a catalytic surface because of convective flows formation from the heated catalytic surface through metal tubes in the structure 1.

The proposed construction of the passive catalytic covnertor can be used on atomic power plants in order to remove hydrogen in rooms and equipment, where it can be generated via radiolysis and other factors.

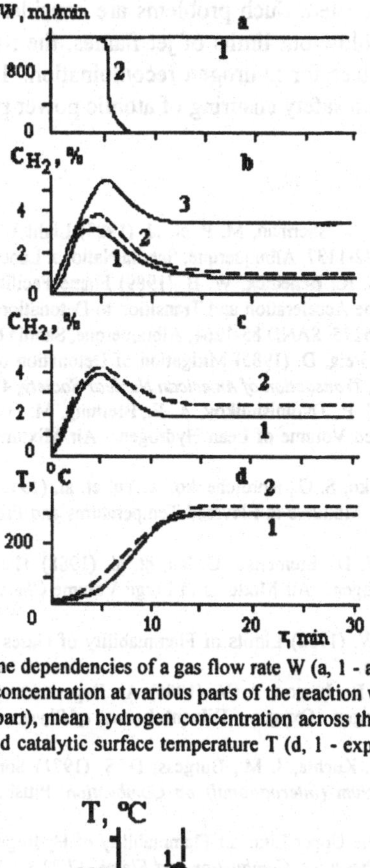

Figure 7. Typical time dependencies of a gas flow rate W (a, 1 - at the reaction vessel entry, 2 - at reaction vessel exit), hydrogen concentration at various parts of the reaction vessel (b,1 - in the upper part, 2 - in the middle, 3 - in the lower part), mean hydrogen concentration across the reaction vessel volume (c, 1 - experiment, 2 - theory [19]), and catalytic surface temperature T (d, 1 - experiment, 2 - theory [19]) for the structure 1.

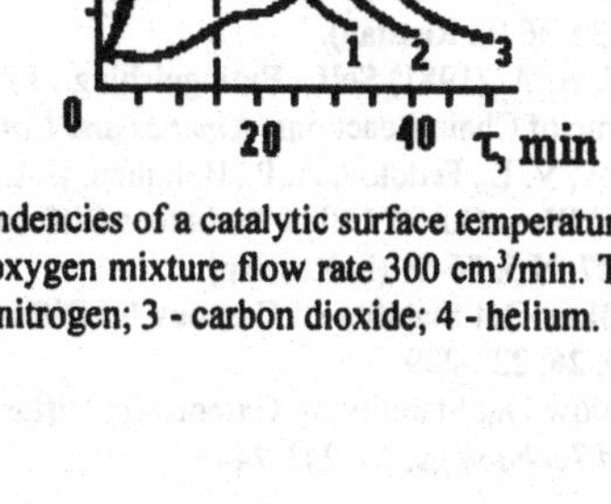

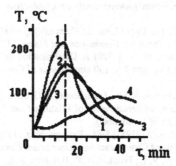

Figure 8. Typical time dependencies of a catalytic surface temperature T at various initial gaseous mixture compositions at the hydrogen-oxygen mixture flow rate 300 cm³/min. The hydrogen-oxygen mixture supply stopped after 15 min. 1 - air; 2 - nitrogen; 3 - carbon dioxide; 4 - helium.

7. Conclusions

In this paper a survey of investigations made in All Russian Scientific Research Institute for Fire Protection last years in the area of hydrogen safety of atomic power plants has been executed. Such problems are considered as hydrogen propagation, flammability limits, blow-out limits of jet flames, the role of initial pressure and temperature, catalytic devices for hydrogen recombination. The obtained results form scientific basis for hydrogen safety ensuring of atomic power plants.

8. References

1. Camp, A. L., Cummings, J. S., Sherman, M. P. et. al. (1983) Light Water Reactor Hydrogen Manual, NUREG /CR-2726. SAND 82-1137, Albuquerque, Sandia National Laboratory.
2. Sherman, M. P., Tieszen, S. R., Benedick, W. B. (1989) Flame Facility. The Effect of Obstacles and Transverse Venting on Flame Acceleration and Transition to Detonation for Hydrogen - Air Mixtures at Large Scale , NUREG /CR-5275. SAND 85-1264, Albuquerque, Sandia National Laboratory,.
3. Kumar, R. K., Skraba, T., Greig, D. (1985) Mitigation of Detonation of Hydrogen - Oxygen - Diluent Mixtures in Large Volumes, *Transaction of American Nuclear Society*, **49**, 255-257.
4. Dorofeev, S. B., Sidorov, V. P., Dvoinishnikov, A. E., Breitung, M. (1996) Deflagration to Detonation Transition in Large Confined Volume of Lean Hydrogen - Air Mixtures,*Combustion and Flame*, **104**, N1/2, 95-110.
5. Shebeko, Yu. N., Tsarichenko, S. G., Korolchenko, A. Yu. et. al. (1995) Burning Velocities and Flammability Limits of Gaseous Mixtures at Elevated Temperatures and Pressures,*Combustion and Flame*, **102**, N3/4, 427-437.
6. Shebeko, Yu. N., Keller, V. D., Eremenko, O. Yu. et. al. (1988) The Regularities of Formation and Combustion of Local Hydrogen - Air Mixtures in Large Volume,*Chemical Industry*, N12, 728-731 (in Russian).
7. Coward, H. F., Jones, G. W. (1952) Limits of Flammability of Gases and Vapours,*Bureau of Mines Bulletin*, N 503, Washington.
8. Shebeko, Yu. N., Iliin, A. B., Ivanov, A. V. (1984) An Experimental Investigation of Flammability Limits in Mixtures of Hydrogen - Oxygen – Diluent,*Journal of Physical Chemistry*, **58**, N4, 862-865 (in Russian).
9. Furno, A. L., Cook, E. V., Kuchta, J. M., Burgess, D. S. (1971) Some Observations of Near Limit Flames, *Thirteenth Symposium (International) on Combustion*, Pittsburgh, The Combustion Institute, 593-599.
10. Holmstedt, G. B. (1971) The Upper Limit of Flammability of Hydrogen in Air, Oxygen and Oxygen - Inert Mixtures at Elevated Pressures, *Combustion and Flame*, **17**, N 3, 229-301.
11. Kogarko, S. M., Ryabikov, O. V. (1970) A Determination of Flammability Limits in Oxyhydrogen Mixtures at Initial Pressures from 1 to 100 atm,*Physics of Combustion and Explosion*,**6**, N 3, 406-407 (in Russian).
12. Shebeko, Yu. N., Korolchenko, A. Yu., Tsarichenko, S. G. et. al. (1989) Influence of Initial Pressure and Temperature on Combustion Characteristics of Mixtures Containing Hydrogen, *Physics of Combustion and Explosion*, **25**, N 2, 32-36 (in Russian).
13. Azatyan, V. V., Shavard, A. A. (1981) Self - Extinguishing of Hydrogen Combustion and Some Aspects of Non-Isothermal Regime of Chain Reactions, *Kinetics and Catalysis*, **22**, N 4, 101-106 (in Russian).
14. Golinevich, G. E., Karpov, V. L., Fedotov, A. P., Bolodian, I. A., Makeev, V. I., Permyakov, A.P. (1991) Natural Stabilization and Blow-Out of Lifted Turbulent Diffusion Gaseous Jet Flame,*Physics of Combustion and Explosion*, **27**, N 5, 76-81 (in Russian).
15. Kalgatgi, G. T. (1981) Blow-Out Stability of Gaseous Jet Diffusion Flame. Part I. Still Air,*Combustion Science and Technology*, **26**, 223-239.
16. Kalgatgi, G. T. (1981) Blow-Out Stability of Gaseous Jet Diffusion Flame. Part II. Effect of Cross Wind, *Combustion Science and Technology*, **26**, 241-244.

17. Vranos, A., Taback, E. D., Shipman, C. W. (1968) An Experimental Study of the Stability of Hydrogen-Air Diffusion Flames, Combustion and Flame, 12, N 3, 253-260.
18. Groomes, E.E. (1966) The Combustion of Hydrocarbons and Fluorosubstituted Hydrocarbons with Nitrogen Trifluoride and Nitrogen Trifluoride-Oxygen Mixtures, Combustion and Flame, 10, N 1, 71-77.
19. Shebeko, Yu.N., Trunev, A.V., Shepelin, V.A. et al. (1995) An Investigation of Non-Flame Combustion of Hydrogen on a Catalytic Surface, Physics of Combustion and Explosion, 31, N 5, 37-43 (in Russian).

HYDROGEN ACCIDENTS AND THEIR HAZARDS

À.À.VASIL'EV, À.I.VALISHEV, V.À.VASIL'EV,
L.V.PANFILOVA, Ì.À.ÒOPCHIAN
Lavrentyev Institute of Hydrodynamics, Siberian Branch of RAS
Novosibirsk State university
630090 Novosibirsk, Russia
gasdet@hydro.nsc.ru

The data on detonation hazards of hydrogen in a wide range of its concentration in mixes with oxygen and air and with water steam are presented. The critical energy of direct initiation of a multifront detonation wave is used as the criterion parameter of detonation hazards of gaseous mixtures. The comparison of experimentally measured and calculation parameters demonstrates their good correlation.

Hydrogen is a unique ecological fuel with highest chemical energy release Q (cal/g) on mass unit (some data for typical gaseous fuels and TNT are presented in Fig.1). H_2 is an explosive hazardous material, therefore the knowledge of main gasdynamic parameters of chemical reaction products is important for estimation of hydrogen hazards in each application, where H_2 is used or may be appeared. The

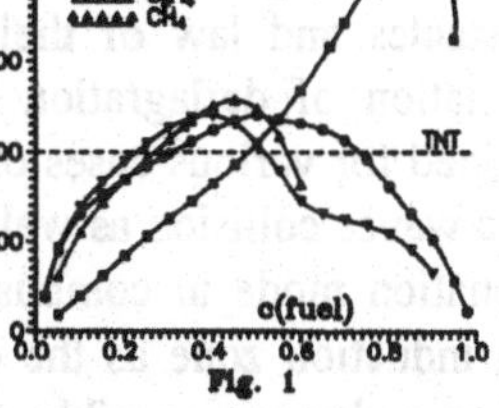

Fig. 1

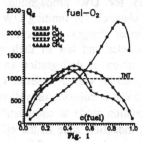

hydrogen accidents may be catastrophic with nuclear reactors on atomic power station, on nuclear submarine and above-water ships, on spacecraft and aviation technique with hydrogen engines, on desalinate water plant, on powerful electrolysis machine, etc. For example, at emergency destruction of a reactor environment at atomic power station (APS) and hit of water in a high-temperature reactor zone a formation of great amount of gaseous hydrogen (Í₂Î decomposition or Li + H_2O reaction) and its subsequent ignition in air is possible. The consequences can be heavier, if the hydrogen combustion occurs in explosion or detonation mode.

V.E. Zarko et al. (eds.), Prevention of Hazardous Fires and Explosions, 151–165.

Nowadays it is well-known, that the detonation wave (DW) in a gaseous mixtures represents quasi-stationary multifront gas-dynamics system that consists of shock and rarefaction waves, contact discontinuous and local zones of chemical reaction. The transverse waves, as the main elements of structure of detonation front, play determining role in initiation and propagation of such multifront detonation. The movement of transverse waves carries periodic character, and their trajectories form the

Fig.2

cellular structure with characteristic scale *a*, named as the cell size (Fig.2). Through *a* the another main parameters of a multifront detonation (with length dimension) may be determined: critical diameters for DW propagation from tube to volume d_{**}; geometrical characteristics of channels for limiting DW propagation d_S, l_{lim} and δ_{lim}; critical diameters of free gaseous charges d_*; linear sizes of gas charges for DW formation L_{form}; sizes of obstacles and law of their space-orientation in turbulent devices for artificial transformation of deflagration to detonation, etc. The critical initiation energy $Å_{*\nu}$ is determined for various cases of symmetry ν through *a* and gas energy $E_{0\nu}$ in area of transverse waves collision as well as diameter d_w of high-velocity bullet for excitation of a detonation mode in combustible mixtures, etc... Through *a* such kinetic parameters of an induction zone as the effective activation energy, pre-exponent and effective reaction order are possible to determine. The technique of calculation of the cell size *a* and the main parameters of a multifront detonation (through *a*) most in detail is stated in [1-5] and is used in computer Program «SAFETY» [6]. The calculations are realized within the framework of ideal gas model and concept of chemical equilibrium of products.

By analogy with low-velocity burning (as the main object of fire safety) the critical initiation energy E_* is used as a basic parameter at comparison of explosive and detonation hazards of combustible mixtures. The E_* value is defined as the minimal energy of the initiator ensuring propagation of that or other process under study in a given mixture. The less the critical initiation energy, the more hazardous the process.

In this paper the part of the most interesting results for various hydrogen mixes is presented for cases of propagation of a detonation wave (DW) and instant explosion, i.e. questions of detonation and explosion safety of hydrogen mixtures. $Ð_0 = 10^5$ Pà (1.0 àtm) and $Ò_0 = 298$ 'Ê are chosen as standard conditions; $c(Í_2)$ is the molar concentration of hydrogen in a mix. On the diagrams the vertical dashed lines (marking st-index) designate the stoichiometric concentrations: $c_* = 0.667$ for hydrogen-oxygen mixture and $c_* = 0.295$ - for mix with air. Other dotted vertical lines mark concentration limits: outside lines - lower and upper limits for flame, nearest to them internal lines - appropriate limits for a detonation.

It is necessary to note that the concentration limits of deflagration and detonation, as a rule, are determined experimentally. Even for well investigated case of ignition and low-velocity laminar combustion under standard conditions ($Ð_0 = 10^5$ Pà and $Ò_0 = 298$ 'Ê) there is a certain scatter in values of limiting concentration. The similar situation is typical for any fuel-oxygen and fuel-air explosive (FOE and FAE) mixtures. For ignition of hydrogen the standard range of concentration of H_2 in oxygen is (0.035÷0.94) and in air - (0.04÷0.75), for example, [7-10]. For detonation in H_2 - O_2 mixtures the following values are accepted - (0.15÷0.93) [9-11]; for H_2 - air mixtures there are disagreements both in lower (0.18 or 0.1) and in upper (0.59÷0.74) concentration limits for detonation [9-11]. At the same time, authors [12] report on hydrogen-air concentration limits for detonation in the range of 0.135÷0.70. It must be noticed that the scatter in experimental detonation limits is caused by two basic reasons: à) insufficient geometrical sizes of the experimental equipment for correct definition of a detonation mode; á) and insufficient initiation energy of a detonation near concentration limits. Near the limits both parameters promptly grow (details below in Figs. 13-16), that makes practically impossible realization of such researches in laboratory conditions, especially for FAE.

It is known, that in kinetic data for any fuel there is a significant disorder, therefore a problem of choice of appropriate kinetic data for description of an induction period is one of the basic at the solution of a task on detonation hazards of given fuel. For hydrogen mixtures in this paper the constants of the Arrhenious equation for an induction period are used from Ref. [13]: A = 5.38 10^{-5} μs mole/l, E = 17150 cal/mole, $k_1 = 0$, $k_2 = 1$, $k_3 = 0$,

$$\tau = \frac{A \cdot \exp(E / RT)}{[f]^{k_1}[o]^{k_2}[in]^{k_3}},$$

or its logarithmic analogue

$$\lg\{[f]^{k_1}[o]^{k_2}[in]^{k_3} \cdot \tau\} = A + B / T,$$

where E is the activation energy of rate control reaction of an induction period, R is the universal gas constant, T is the mixture temperature in an induction zone. The quantities in square brackets are concentrations of a mixture component (f - fuel, o - oxidizer, in - inert additive), A and k_i are the numerical coefficients.

The DW chemical energy-release Q (cal/g) is illustrated on Fig.1 showing of exceptional energy supply sources of a hydrogen-oxygen mixture in comparison with other fuels: H_2 - maximal Q-value equals 2255, CH_4 - 1285, C_2H_2 - 1206, C_2H_4 - 1205, etc. while TNT - 1000 cal/g. For FAE the maximal Q-values are lower: H_2 - 752, C_2H_2 - 644, C_2H_4 - 590, CH_4 - 571...

In Figs. 3-4 the values of detonation velocity D_0 (m/s) and DW characteristic temperatures (in Ê) are presented: T_2 is the gas temperature in an induction zone, Ò is that for reaction products and $Ò_3$ is that for reflected gas in induction zone. The DW velocity monotonously grows with increasing hydrogen molar concentration and only close to the upper limit begins to decrease. For FOE maximal D_0 ~ 3800 m/s is observed

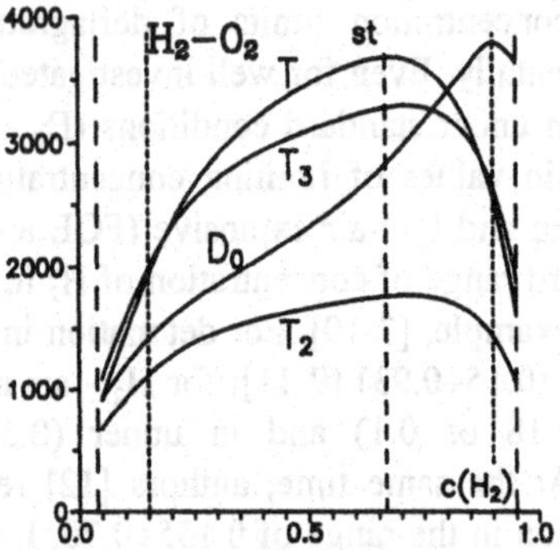

Fig. 3

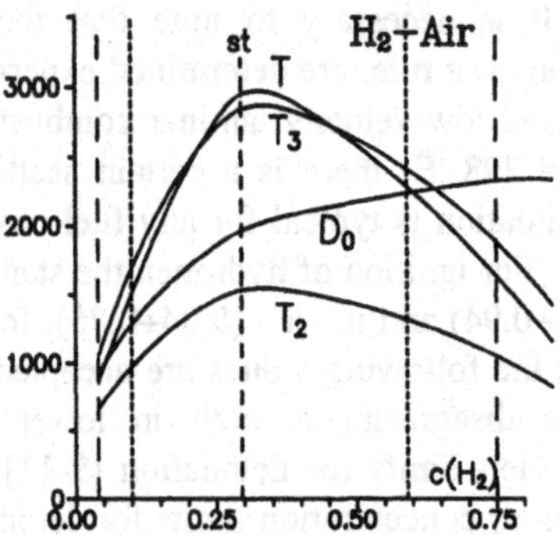

Fig. 4

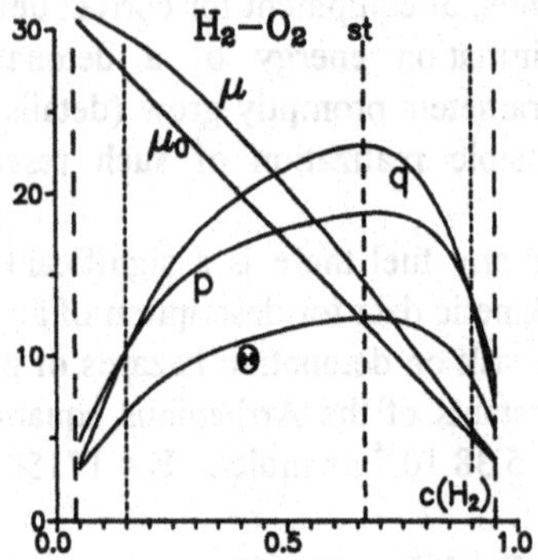

Fig. 5

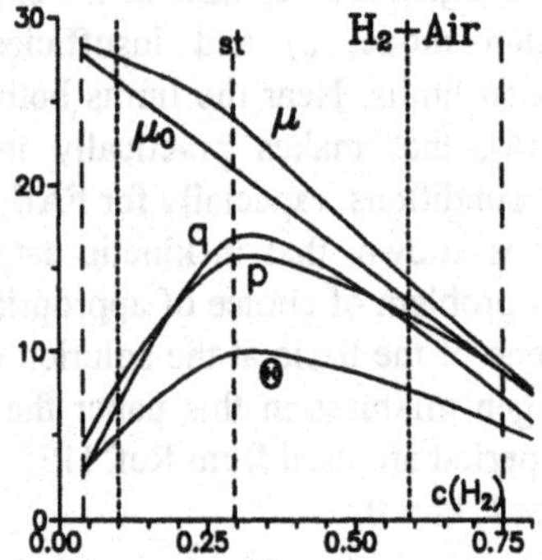

Fig. 6

in a mixture with $c = 0.9$, for FAE the maximum D_0 ~ 2345 m/s is located with $c = 0.8$. For stoichiometric FOE and FAE D_0 are equal 2837 and 1966 m/ñ, accordingly. Maximal temperature of products Ò ~ 3682 °K for FOE is reached at stoichiometric ratio. The maximum Ò ~ 2956 °K for FAE is reached with $c = 0.35$ and differs slightly from a stoichiometry, where Ò = 2947 °Ê. Maximums of D_0 and Ò are strongly displaced from each other. Typically, the characteristic temperatures (and D_0) decrease when approaching limits. It must be noted that the equations system for detonation has solution even besides the experimentally measured concentration limits, that once again indicates conventional meaning of limiting values and necessity of development of the accurate theory for concentration limits.

In Figs. 5-6 the molecular weights of an initial mixture μ_0 (g/mole) and of detonation products μ, ratios of pressure p = Ð/$Ð_0$ and temperature θ = Ò/$Ò_0$ in products, dimensionless chemical energy-release q = Q/c_0^2 ($ñ_0$ is the sound speed in initial mixture) are presented. For stoichiometric mixtures ð = 18.78, θ = 12.35 and q = 22.93 (FOE) and ð = 15.6, θ = 9.88 and q = 16.8 (FAE). Note, that $\mu > \mu_0$ for every mixture. The maximum p = 18.82 for FOE is observed for a mixture with c = 0.7; for FAE p = 15.6 with c = 0.295÷0.35 (and here the maximum q is located). When approaching the limits, the values of ð, θ, q decrease.

The dynamic pressures ρu^2 of detonation products and of gas in an induction zone give the measure for dynamic loading: with stoichiometric concentration the dynamic pressures are equal 15 and 148 atm accordingly for FOE (static pressures equal to 18.8 and 33.0 atm) and 12 and 118 atm for FAE (static pressures equal to 15.6 and 27.7 àtm). The dynamic pressure in a shock wave is much higher, than in DW (with identical wave speed). The maximum ρu^2 values are observed at stoichiometric concentrations and reduced at limits approaching (Fig.7, FOE for example). Let's note that the powerful dynamic pressure in DW can be used for acceleration of particles (for example, in installations for detonation dusting of powder coverings on various details) or for effective cleaning the technological equipment from dust sediments.

With estimations of temperature loading the important role belongs to specific heat of detonation products (Fig. 8 for FOE). Its maximal values correspond to stoichiometric mixtures: for FOE equilibrium C_P and C_V of detonation products are equal 56.4 and 46.2 cal/mole K, accordingly, and frozen C_P = 11.4 cal/mole K (frozen $C_V = C_P - R$). For initial stoichoimetric mixture the frozen specific heat C_P = 7 cal/mole K; for FAE equilibrium C_P and C_V of detonation products are equal 19.2 and 16.4 cal/mole K, accordingly, and frozen C_P = 10.2 cal/mole K, while for initial stoichiometric mixture the frozen C_P = 6.8 cal/mole K.

In Figs. 9-10 there are presented the pressure ratio in a detonation wave $ð_d$ and at instant explosion $ð_v$ (the maximum $ð_v$ = 9.6 for FOE is achieved at c = 0.7, for FAE $ð_v$ = 8.0 with c = 0.295÷0.35), dimensionless chemical energy-release of instant explosion $q_v = Q_v/c_0^2$ and appropriate temperature ratio θ_v. For estimations it is possible to assume $P_d \sim 2P_v$. The temperatures of reaction products for detonation $Ò_d$ and for instant explosion in closed volume $Ò_v$ are presented in Fig. 11-12 along with chemical energy-release of reaction Q (cal/g) for these processes. For FOE maximal Q ~ 2370 cal/g is characteristic of a mixture with c = 0.85 for V=const (for detonation with above c the value Q ~ 2255 cal/g), for FAE maximal Q_v ~ 770 cal/g with c = 0.45.

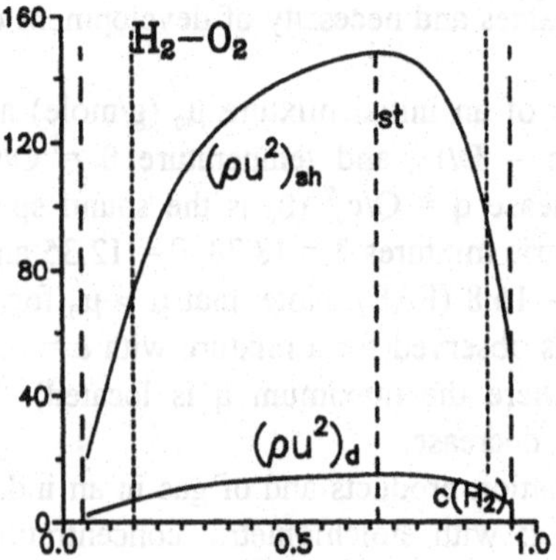

Fig. 7

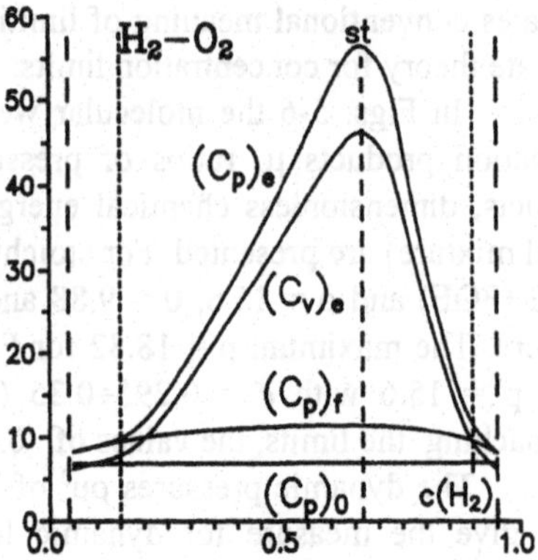

Fig. 8

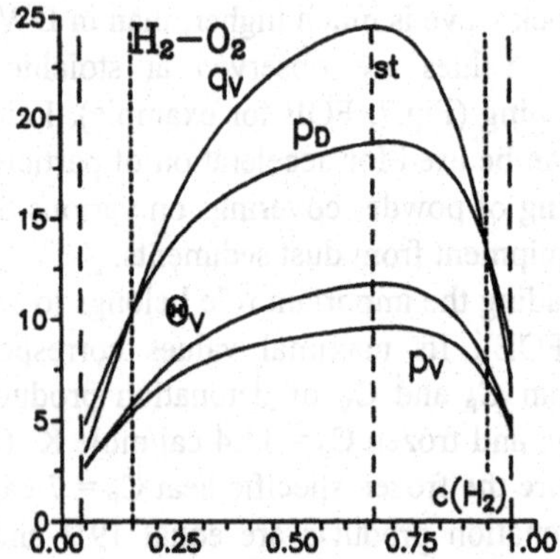

Fig. 9

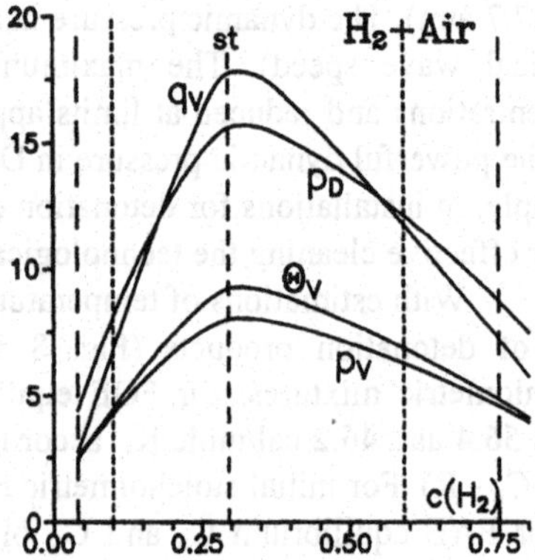

Fig. 10

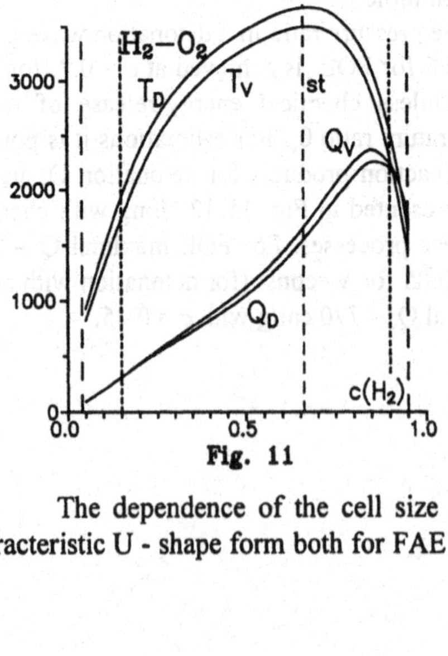

Fig. 11

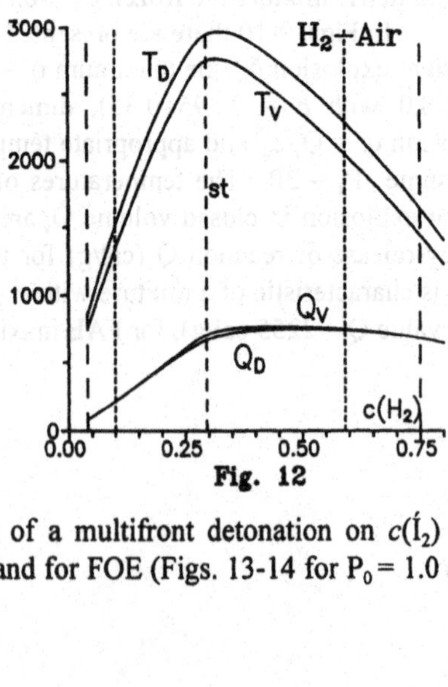

Fig. 12

The dependence of the cell size a of a multifront detonation on $c(Í_2)$ have characteristic U - shape form both for FAE and for FOE (Figs. 13-14 for $P_0 = 1.0$ atm).

The minimal $a \sim 1.0$ mm for FOE and $a \sim 11.0$ mm for FAE also is close to values of a in stoichiometric mixtures (for $2H_2 + O_2$ $a = 1.6$ mm).

The dependence of critical energy of detonation initiation in hydrogen mixtures for plane ($Å_1$ in J/sm^2), cylindrical ($Å_2$ in J/sm) and spherical ($Å_3$ in J) cases of symmetry have U - shape form similar $a(\tilde{n})$ ($Å_3$ in Figs.15-16 as an example). The minimal energies for FOE are equal 0.7 J/sm^2, 0.3 J/sm and 1.9 J, for FAE - 8.4 J/sm^2, 37.3 J/sm and 3354 J (about ~ 0.8 g TNT). Let's note extremely fast increase of all these values in the vicinity of limits (in Figs.13-16 a vertical logarithmic scale is used). It is the basic reason for large scatter of concentration limit values reported by various authors (because of use of the equipment with the insufficient geometrical and power characteristics for correct definition of combustion and detonation limits).Figure17 demonstrates change of critical initiation energies $Å_{*\nu}$ at variation of initial pressure for $2H_2 + O_2$. It is seen, that the efficiency of initiation by a charge of the certain symmetry ν changes with $Ð_0$: at elevated pressures the point (spherical) initiation is most dangerous, but at low pressure the most dangerous is plane initiation.

Figure 18 shows variation with $c(H_2)$ of ratio a/l_{10}, where $l_{10} = (D_0 - u_2)\,\tau_{10}$ is the induction zone size for Chapman-Jouguet DW (τ_{10} is the ignition delay of a mixture). An inconstancy of this ratio is seen in contradiction to the assumption [2-3], where $a/l_{10}=const=29$ (upper dotted horizontal line). Also the ratios d_{**}/a and l_{**}/a are changeable for various mixtures, where l_{**} is the channel width at which the reinitiation of a cylindrical multifront detonation with transition of a wave from the narrow channel in wide is observed (ratios $l_{**}/a=const \approx 10$ for cylindrical and $d_{**}/a=const \approx 13$ for spherical cases (lower dotted horizontal line) of initiation during long time were used as universal criteria for estimations l_{**} and d_{**} according to the assumptions [2-3,14-18]).

The hydrogen mixtures with the water steam are of special interest at the hazards analysis. Some aspects of such systems were analyzed in [21-27] in the assumption of a homogeneous mixture (steam in a gas phase). Both homogeneous and heterogeneous systems were examined in [28]. At calculation an initial pressure of a mixture varied in a range 0,1÷50 atm, initial temperature - up to 500 $^{\circ}$Ñ, steam concentration was increased up to values when the extremely fast increase of the cell size and critical initiation energy was already fixed (analogue of a concentration limit for detonation). Up to saturation condition (at initial temperature) steam was considered as gaseous, and at higher concentration it was treated as heterogeneous gaseous-droplets system.

The detonation velocities D_0 as a function of molar steam concentration $c(H_2O)$ at some fixed initial temperatures are demonstrated in Fig.19 for a case when common pressure of a FAE mix $Ð_0$ is constant ($P_0 = 1.0$ atm in Fig.19). D_0 diminishes with c increasing, such behavior is typical also for detonation temperature T. For example, at $T_0 = 348$ K the velocity decreases for FAE from 1963 m/s at $c = 0$ up to 1328 m/s at $c = 0.54$ and for FOE - from 2821 m/s at $c = 0$ up to 1515 m/s at $c = 0.70$. At the same time T decreases from 2951 °K up to 1278 °K for FAE and from 3658 °K up to 1431 °K for FOE. Its may be noted that at higher c only H_2O is presented in detonation products.

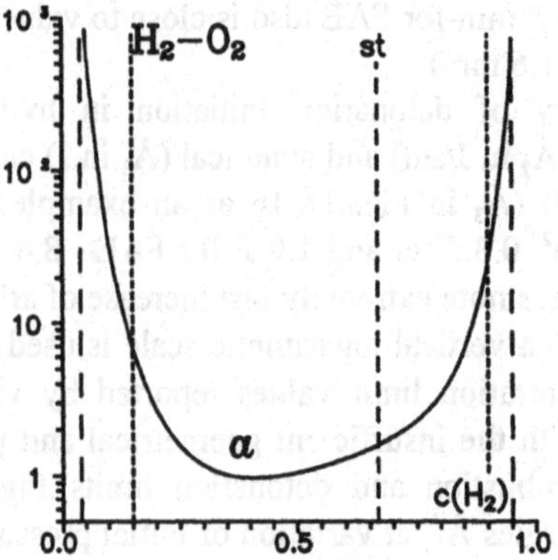

Fig. 13

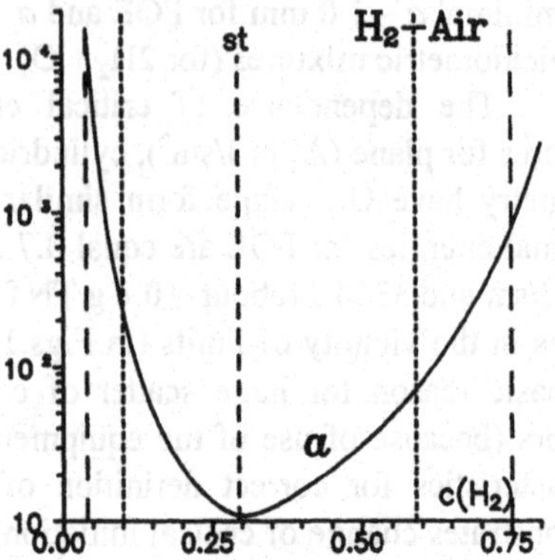

Fig. 14

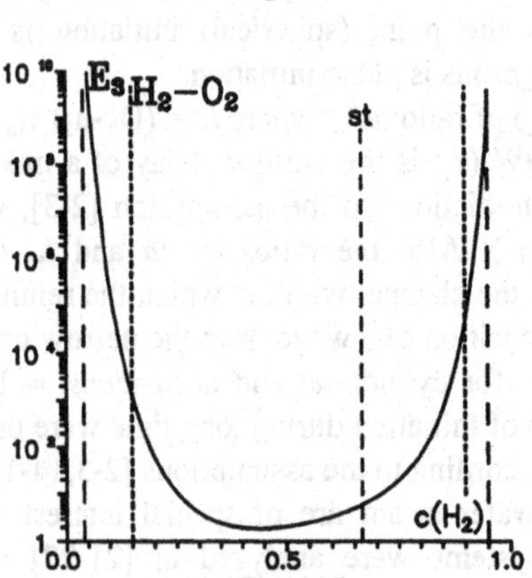

Fig. 15

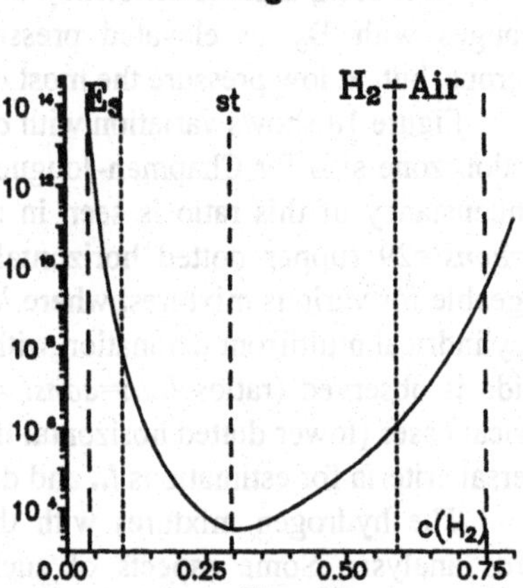

Fig. 16

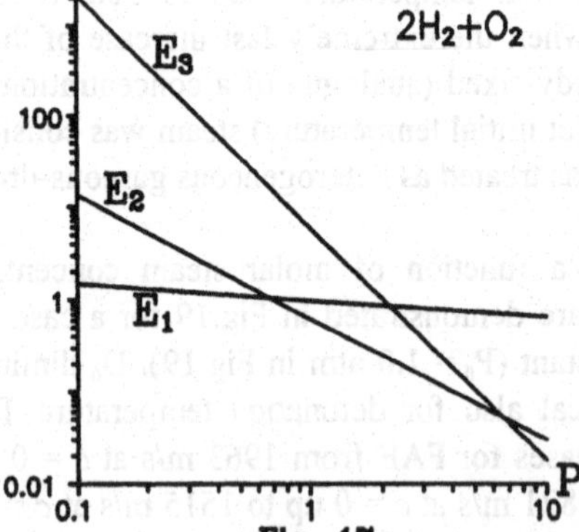

Fig. 17

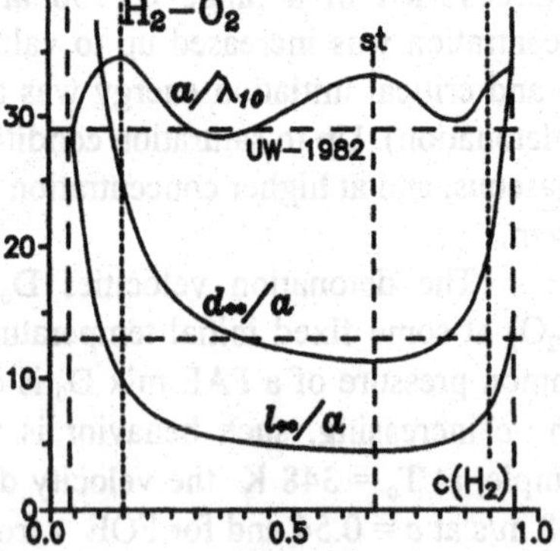

Fig. 18

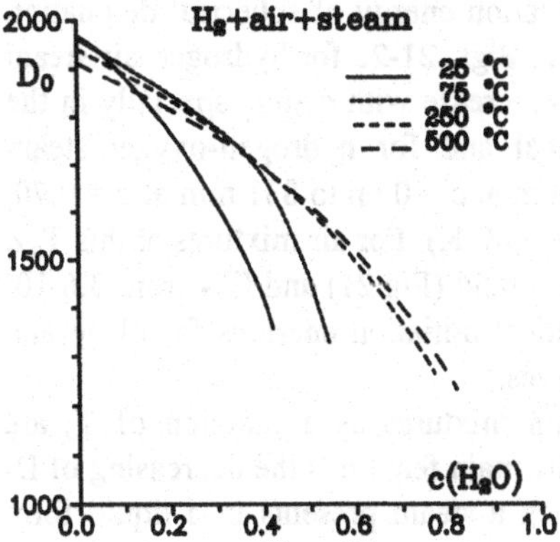

Fig. 19

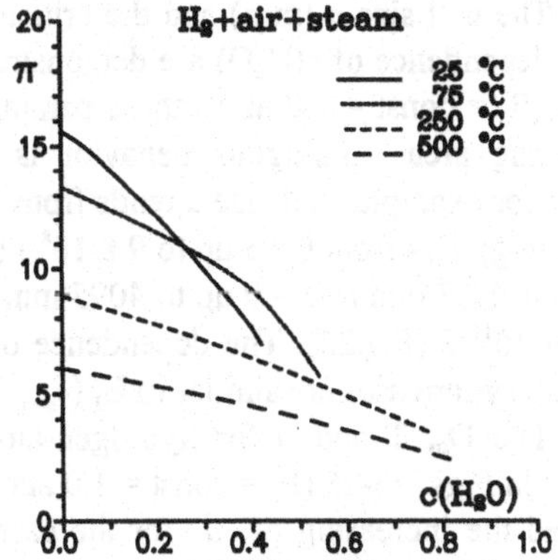

Fig. 20

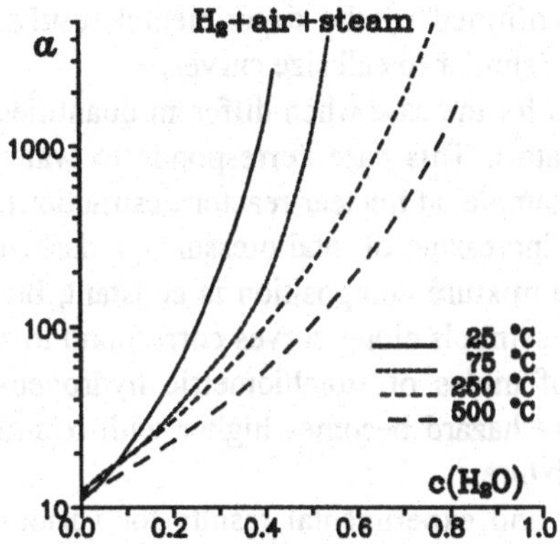

Fig. 21

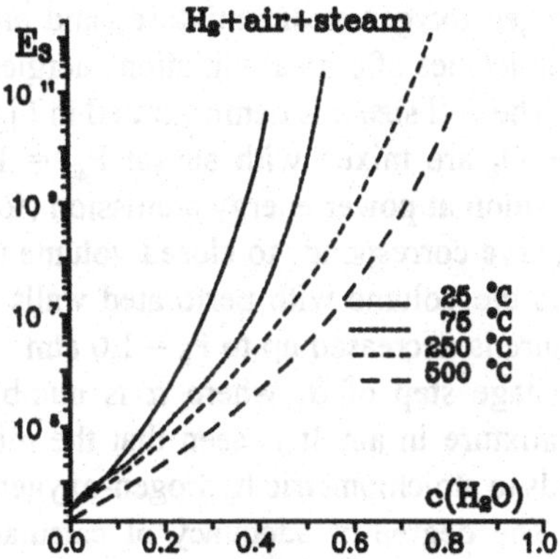

Fig. 22

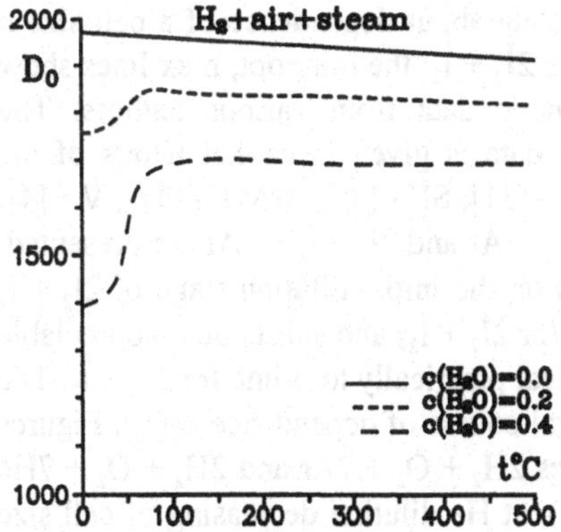

Fig. 23

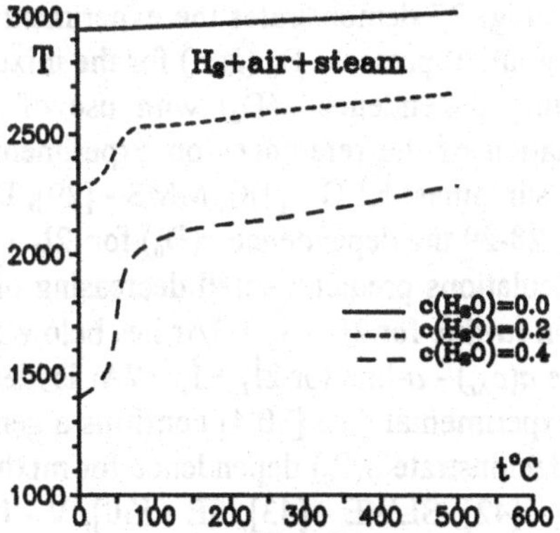

Fig. 24

The dimensionless pressure π=P/P_0 of detonation products in dependence of $c(H_2O)$ is demonstrated in Fig.20 (P_0 = const = 1.0 atm): the values of π monotonously decrease not only with c, but also with T_0 (the curves for T/T_0 are similar).

The cell size *a* (mm) and the critical initiation energy of spherical detonation E_3 (J) in dependence of $c(H_2O)$ are demonstrated in Figs. 21-22 for hydrogen-air-steam mixtures (P_0 = const = 1.0 atm): these parameters increase with *c* and especially in the near-limiting area. Analogous behavior is typical and for hydrogen-oxygen-steam mixtures: for example, cell size *a* tends from 1.7 mm at *c* = 0 up to 831 mm at *c* = 0.70, critical energy E_3 - from 6,9 J up to 9.8 10^8 J (T_0 = 348 °K). For air mixtures at this T_0 *a* varies from 11.7 mm at *c* = 0 up to 4095 mm at *c* = 0.54 (Fig.21) and E_3 - from 3.6 10^3 up to 1.9 10^{11} J (Fig.22). The dependence of critical initiation energies for plane and cylindrical symmetries are similar to E_3 (c_{steam}) curves.

The D_0, T and *a* for hydrogen-air-steam mixtures as a function of T_0 are presented in Figs. 23-25 (P_0 = const = 1.0 atm). The main feature is the decreasing of D_0 and T and the increasing of *a* for the case when a steam presents in a liquid form (t_0<100 °C at P_0 = 1.0 atm). This effect is caused by the heat losses on an evaporation of liquid steam, which decrease the chemical energy release. This effects are typical also for hydrogen-oxygen-steam mixtures and are confirmed by the experimental results. The dependencies of critical initiation energies are similar to cell size curves.

The cell size *a* is demonstrated in Fig. 26 for the case when different quantities of $2H_2 + O_2$ are mixed with air (at P_{air} = 1.0 atm). This case corresponds to water decomposition at power energy admission (for example, at nuclear reactor destruction). A solid curve corresponds to closed volume (the increasing of total pressure), a dashed curve - to the volume with perforated walls (the mixture composition is constant, but the pressure is decreased up to P_0 = 1.0 atm. The symbols along curves correspond to a 10-percentage step of α, where α is number of moles of stoichiometric hydrogen-oxygen mixture in air. It is seen that the mixture hazard becomes higher with α and FAE tends to stoichiometric hydrogen-oxygen mixture.

The degree of adequacy of calculating and experimental results for various FOE and FAE is thoroughly analyzed in [4], here some data for hydrogen mixes are given as an example only.

Fig. 27 demonstrates the experimental data about dependence of a cell size *a* (mm) on initial pressure $Ð_0$ (àtm) for the mixture 2Í$_2$ + Î$_2$: the min, opt, max lines show calculating dependence *a*($Ð_0$) with use of kinetic data from various authors. The abbreviation of the references on experimental data is given in capital letters of the authors surnames: KLG - [18], MMS - [29], DT - [31], SE - [30], WMT - [32], V - [4]. In Figs. 28-29 the dependence *a*($Ð_0$) for 2Í$_2$ + Î$_2$ + 3Ar and 2Í$_2$ + Î$_2$ + 7Ar are presented. The calculations predict a small decreasing of *a* on the initial dilution stage of 2Í$_2$ + Î$_2$ by argon (*a*-line for 2Í$_2$ + Î$_2$ + 3Ar lies below as for 2Í$_2$ + Î$_2$) and subsequent appreciable increase *a*(c_{Ar}) - *a*-line for 2Í$_2$ + Î$_2$ + 7Ar is identical practically to *a*-line for 2Í$_2$ + Î$_2$. The set of experimental data [30,4] confirms a general course of dependence *a*($Ð_0$). Figures 29-30 demonstrate *a*(P_0) dependence for mixtures $2H_2 + O_2$ + 7Ar and $2H_2 + O_2$ + 7He (PBGD - [42], SLWE - [43], SE - [30], V - [4]), at He dilution decreasing of cell size not observed.

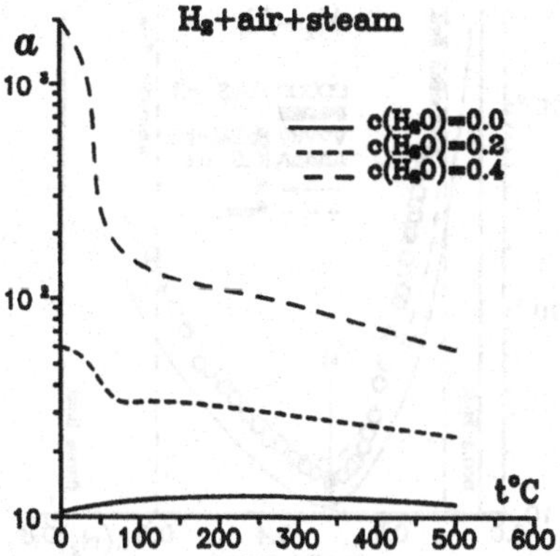

Fig. 25

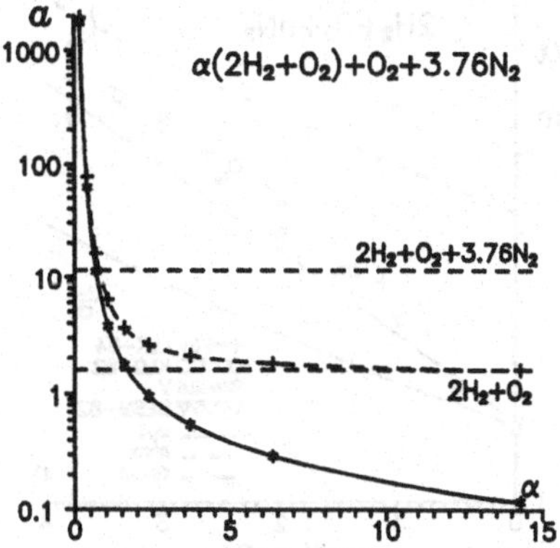

Fig. 26

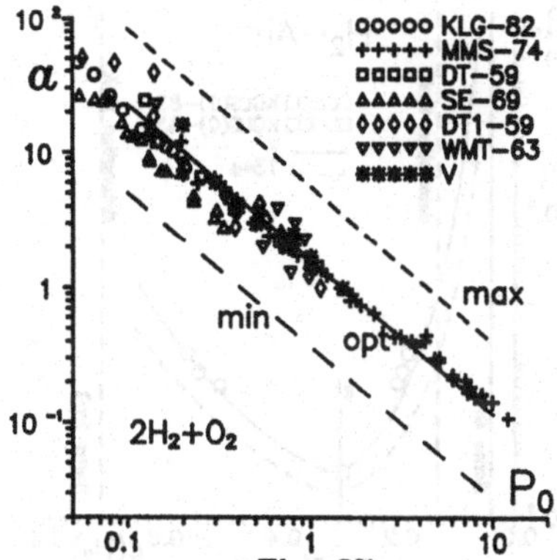

Fig. 27

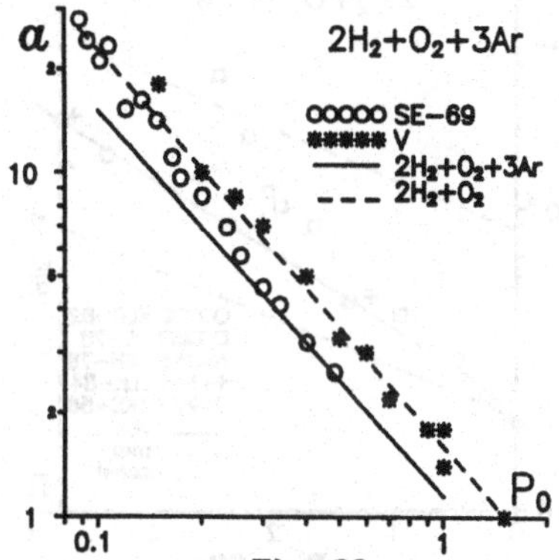

Fig. 28

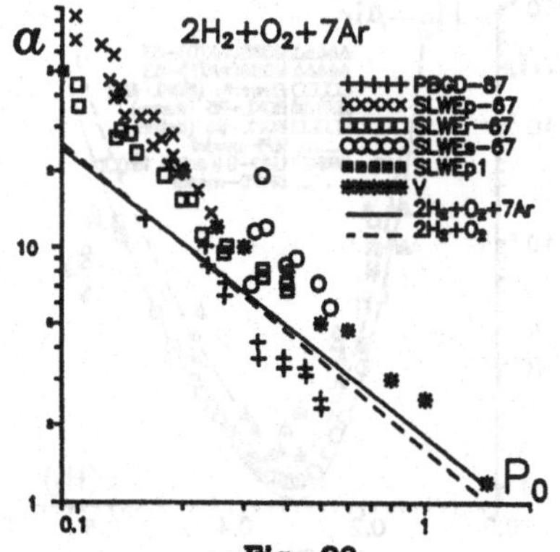

Fig. 29

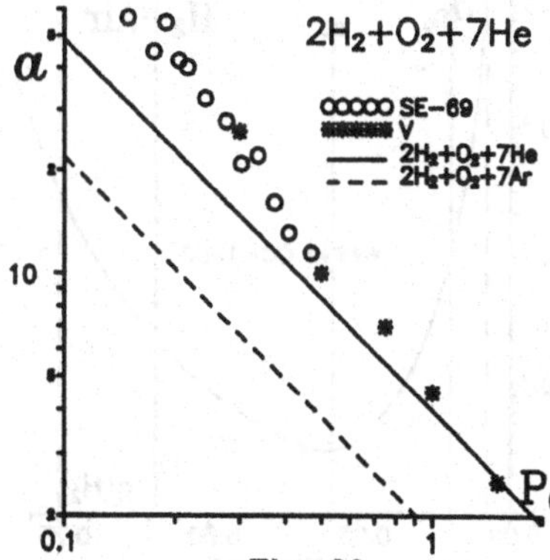

Fig. 30

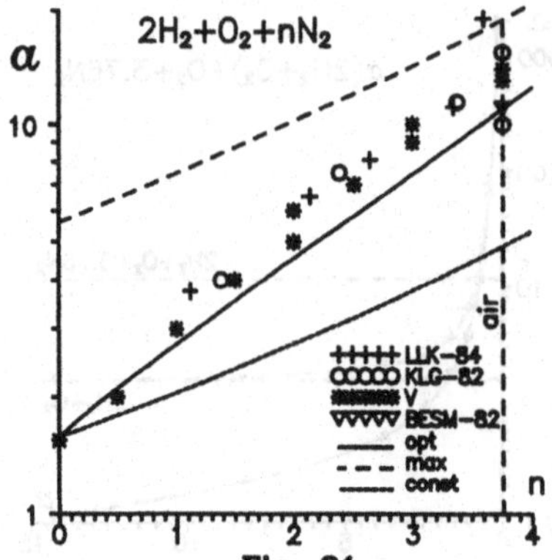

Fig. 31

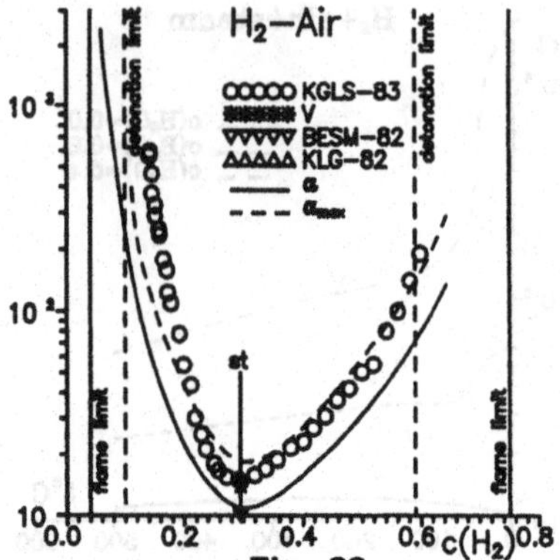

Fig. 32

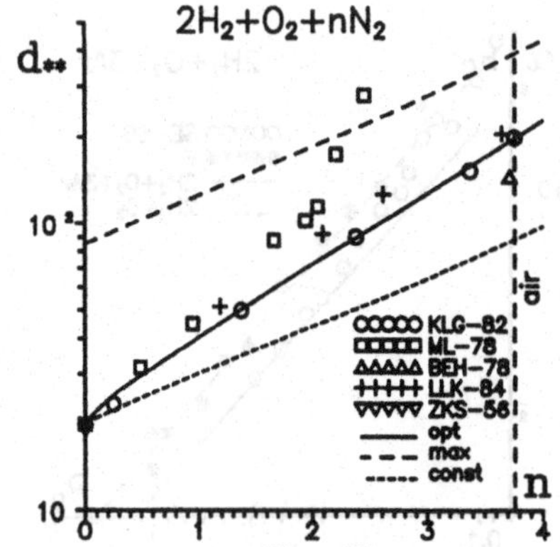

Fig. 33

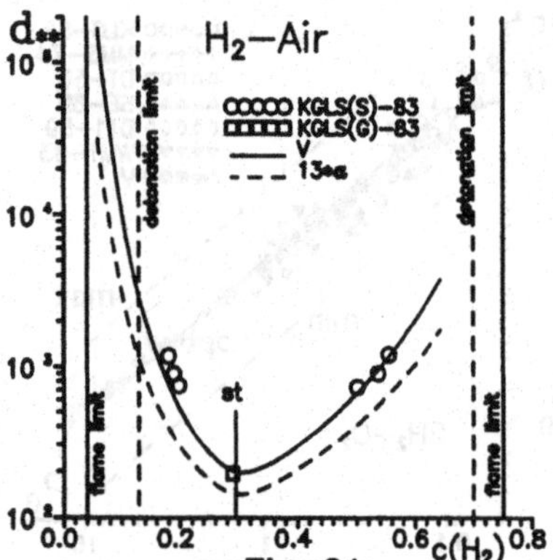

Fig. 34

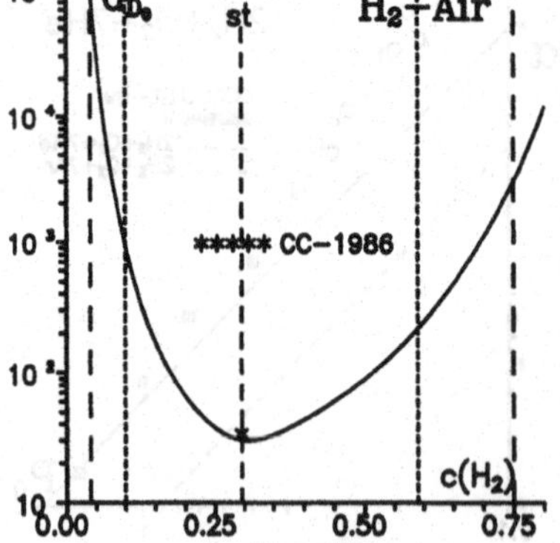

Fig. 35

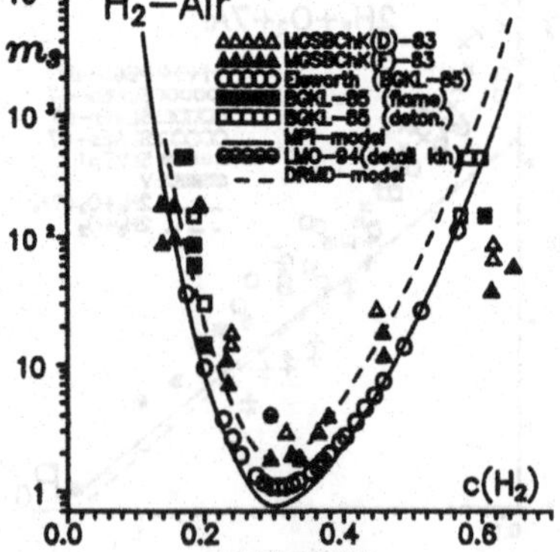

Fig. 36

The experimental and calculating cell sizes in mixtures with nitrogen added to $2H_2 + O_2$ are presented in Fig.31 for total initial pressure $P_0 = 1.0$ atm (LLK - [36],

KLG - [18], V - [4], BESM - [41]), *a*-dependence on molar concentration of hydrogen in air demonstrates in Fig.32 (KGLS - [3], BESM - [41], KLG - [18], V - [4]).

In Figs. 33-34 the data about diffraction diameter d.. (mm) are presented. When d=d.. a detonation initiation of an explosive mixture is realized with DW passing from a pipe into free volume. Figure 33 shows dependence of d.. on dilution degree of mixture $2Î_2 + Î_2$ with N_2 (KLG - [18], ML - [33], BEH - [35], LLK - [36], ZKS - [34]), while Fig.34 - on molar concentration of H_2 in air (KGLS - [3]) at $Đ_0$ = 1.0 àtm.

In Fig. 35 the data are presented on diameter of sphere capable to excite detonation in FAE if sphere velocity w = D_0: the minimal diameter ~ 2.5 mm for FOE and ~ 30 mm for FAE [19-20].

In Fig. 36 calculating and experimental data are given on critical weight m. (g) of TNT charge for initiation of a spherical detonation in hydrogen-air mixtures (ÌGSBChK - [37], Elsworth - from [38], BGKL - [38], ABS - [39], LÌÎ - calculations of [40] with use of model with detailed kinetic for an induction period). For stoichiometric hydrogen-air mixture the experimental value of minimal TNT charge weight m. for initiation of a detonation is about ~ 1.1 g [39], the calculation predicts about 0.8 g.

The comparison of calculating and experimental data shows that the Program "SAFETY" describes satisfactory the basic parameters of multifront DW needed for estimations of explosion hazard of combustible systems. Use of the Program for estimations of explosion hazard of hydrogen mixtures can appear productive for development of necessary arrangements on decrease of an accident probability.

References.

1. Vasil'ev A.A., Nikolaev Yu. A. (1978) *Closed theoretical model of a detonation cell*,Acta Astr., 5. p.983-996.
2. Westbrook C.K., Urtiew P.A. (1982) *Chemical kinetic prediction of critical parameters in gaseous detonations*, 19-th Symp. (International) on Combust., -p.615-623.
3. Knystautas R., Guirao C., Lee J.H., Sulmistras A. (1983) Measurement of cell size in hydro-carbon-air mixtures and predictions of critical tube diameter, critical initiation energy and detonation limits, in J.Bowen, N.Manson, A.Oppenheim and R.Soloukhin (eds), *"Dynamics of Shock Waves, Explosions and Detonations"*, v.94 of "Progress in Astronautics and Aeronautics". New-York, -p.23-37.
4. Vasil'ev À.À. (1995) *Near critical regimes of a gas detonation* (in Russian), Novosibirsk.
5. Vasil'ev A.A. (1997) *Detonation hazard of gaseous mixtures*, in Proc. 28th Intern. Conference of Fraunhofer Instittte of Chemical Technologies. Germany, P.50/1-50/14.
6. Vasil'ev À.À., Valishev À.I., Vasil'ev V.À., Panfilova L.V., Òîpchian Ì.À. (1997) *Detonation waves parameters at increased pressure and temperatures* (in Russian),Khimicheskaja Fizika, 16,11. See also: Chem. Phys. Reports, 1997, 16 (9), p. 1659-1666.
7. Lewis B., Elbe G. (1984) *Combustion, flame and explosions in gases* (in Russian), «Mir», Ìoscow.
8. *Hydrogen. Properties, production, storage, transportation, application* (1989) (in Russian), Reference book (D.Ju.Gamburg and N.F.Dubovkin (eds)), «Chemistry», Ìoscow.
9. *Chemical encyclopedia* (1988) (in 5 volumes) (in Russian) ,I.L.Ênunjants (eds). "Soviet encyclopedia ", Ìoscow.
10. Àndreev Ê.Ê., Beljaev À.F. (1960) *The theory of explosive substances* (in Russian), Gostechizdat, Ìoscow.
11. Nettleton Ì. (1989) *Detonation in gases* (in Russian), «Mir», Ìoscow.
12. Tieszen S.R., Sherman M.P., Benedick W.B., Shepherd J.E., Knystautas R., Lee J.H. (1986) Detonation cell size measurements in hydrogen-air-steam mixtures,in Bowen, Leyer and Soloukhin (eds); *"Dynamics of Explosion"*, v.106 of "Progress in Astronautics and Aeronautics". -N.Y. -p.205-219.

13. Strehlow R.A., Crooker A.J., Cusey R.E. (1967) *Detonation initiation behind an accelerating shock wave*, Comb. Flame, 11,4. -p.339-351.
14. Ìitrofanov V.V., Soloukhin R.I. (1964) *About diffraction of a multifront detonation wave* (in Russian), Doklady USSR Akademii Nauk, 159,5. -ñ.1003-1006.
15. Edwards D.H., Thomas G.O., Nettleton M.A. (1979) *The diffraction of a planar detonation wave at an abrupt area change*, J. Fluid Mech., 95,1. -p.79-96.
16. Edwards D.H., Thomas G.O., Nettleton M.A. (1981) Diffraction of a planar detonation in various fuel-oxygen mixtures at an area change, in J.Bowen, N.Manson, A.Oppenheim and R.Soloukhin (eds), *"Gasdynamics of detonation and explosions"*; v.75 of "Progress inAstronautics and Aeronautics". -N.Y. -p.341-357.
17. Lee J.H., Knystautas R., Guirao C. (1982) The link between cell size, critical tube diameter, initiation energy and detonability limits, in J.Lee and C.Guirao (eds), *"Fuel-Air Explosions"*, University of Waterloo press. -p.157-189.
18. Knystautas R., Lee J.H., Guirao C.M. (1982) *The critical tube diameter for detonation failure in hydrocarbon-air mixtures*, Comb. Flame, 48. -p.63-83.
19. Vasil'ev A.A. (1994) *Initiation of gaseous detonation by a high speed body*, Shock Waves, 3,4. -p.321-326.
20. Vasil'ev À.À., Kulakov B.I., Ìitrofanov V.V., Silvestrov V.V., Òitov V.Ì. (1994) *Initiation of explosive gas mixes by high-velocity body* (in Russian), Dokladi Akademii Nauk. 338,2. -p.188-190.
21. Vasil'ev A.A., Topchian M.E., Ul'yanitsky V.Yu. (1979) *The influence of initial temperature on the parameters of gaseous detonation* (in Russian), Fizika Gorenija i Vzriva, 14,6. -p.149-152.
22. Ciccarelli G., Ginsberg T., Boccio J., Economos C., Sato K., Kinoshita M. (1994) *Detonation cell size measurements and predictions in hydrogen-air-steam mixtures at elevated temperatures*, Combust. Flame, 99. -p.212-220.
23. Kumar R.K. (1990) *Detonation cell widths in hydrogen-oxygen-diluent mixtures*, Combust. Flame, -80.-p.157-169.
24. Shepherd J.E. (1986) Chemical kinetics of hydrogen-air-diluent detonations, in Bowen,Leyer and Soloukhin (eds), *"Dynamics of Explosion"*; v.106 of "Progress inAstronautics and Aeronautics". -N.Y. -p.263-293.
25. Stamps D.W., Tieszen S.R. (1991) *The influence of initial pressure and temperature on hydrogen-air-diluent detonations*, Combust. Flame, -83. -p.353-364.
26. Tieszen S.R., Sherman M.P., Benedick W.B., Shepherd J.E., Knystautas R., Lee J.H. (1986) Detonation cell size measurements in hydrogen-air-steam mixtures, in Bowen, Leyer and Soloukhin (eds), *"Dynamics of Explosion"*; v.106 of "Progress in Astronautics and Aeronautics". -N.Y. -p.205-219.
27. Tieszen S.R., Stamps D.W., Westbrook C.K., Pitz W.J. (1991) *Gaseous hydrocarbon-air detonations*, Combust. Flame,-84,33.-p.376- 390.
28. Vasil'ev A.A., Vasil'ev V.A. (1997) The steam influence on hydrogen-oxygen and hydrogen-air detonation, in *Proc. 16th ICDERS*, University of Mining and Metallurgy, AGH,Cracow, Poland. -P.385-388.
29. Ìànzalej V.I., Ìitrofanov V.V., Subbotin V.À. (1974) *Measurement of nonuniformities of detonation front in gas mixes with the increased pressure* (in Russian), Fizika Gorenija i Vzriva, -10,1. -p.102-110.
30. Strehlow R.A., Engel C.D. (1969) *Transverse waves in detonation: II. Structure andspacings in H_2 - O_2, C_2H_2 - O_2, C_2H_4 - O_2 and CH_4 - O_2 systems*, AIAA J., -7,3. -p.492-496.
31. Denisov Ju.N., Òroshin Ja.Ê. (1959) *A pulsing and spin detonation of gas mixes in pipes* (in Russian), Dokladi USSR Akademii Nauk, -125,1. -ñ.110-113.
32. Voitsehovsky B.V., Ìitrofanov V.V., Òîpchian Ì.À. (1963) *Structure of detonation front in gases* (in Russian), Izdatelstvo SB USSR Academy of Sci., Novosibirsk.
33. Matsui H., Lee J.H. (1978) *On the measure of the relative detonation hazards of gaseous fuel-oxygen and air mixtures*, 17-th Symp. (International) on Combust., -p.1269-1280.
34. Zeldovitch Ja.B., Êîgàrkî S.Ì., Simînîv N.N. (1956) *An experimental research of a spherical gas detonation (in Russian)*, Journal of Technic. Phys., 26,8. p.1744-1768.
35. Bull D.C., Elsworth I.E., Hooper G. (1978) *Initiation of spherical detonation in hydrocarbon-air mixtures*, Acta Astr., -5. P.997-1008.
36. Lui Y.K., Lee J.H., Knystautas R. (1984) *Effect of geometry on the transmission of detonation through an orifice* // Combust. Flame, -56. -p.215-225.

37. Ìàkååv V.I., Gîstintsåv Ju.À., Strogànîv V.V., Bohon Ju.À., Chernushkin Ju.N., Êulikov V.N. (1983) *Burning and detonation of hydrogen-air mixes in free areas* (in Russian), Fizika Gorenija i Vzriva. - 19,5. -p.16-18.
38. Benedick W.B., Guirao C.M., Knystautas R., Lee J.H. (1986) Critical charge for direct initiation of detonation in gaseous fuel-air mixtures , in Bowen,Leyer and Soloukhin (eds), *"Dynamics of Explosion"*; v.106 of "Progress in Astronautics and Aeronautics". -N.Y. -p.181-202.
39. Atkinson R., Bull D.C., Shuff R.I. (1980) *Initiation of spherical detonation in hydrogen-air*, Comb. Flame, -39. -p.287- 300.
40. Levin V.A., Osinkin S.F., Markov V.V.(1994) Direct initiation of detonation in a hydrogen-air mixture, in *"Combustion, Detonation, Shock Waves"*, v.2. Proceedings of theZeldovitch Memorial. Russian Section of the Combustion Institute, Moscow, -p.363-365.
41. Bull D.C., Elsworth J.E., Shuff P.J., Metcalfe E. (1982) *Detonation cell structures in fuel-air mixtures*, Combust. Flame, -45,1. -p.7-22.
42. Presles H.N., Bauer P., Guerraud C., Desbordes D. (1987) *Study of the head on detonation wave structure in gaseous explosives*, J. de Physique, -48. -p.(C4-119)-(C4-124).
43. Strehlow R.A., Liaugminas R., Watson R.H., Eyman J.R. (1967) *Transverse wave structure in detonations, in* 11-th Symp. (International) on Combust., -p.683-692.

THIN LAYER BOILOVER OF PURE OR MULTICOMPONENT FUELS

J.P. GARO, J.P. VANTELON
Laboratoire de Combustion et de Détonique
Ecole Nationale Supérieure de Mécanique et d'Aérotechnique
University of Poitiers. Téléport 2 - BP 109
86960 Futuroscope Cedex France

1. Abstract

In situ burning of fuels spilled on water is of interest as a means for cleaning up these spills. Results of small-scale experiments that investigate the combustion of this layers of fuel on water are reported. They allow to analyze the events that take place in this complex combustion process, which can lead to the occurrence of explosive burning, normally referred to as boilover. The work concerns primarily the influence of the initial fuel layer thickness, pool diameter and fuel boiling point, on the burning rate, time to the start of boilover, burned mass ratio, boilover intensity and temperature history of fuel and water. The temperature measurements show that the phenomenon may be due to boiling nucleation near the fuel/water interface, in sublayer water that has been superheated. A simple heat transfer analysis of the fuel and water heating provides information about the characteristics of boilover. Thicker fuel and superheated water layers result in a stronger and faster ejection of the fuel from the pan toward the flame, and consequently in a more explosive and hazardous boilover event.

2. Introduction

The burning of a liquid fuel floating on water is a potential important hazard in unwanted fires. Although in such a case, the fuel burning itself is similar to that of a single fuel, the presence of the water introduces a number of effects that are caused by the transfer of heat from the fuel to the water underneath. The heat transferred to the water, together with surface temperature and surface tension effects, is the major reason why the fuel layer must have a minimum thickness to be ignited. But this heat transferred in depth may also induce water boiling and splashing, a phenomenon referred to as boilover. The fuel burns in a disruptive fashion which is caused by the boiling of the water underneath and results in a sharp increase in burning rate and external radiation, and often in the explosive burning of the fuel.

In general, boilover occurs with fuels whose boiling temperature is higher than that of the water. This phenomenon is often encountered with fires involving large storage

V.E. Zarko et al. (eds.), Prevention of Hazardous Fires and Explosions, 167–182.

tanks containing multicomponent fuels (particularly crude oils or other heavy oils), leading to explosive vaporization of the water often present on the bottom of the tank. The heat transfer through the fuel arises from the phenomenon of the hot-zone formation, a zone of practically uniform temperature and composition that propagates through the interior of the fuel. The hot-zone formation in large storage tanks and its extent are normally explained by the generation of continuous and vigorous convective mixing within the liquid, stimulated by selective evaporation of the light ends or by the generation, ascent, and growth of vapors released throughout most of this zone (1-3).

However, boilover can also occur with the burning of thin layers of these liquids floating on water. Studies specifically concerning thin-layer boilover of pure or multicomponent fuels are, on the whole, rather recent. Most of these studies have been conducted on small or medium-scale fires in a laboratory. Although test pans of finite diameter cannot perfectly model real situations, they ensure calm external conditions, stable flames and nearly uniform heat transfer through the fuel and the fuel/water interface which help the onset of nearly uniform boiling at this interface. This facilitates the experiments, while it is anticipated that the results will give a better understanding of the boilover process.

Ito et al. (4), using n-decane as fuel, provided detailed physical structure and temperature measurements within the fuel layer and the supporting water sublayer at the onset of boilover. They showed that, at this onset, the water layer is superheated and dynamically unstable. Petty (5) conducted a series of experiments to simulate large-scale crude oil fires (2 m in diameter) and concluded that fuel surface temperature remains unchanged throughout the burning process, in contradiction to the interpretation of fuel burning as a distillation process. Other researchers (6-9) have begun to investigate the mechanism leading to boilover. Koseki and Mulholland (6) and Koseki et al. (7) reported that fuel thickness is an important factor in regard to boilover. Their results show that the thicker the fuel layer the more violent boilover is, and the higher the burning rate and external radiation. Their temperature measurements also indicate probable superheating of the water when boilover occurs. Arai et al. (8), in studies with simple and multicomponent fuels, brought out some fundamental aspects of the effect of a boiling water sublayer on the behavior of the pool fires. For fuels whose boiling point temperature is higher than that of water, they observed that the higher the fuel's boiling point, the more intense the boilover process was. Data obtained include temperature and mass loss history of the liquid fuel and water. Based on these data, Inamura et al. (9) described a one-dimensional model to predict the time required for the water sublayer to start to boil.

Despite some overlaps, these studies are often quite different and comparative analysis is not always obvious. In fact, it appears that a complete understanding of boilover is still lacking, most likely because of the complexity of the mechanisms involved. For this reason, the present authors, a few years ago, began a systematic and comprehensive study of the thin-layer boilover phenomena (10) (11). The work was concerned primarily with the influence of two parameters, initial fuel layer thickness and pool diameter, on the burning rate, time to the start of boilover, burned mass ratio, boilover intensity and temperature history of the fuel and water. Experiments were also conducted to determine the effect of fuel boiling point on the onset and characteristics of the boilover. On the basis of the results, the mechanisms leading to boilover, and its intensity, are discussed.

3. Experiment

The pool burning tests of a fuel layer floating on water were conducted in a large-scale test cell vented by natural convection. Stainless-steel pans 6-cm deep and of inner diameters 15, 23, 30 and 50 cm were used in the experiments. The pans were placed on a load cell to measure the consumption of fuel as a function of time. The load cell had a response time of 60 ms and an accuracy within ± 0.5 g. It was frequently calibrated to insure reliable measurements.

For each pan diameter, different initial fuel layer thicknesses were tested (ranging from 2 to 20 mm). Before each test, water was first poured on the pan and next the fuel until it reached 1 mm below the pan lip. During the combustion, the location of the fuel/water interface remained fixed. Therefore, since the height of the burning fuel decreases as combustion progresses, the freeboard length increases during the experiment. However, this freeboard length increase, appeared to have only a minor effect on the measured combustion rate.

Fuel and water temperature were measured with an array of four, stainless-steel sheeted, chromel-alumel thermocouples of 0.5 mm diameter inserted horizontally through the side wall of the pan with their junction located along the centerline. After a short period of time from ignition, the burning rate reached steady-state, defined here as the pre-boilover burning rate. At the onset of thin-layer boilover, the burning rate increased widely, with intense splashing of water and fuel.

Special attention was given to the accurate measurement of the fuel-water interface temperature (within ± 2°C). Some characteristic experiments were conducted to video record the evolution of the fuel-water interface at the onset of boilover. A pyrex pan of the same dimensions as the steel pan, together with an argon-ion laser sheet and a video camera were used for these experiments. The laser sheet, passing through the center of the pan, was used to enhance the illumination of the fuel layer.

To reduce possible experimental complications peculiar to multicomponent fuels, heating oil has been selected for our experiments on initial fuel layer thickness and pool diameter effects. Since it is composed of a mixture of $C_{14} - C_{21}$ hydrocarbons, it offers a relatively narrow range of boiling points (≈ 250 – 350°C), while its burning temperature is high enough to heat the water underneath to its boiling point. Thus, even if it presents a more restricted range of hydrocarbons than the crude oils, it appears well suited for a study of thin-layer boilover.

To determine the effect of the fuel boiling point, five single component fuels were used : toluene, n-octane, xylene, n-decane and hexadecane, which had boiling points ranging from 383 K to 560 K. The heating oil previously used (components with a narrow range of volatility) and a crude oil (component with large range of volatility) were also used. Relevant physical properties for all these fuels are given in Table 1.

Table 1 : Thermophysical properties of the fuels and water. [a] Value obtained with water used.

	Crude oil	Heating oil	Hexdecane	n-Decane	Xylene	n-Octane	Toluene	Water
Boiling point (K)	478 (mean value)	518 (mean value)	560.2	433.2	412	398.2	383.2	373
Surface tension $\times 10^3$ ($N.m^{-1}$) (at 293 K)	27.88	29.54	27.41	23.82	28.75	21.62	28.52	64.62[a]
Viscosity $\times 10^6$.($m^2.s^{-1}$) (at 293 K)	9.38	5.31	4.32	2.62	0.76	0.77	0.69	1.01[a]

4. Results

4.1. BURNING RATE

Figure 1 shows the surface regression rate as a function of the initial heating-oil layer thickness, for the different pool diameters tested. It can be observed that the regression rate increases first with increasing initial fuel layer thickness, and then reaches a constant limiting value that is characteristic of each pan size. This limiting burning rate increases with pool diameter, as is observed for this range of pool sizes. The present results seem to indicate that these limiting values are reached for layer thicknesses around 1 cm or larger. The variation of the combustion rate with the initial fuel layer thickness is due to heat losses to the water layer underneath. When the fuel thickness is small, the water acts as a heat sink, and the burning rate is reduced. In some specific test, it was observed that, if the fuel layer becomes thin enough (around 0.5 mm), the combustion can be quenched by heat loss to the water. When the fuel thickness is increased, this influence lessens, and there is time for the fuel to reach steady-state burning.

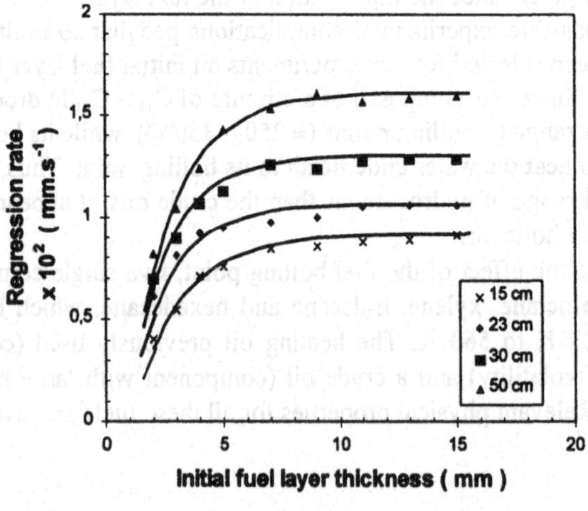

Figure 1. Regression rate as a function of initial heating-oil layer thickness for different pool diameters.

4.2. TIME TO THE START OF BOILOVER

Figure 2 shows the time to the onset of boilover as a function of the initial heating-oil layer thickness. It is seen that the dependence appears practically linear. If, as explained below, it is assumed that thin-layer boilover starts when the temperature at the heating oil/water interface reaches a given value, then these straight lines can be considered to be representative of a constant, average, apparent thermal penetration rate (actually, this interface is not really calm and well defined because of the occurrence of some frothing). The larger the pool size, the higher the penetration rate, which is consistent with the increase of burning rate with the pool size. It can also be observed that these lines cut the thickness coordinate, showing that there is a minimum thickness for the combustion to be sustained. Similar experiments are reported by Koseki et al. (7) with crude oil and a larger range of pan diameters (0.3-2.7 m). However, as a result of large scatter in their test results, they only deduced an average thermal penetration rate from a linear fit to the data.

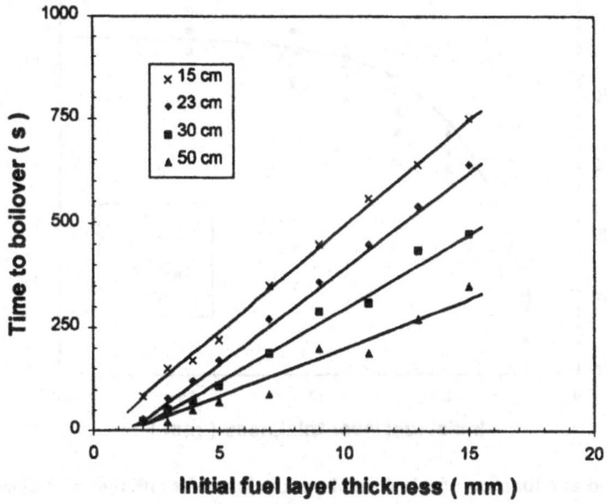

Figure 2. Pre-boilover time as a function of initial fuel layer thickness for different pool diameters.

If the regression rate of the fuel surface is known, then it is possible to deduce, by difference from the fitted slope, the effective thermal penetration rate responsible for boilover. The calculated values are reported in Table 2. It is noticed that the termal penetration rate is of the same order of magnitude as the heating-oil surface regression rate.

Table 2 : Thermal penetration rate results.

Pan diameter (cm)	15	23	30	50
Regression rate × 10^2 ($mm.s^{-1}$)	1	1.1	1.3	1.7
Slope of a linear fit to the data × 10^2 ($mm.s^{-1}$)	1.9	2.4	2.7	3.5
Thermal penetration rate × 10^2 ($mm.s^{-1}$)	0.9	1.3	1.4	1.8

4.3. BURNED MASS RATIO

The burned mass ratio is defined as the ratio between the amount of heating oil burnt before occurrence of boilover and the initial amount of fuel. This ratio can be determined either by comparison of the fuel thicknesses or from the mass loss. Figure 3 shows this ratio as a function of the initial fuel layer thickness. Independent of the diameter used, a limiting value of about 50% is reached when the initial layer thickness exceeds about 1 cm. This limiting value is consistent with the values of thermal penetration rate, responsible for boilover, and the fuel regression rate, respectively. It is also consistent with the onset of boilover at a constant temperature at the fuel/water interface. In the case of initial fuel layer thicknesses smaller than 1 cm, the comparison between regression rate and thermal penetration rate permits one to asses quite well the measured values of burned mass ratio reported in Fig. 3.

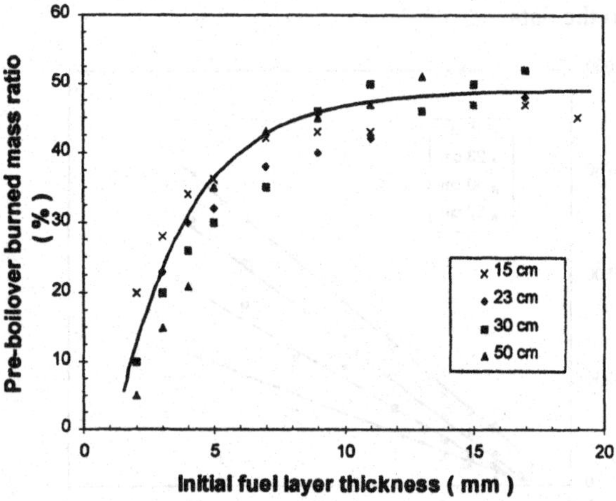

Figure 3. Burned mass ratio as a function of initial fuel layer thickness for different pool diameters.

4.4. BOILOVER INTENSITY

The intensity of boilover is defined here as the ratio between the mass loss rate of heating-oil during the boilover period and the maximum heating-oil combustion rate during the pre-boilover period. Note that the load cell signal is very noisy and that the associated uncertainly is within 20%. Moreover, one must be well aware that this corresponds to fuel burnt during eruptive vaporization and also to burning droplets randomly ejected outside the pan, together with an appreciable amount of water. An estimation of the amount of lost water can be made from the difference between the total mass loss recorded and the initial mass of fuel. In any case, the estimation of the boilover intensity is very approximate and only can be viewed as indicative of the intensity of the phenomenon. Figure 4 shows the boilover intensity thus estimated as a function of the initial fuel layer thickness, for different pool sizes. It is seen that the intensity increases with the thickness but decreases strongly with the pool diameter. The last observation was previously reported by Koseki et al. (7) with crude oil.

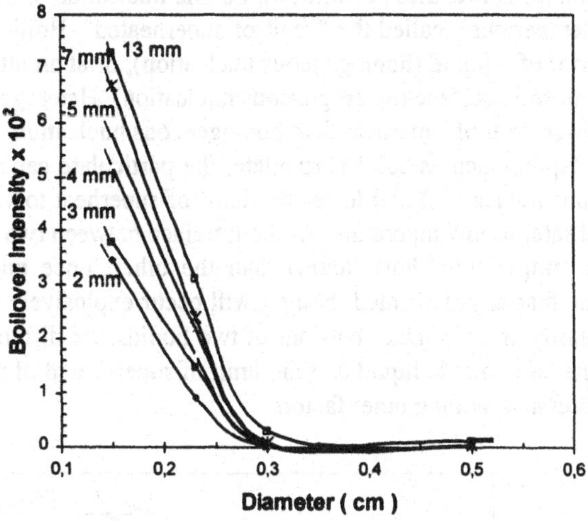

Figure 4. Boilover intensity as a function of initial fuel layer thickness for different pool diameters.

4.5. LIQUID TEMPERATURE HISTORY AND BOILOVER GENERAL CHARACTERISTICS

Figure 5 shows the variation of the temperature with the distance from the heating-oil/water interface, for different times after the start of the test (boilover occurs at 630 s), for the case of an initial heating-oil layer thickness of 11 mm. Also presented in the figure is the evolution of the fuel surface level for the different periods of time considered. All the temperature measurements presented were made with the smallest pan (15 cm in diameter). Tests conducted to verify the radial temperature variation showed that the variations were not significant whatever the initial fuel layer thickness. Figure 5 also illustrates the variation of the fuel surface temperature with time in the pre-boilover stage. Contrary to the constant value reported by Petty (5) with crude oils combustion, an appreciable increase in temperature with time is observable here.

Temperature histories, particularly at the fuel water interface, provided interesting information about the event taking place during the onset of boilover. An interesting result is that boilover appears to occur, in all cases, when the temperature at this interface reaches a value of approximately 120 °C (see Fig. 5 for an initial heating-oil layer thickness of 11 mm).

The experimental observation that there is a rapid transition from normal pool burning to disruptive burning, together with the observation that this transition occurs at an approximately fixed temperature that is above the saturation temperature of the water, indicates that the phenomena may be caused by the boiling nucleation of the water at the water/fuel interface.

It is well known (12) that a liquid that is not in contact with a gas phase can be superheated, at constant pressure, to temperatures that are well above the liquid saturation temperature. Under these conditions, bubble nucleation will occur within the liquid at a fixed temperature, called the "limit of superheated". Boiling nucleation can occcur at the interior of a liquid (homogeneous nucleation), or at an interface between a liquid and a smooth solid surface (heterogeneous nucleation). Heterogeneous nucleation generally has a lower limit of superheat than homogeneous nucleation. Also, if there are impurities in the liquid, such as solid particulate, the particulate can act as nucleation sites (heterogeneous nucleation) and lower the limit of superheat to values that can be close to the liquid saturation temperature. At the interface between two liquids, one with higher saturation temperature (host liquid) than the other, once nucleate boiling is initiated in the one that is superheated, boiling will occur explosively with an intensity that depends primarily on the surface tensions of two liquids, the difference between the boiling point of the less volatile liquid and the limit of superheated of the more volatile, and the ambiant pressure, among other factors.

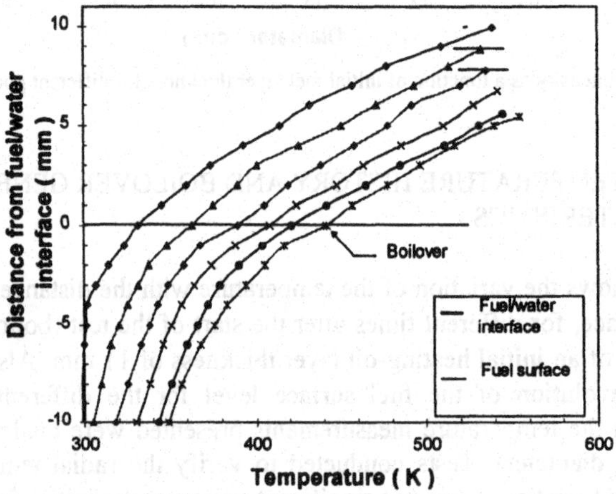

Figure 5. Development of vertical temperature profile (pan of 15 cm in diameter and initial fuel layer thickness of 11 mm).

Thus, heat is transferred from the surface to the liquid interior, causing the temperature at the fuel/water interface, and at the water interior, to increase to values well above the water saturation temperature at the corresponding pressure (approximately 100°C); i.e., the water becomes superheated. Under these conditions, bubbles in the superheated water could nucleate, most likely heterogeneously at the fuel/water interface, and grow explosively. The violent eruption of the water bubbles will cause the breakup, and possibly the atomization, of the fuel layer above, projecting the fuel outward. The projected fuel burns with the ambient air, increasing the flame extension and, if the explosive nucleation is sufficiently strong, generating a flame-ball-type burning.

The evolution of the heating-oil/water interface temperature with time is shown in Fig. 6, for different initial thicknesses of the oil layer. Although this temperature is difficult to measure accurately because of liquid motion and, sometimes, foaming at the interface, it appears that the value increases slightly as the heating-oil layer thickness increases. As described above, it is suggested that the event is caused by the onset of boiling nucleation at the oil/water interface. Also consistent with this the process is the observed rapid drop in temperature at the interface, which is due to the liquid motion at the vicinity of the nucleation site, and the rapid exchange of liquid as colder water fills the void left by the water bubbles as they leave the site, or explode. It is noteworthy that the low value of the levels of superheat (≈ 20°C) is smaller than that expected from experiments of the nucleation of water in hydrocarbons (13). It is plausible to attribute this difference to changes in surface and interfacial tensions due to the adsorption of impurities at the interface between liquids, which may lead to the heterogeneous nucleation of the water rather than to its homogeneous nucleation. Unfortunately, no experimental evidence is available to confirm this statement.

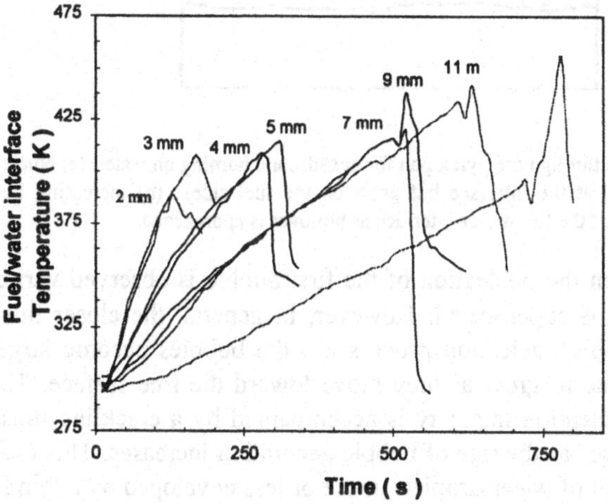

Figure 6. Fuel/water interface temperature as a function of time for different initial fuel layer thickness.

Additional information about the evolution of the fuel and water phases and of the interface during the onset of boilover can be obtained from the simultaneous video recording of the fuel-water interface, as shown in Fig. 7. The formation of the first bubbles is easily discernible, as well as how the bubbles grow, break away from the interface, rise, and finally reach the free surface. The most important information, however, is the location of the bubble initiation with respect to the interface and their subsequent development. Figure 7a shows that the bubbles are initiated at the interface but grow on the fuel side. Theoretical analysis of the problem indicate that the characteristics of the bubble growth, when a liquid is superheated in contact with another liquid, depends on the relative magnitude of the interfacial tensions (14). The present case clearly corresponds to the "bubble blowing" regime (14); in other words, the surface tension of water is larger than the sum of the surface tension of fuel and the interfacial tension. Although this inequality is difficult to quantify because interfacial tensions of fuels are generally not available, the surface tension of water is large enough to ensure it (see Table 1).

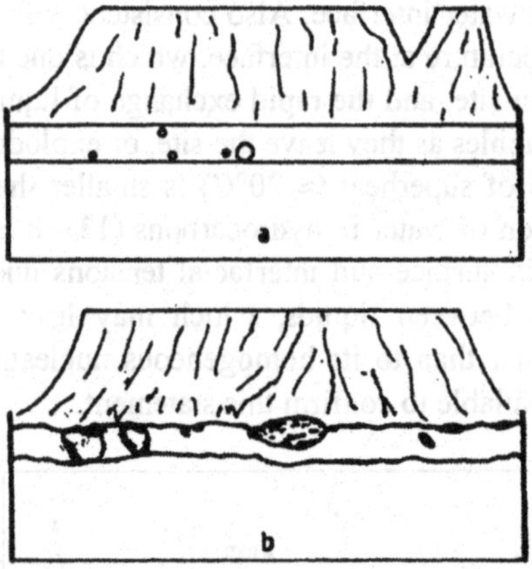

Figure 7. Side photographs through the pyrex pan for hexadecane burning on water. (a) Nucleation of the fisrt bubbles (they are initiated at the interface but grow on the fuel side) ; (b) increasing bubble nucleation intensity and disturbances of the fuel-water interface as boilover is approached.

The moment when the nucleation of the first bubble is observed varies during the period that the water is superheated. However, in general, the closer this period, the more intense the bubble nucleation process is ; the bubbles become larger and more numerous and continue to grow as they move toward the free surface. This period of increasing bubble nucleation intensity is accompanied by a crackling noise that has a frequency that increases as the rate of bubble generation increases. This crackling noise appears to be the result of water droplets, more or less enveloped by a thin layer of fuel, that are projected into the flame zone. These small droplets explode due to the nucleation of the water (15) and cause the characteristic crackling noise. The increase in the crackling noise intensity and frequency is generally the precursor of the boilover, and can be used to characterize its onset.

The simultaneous recording of temperature and fuel mass loss, as the boilover is approached, provided also interesting information about the events taking place during the onset of the phenomenon. At this onset, a sudden increase in fuel weight is observed. This apparent change of weight is due to the sudden expansion of water vapor at the nucleation site. The effect is naturally bigger, the higher is the internal pressure. The effect of pool diameter on the weight increase and, for that matter, on the intensity of the fire after boilover, is probably due to surface tension effects at the fuel surface (fuel/air interface). This surface tension will tend to deter the growth of the bubbles underneath, and the subsequent breakup of the fuel layer. The smaller the pan diameter, the stronger will be the fuel surface tension effect because of the capillary forces at the contact between the fuel and the pan edge.

It is found that the sudden increase in fuel weight, often coincides with a brief drop in temperature, which is immediately followed by a rapid and large temperature increase. It should be pointed out that the peaks of temperature and weight loss observed are, in fact, the envelopes of irregular fluctuations resulting from the tumultuous and violent character of the phenomenon. The extent in time of these peaks does not necessarily correspond to the actual extent of time of the eruptive vaporization observed during the boilover phenomena, which in general is much shorter. The actual time extent of the explosive burning is always very difficult to determine with reference to the total extent in time of the peaks. The sudden changes in weight loss and interface temperature coincide with significant disturbances of the water-fuel interface, as can be seen in Fig. 7b, and a corresponding increase in vaporization intensity. As a result, the thermocouple at the interface, which is fixed, can at any instant be in contact with either cold water or hot fuel that is rushing to fill up the voids left by the rising bubbles. Moreover, the vaporization acts as a heat sink that sometimes is also evidenced in the thermocouple reading. However, in general, the fuel is so intensively agitated that the thermocouple gives only an average temperature around the interface. This stirring effect has already been noted by Arai et al. (8).

The violent vaporization (i.e., the actual boilover) generally occurs when the rate of bubble nucleation increases so rapidly that bubbles cannot be evacuated toward the fuel surface. The large volume of water vapor generated at the interface suddenly breaks through the fuel layer above, ejecting fuel drops and columns toward the flame. The result is often spectacular, producing a column or ball of fire of very large proportions. The fuel ejection can deplete the fuel in the pan and, consequently, causes the sudden termination of the pool fuel burning. Occasionally, the boilover is not too intense and the disruptive burning can become repetitive.

From the above description, it can be seen that boilover is a rather complicated phenomenon that is difficult to model. Experimentally, it is possible to vary some of the parameters of the process to observe their effect on the boilover characteristics and, through them, infer some of its controlling mechanisms. Parameters that affect boilover strongly are the initial thickness of the fuel layer, the pan diameter, and the fuel type. The effect of the former two has been addressed above with heating oil as fuel. In the following paragraph, we are analyzing the effect of the latter.

4.6. INFLUENCE OF THE FUEL BOILING POINT

The influence of the fuel type on the boilover intensity, as a function of the difference between the fuel and water boiling point, is presented in Fig. 8 for the fuels tested. An average vaporization temperature is used for the crude and heating oils. The boilover intensity, although a rough measurement, provides useful, comparative information about the explosive character of the boilover. Although, considering that the boilover for toluene is rather weak, it appears more appropriate to state that the boiling point of the fuel has to be above 120°C, for significant boilover to occur. The data in Fig. 8 also show that the boilover intensity increases as the difference between the fuel and water boiling points increases, in agreement with previous observation (8). The information in Fig. 8 is complemented with the data in Fig. 9 on the effect of fuel boiling point on the evolution of the temperature of fuel/water interface, particularly since the magnitude of the temperature increase at the onset of boilover also provides qualitative information about the intensity of the boilover phenomenon. From the results of Fig. 9, it is seen that the amplitude of the temperature increase at the onset of boilover increases as the difference between the fuel and water boiling points increases, corroborating that the intensity of the boilover process is strongly dependent on the boiling point of the fuel.

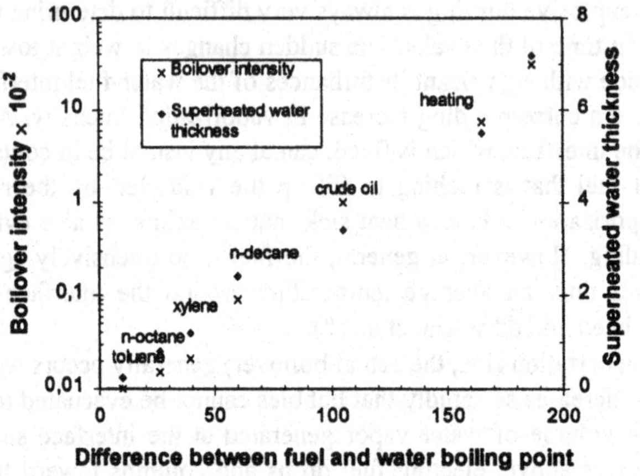

Figure 8. Boilover intensity and superheated water thickness as a function of the difference between fuel and water boiling point (initial fuel layer thickness : 13 mm ; pan diameter : 15 cm).

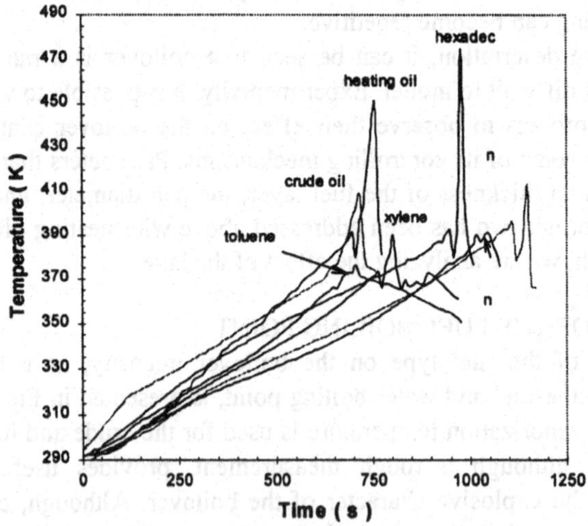

Figure 9. Fuel/water interface temperature as a function of time (initial fuel layer thickness : 13 mm ; pan diameter : 15 cm).

Also informative are the measurements presented in Fig. 10 concerning the effect of the fuel type on the preboilover burned mass ratio (burning efficiency). The burning efficiency gives information about the amount of fuel left at the time of boilover and, consequently, of the potential fire-ball size from the ejection of fuel caused by the boilover process. The results of Fig. 10 show that the percentage of fuel consumed before boilover decreases as the fuel boiling point is increased or, equivalently, that the thickness of the fuel layer at the time of boilover increases with the fuel boiling point. Consequently, the quantity of fuel ejected into the flame, and the resulting size of the fire-ball, is larger with fuels of higher boiling point, i.e., the overall intensity of boilover increases as the fuel boiling point is increased. Despite the difficulty of measuring accurately the level of superheat at the time of boilover (around 20 °C) because of the liquid motion and the random character of the phenomenon, it seems that there is a weak trend toward an increase in this level as the difference between the boiling points of water and fuel increases. This is probably due to the need for a larger pressure in the bubbles to overcome the higher pressure that results from the greater fuel layer thickness.

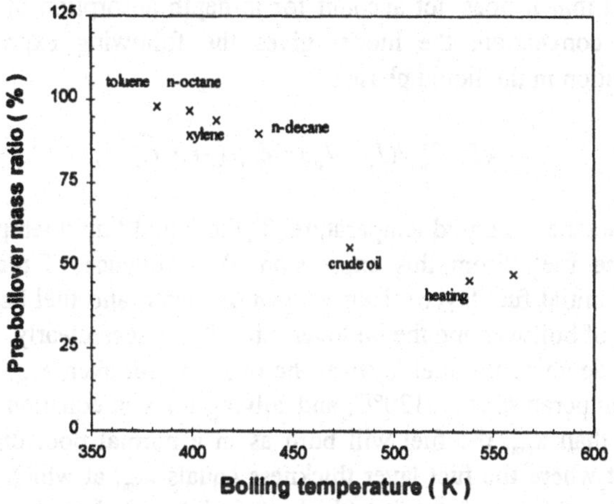

Figure 10. Pre-boilover burned mass ratio as a function of fuel boiling point (initial fuel layer thickness : 13 mm).

The effect of the fuel boiling point on the thickness of the layer of superheated water (considered to be between 100°C and 120°C) at the time of onset of bubble nucleation is shown in Fig. 8. It is seen that superheated water layer thickness increases as the fuel boiling point increases. This information is important because a thicker layer of superheated water and, consequently, a larger mass of evaporated water at boilover, contributes to a more intense boilover process by enhancing the expansive effect of the water vapor on the ejection of the fuel toward the flame.

4.7. MODELING OF THE FUEL AND WATER HEATING

The above results indicate that the characteristics of the boilover phenomena are determined primarily by the onset of water bubble nucleation at the interface, and the thickness of the fuel and superheated water layers at onset of boilover. The characteristics of the latter variables are determined primarily by the heat transfer through the liquid phase and, therefore, the effect of the fuel boiling point on the boilover characteristics should be reflected in the liquid heating process.

If heat transfer from the fuel surface to the liquid phase is assumed to be limited to conduction, and assuming a constant regression rate and that both the fuel and water have approximately the same thermal diffusivity α, the following simple, one-dimensional, quasi-steady, heat conduction equation may support the description of the spacial evolution of the temperature :

$$\frac{d^2T}{dx^2} = \frac{r}{\alpha}\frac{dT}{dx} \qquad [1]$$

where time has been replaced by x/r (where x is the depth from the fuel surface and and r the fuel surface regression rate (assumed to be steady)).

Despite the fact that it does not account for in-depth absorption of radiation, and possible effect of convection, the model gives the following expression for the temperature distribution in the liquid phase :

$$(T - T_o)/(T_b - T_o) = \exp(-r.x/\alpha) \qquad [2]$$

where T is the instantaneous liquid temperature, T_o the initial liquid temperature, T_b the boiling point of the fuel. From this expression, the influence of main parameters observed above, as initial fuel layer thickness, pan diameter, and fuel boiling point, on the time to the start of boilover and the boilover intensity, is seen clearly.

The maximum depth of the fuel layer at the onset of boilover, x_{bm}, is obtained by setting the liquid temperature at T=120°C, and solving for x in equation [2]. For initial fuel layers thicker than x_{bm}, the fuel will burn as in a normal pool until the surface regresses to a point where the fuel layer thickness equals x_{bm}, at which point boilover will occur with its maximum intensity (thickest fuel layer). For thinner initial fuel layers, the fuel will burn as in a pool until the water temperature reaches 120°C , at which point boilover will occur but with a lesser intensity, depending on the final fuel layer thickness.

A matter of concern is the boilover intensity. It appears that the determining factors are the thickness of the fuel layer at the time that nucleation of the water starts and the thickness of the layer of superheated water (assumed to be where the water is between 100 and approximately 120°C and may gasify). The thicker these layers, the more intense the boilover. These statements appeared verified when studying the influence of the aforementioned three main parameters. One aspect related to the effect of the initial fuel layer thickness is that boilover intensity increases with the final thickness of fuel when the water reaches the nucleation temperature. Another aspect results from the thermal penetration through the liquid. Since the thicker the initial fuel layer, the longer it takes to reach the critical fuel thickness at the time of nucleation, the deeper the thermal wave penetrates. The result is a thicker layer of superheated water which also results, as indicated above, in a more intense boilover.

As regard the effect of the fuel layer thickness on the boilover intensity, it should be pointed out that there may be additional effects other than simply the amount of fuel ejected into the flame, as indicated above. A possible additional effect is the delay of the onset of water nucleation due to the increased hydrostatic pressure, mentioned previously, at the fuel/water interface as the fuel thickness increases. A higher pressure will require a higher degree of superheat for the bubble to grow, and therefore, a thicker layer of superheated water. Futhermore, the thicker fuel layer may also initially restrain the expansion of the vaporized water until enough pressure in the vapor is built up to eject the fuel above. In fact, from the data in Table 1, it can be seen that, in general, the fuels with higher viscosity tend to experience a more intense boilover. In those cases, it is observed experimentally that the onset of boilover is characterized by the formation of a vapor film at the fuel/water interface, rather than individual bubbles.

The effect of the pan diameter was shown as resulting from a larger surface heat flux at the fuel surface (pan larger and sootier and more radiative flames). As the pan diameter is increased, the regression rate increases and the thickness of the superheated water layer decrease and, consequently, the boilover intensity decreases. An additional aspect was related to transient effects. As the pan diameter is increased, the surface heat flux increases and then the liquid phase is heated faster and the water reaches the nucleation temperature sooner. As a consequence, the penetration of the thermal wave is smaller and also the thickness of the superheated water and the intensity of boilover.

As for the effect of the boiling point of the fuel, a higher value results in an increase of the fuel layer thickness at the time of nucleation and in an increase of the superheated water layer thickness. This is likewise clearly shown by Eq. [2] and agrees with the observations.

The effect of the fuel boiling point on the thickness of the layer of superheated water (considered to be between 100 and 120°C) at the time of onset of bubble nucleation is shown in Fig. 8. It is seen that the superheated water layer thickness increases as the fuel boiling point increases. This information is important because a thicker layer of superheated water and, consequently, a larger mass of evaporated water at boilover contributes to a more intense boilover process by enhancing the expansive effect of the water vapor on the ejection of the fuel toward the flame.

However, if the Eq.[2] helps in understanding how the different problem parameters affect boilover, it presents some limitations. To apply it throughout the whole liquid phase, it is assumed that the thermal diffusivity of the fuel and water are similar. On the other hand, steady-state is assumed but, in reality, there are transient effects related to the time needed to the regression rate and temperature profiles to become steady. Then, it appears that the viability of this simple temperature distribution model is restricted to the thickest initial fuel layers (more than approximately 0,8-1 cm). For such conditions, where the above assumptions are realistic, the predictions of the analysis agree well (≈ 10%) with the experimental observations but the calculated time to the the start of boilover is larger than the measured time, indicating that radiative heat transfer in depth is neglected. The model is only valid for highly viscous fuels, such as multicomponent fuels (crude oils, heating oil, etc.). As the viscosity drops, correction due to the increasing in-depth absorption of radiation may take an increased role.

5. Conclusion

The present study of the combustion of a fuel spilled on water emphasizes the importance of heat transfer in the direction normal to the fuel and sublayer surfaces. It is shown that the fuel-layer thickness, the pool dimension, the fuel boiling point, are important factors in regard to the surface layer energy balance, the combustion of the fuel and the possible onset of boilover. The characteristics of the boilover intensity lead to consideration of the fundamental aspects that cause this phenomenon to occur. The observations that the transition to boilover is fast and has a violent explosive character, together with the observation that the water temperature reaches a temperature which is above the saturation temperature of the water, suggest that the event is due to the onset of boiling nucleation in the superheated water at the fuel/water interface.

Nevertheless, although a simple heat transfer analysis of the fuel and water heating (Eq. [1]) provides information about the limiting conditions for boilover and helps understanding how the different problem parameters affect it, it presents some limitations due to its simplifying assumptions, in particular those of quasi-steady heat tranfer, equal thermal diffusivities of the fuel and water, and negligible in-depth radiation absorption. Thus, there is still a need to develop models to predict accurately temperature histories in the liquid and time to the onset of boilover and to describe the water nucleation at the fuel-vapor interface, the bubble dynamics, and the fuel ejection. Of particular importance should be the development of a model that would accurately predict the temperature at which heterogeneous nucleation of the water occurs under the conditions encountered in liquid fuels spilled on water.

6. Acknowledgements

The authors express their thanks to Pr A. C. Fernadez-Pello of the University of California, Berkeley, for its interest and help in this study.

7. References

1. Hall, H., Mech. Eng., 47 (7):540-544 (1925).
2. Burgoyne, J. H., and Katan, L. L., J. Instit. Pet. 33:158-191 (1947).
3. Hasegawa, K., Fire Safety Science. Proceedings of the Second International Symposium, Hemisphere, New-York, 1989, pp.221-230.
4. Ito, A., Inamura, T., and Saito, K., ASME/JSME, Thermal Engineering Proceedings, 5: 277-282 (1991).
5. Petty, S. E., Fire Safety J. 5:123-134 (1982).
6. Koseki, H. and Mulholland, G.W., Fire Technol. 27 (1):54-65 (1991).
7. Koseki, H., Kokkala, M., and Mulholland, G.W., Fire Safety Science. Proceedings of the Third International Symposium, Elsevier, London and New-York, 1991, pp. 865-874.
8. Arai, M., Saito, K., and Altenkirch, R., Combust. Sci. Technol. 71:25-90 (1990).
9. Inamura, T., Saito, K., and Tagavi, K.A., Combust. Sci. Technol. 86:105-109 (1992).
10. Garo, J.P., Vantelon, J.P., and Fernandez-Pello, A.C., Twenty-Fifth Symposium (International) on Combustion, The combustion Institute, Pittsburgh, 1994, pp 1481-1488.
11. Garo, J.P., Vantelon, J.P., and Fernandez-Pello, A.C., Twenty-Sixth Symposium (International) on Combustion, The combustion Institute, Pittsburgh, 1996, pp. 1461-1467.
12. Blander, M., and Katz, J. L., AICHE J. 21(5) :834-848.
13. Lasheras, J. C., Fernandez-Pello, A. C., and Dryer, F. L., Combust. Sci. Technol. 21:1-14 (1979).
14. Jarvis, T. J., Donohue, M. D., and Katz, J. L., J. Colloid Interface Sci. 50:359-368 (1975).
15. Collier, J. G., Convection Boiling and Condensation, Mc Graw Hill, Maindenhead, Berkshire, England, 1981, p.171.

INDUSTRIAL ACCIDENT MODELLING: CONSEQUENCES AND RISK

E.A.GRANOVSKY, V.A.LYFAR', E.V.VASILYUK
Scientific Centre of Risk Investigation RIZIKON,
Severodonetsk -11, Lugansk region 349940, Ukraine

1. Introduction

The industrial development is accompanied by growth of scale and frequency of repetition of accidents associated with fires, explosions, and dangerous substance exhausts. The major accidents lead to the people death, substantial material losses, and serious ecological consequences [1, 2]. Prevention of the large scale accidents and limitation of their consequences at the industrial objects where dangerous substances are present take elaboration of the method for analysis of a hazard and risk. The analysis representing an 'illness diagnosis' is based on the complex investigation of a production plant. The algorithm is given in Fig. 1.

2. Hazard analysis

The hazard analysis is the most important stage of the investigation with a task to reveal all possible characteristic hazards of an industrial object which may lead to initiating and developing an accident. It's impossible to determine a risk of an accident if hazards that can lead to this accident are unknown.

The hazard analysis associated with violating conditions of the safe object operation consists of:

- revealing (identifying) dangerous substances and their specific properties;
- determining the boundary conditions under which a manifestation of the dangerous properties of the substances and an accident origination become possible;
- analysing deviations from a production process which lead to an accident origination.

To generalise the investigations of operational hazards the analysed object is considered as a chemical engineering system (CES) that deals with stages:

- preparation of raw materials and products to chemical conversion;
- chemical conversion;
- purification and separation of target products and by-products.

Each of the stages includes the characteristic elements in this CES. Thereby a typical reactor in which a typical physicochemical process runs is regarded as CES.

V.E. Zarko et al. (eds.), Prevention of Hazardous Fires and Explosions, 183–197.

A reactive medium and a presence of dangerous substances in it are analysed within every specified element. Thereby not only possibility of manifestation of the dangerous properties of the substances when leaving the reactor and contacting with atmosphere but possibility of a dangerous process within reactor or pipeline and possibility of an uncontrolled reaction are considered.

The boundary conditions are determined as a set of parameter values for production process run within the specified CES element under which a manifestation of dangerous properties of substances and an accident origination become possible. Conditions under which there are possible:

- spontaneous acceleration of a purposed and/or side exothermic reaction,
- generation of an explosive or combustible medium within production apparatus,
- exceeding the allowable value of the corrosion rate,
- seal failure of a detachable or packed joint,
- loss of integrity of reactor or pipeline walls,
- other violation by which the process can go out of control;

are regarded as boundary ones.

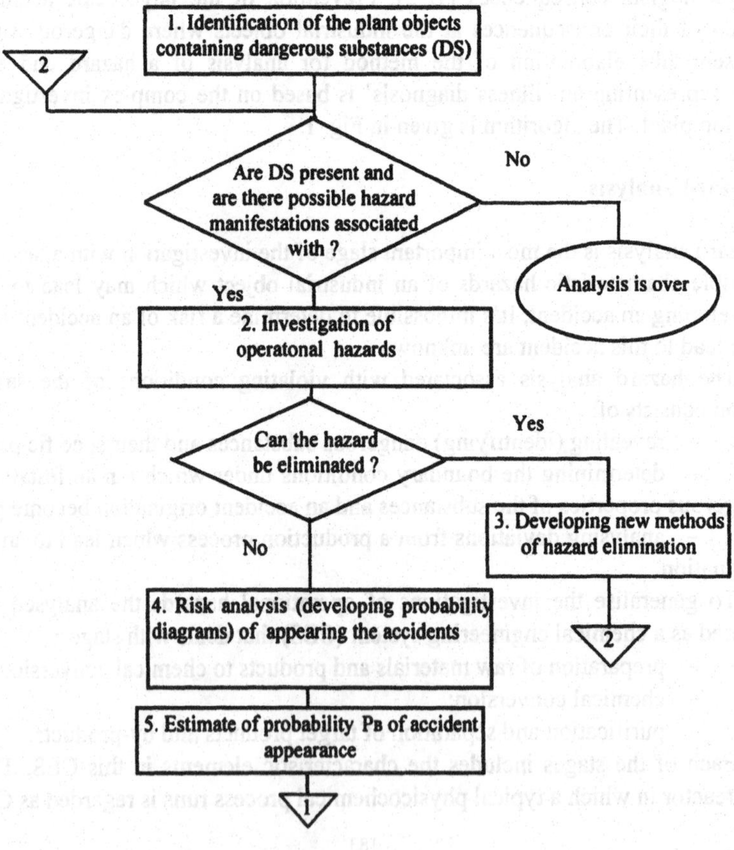

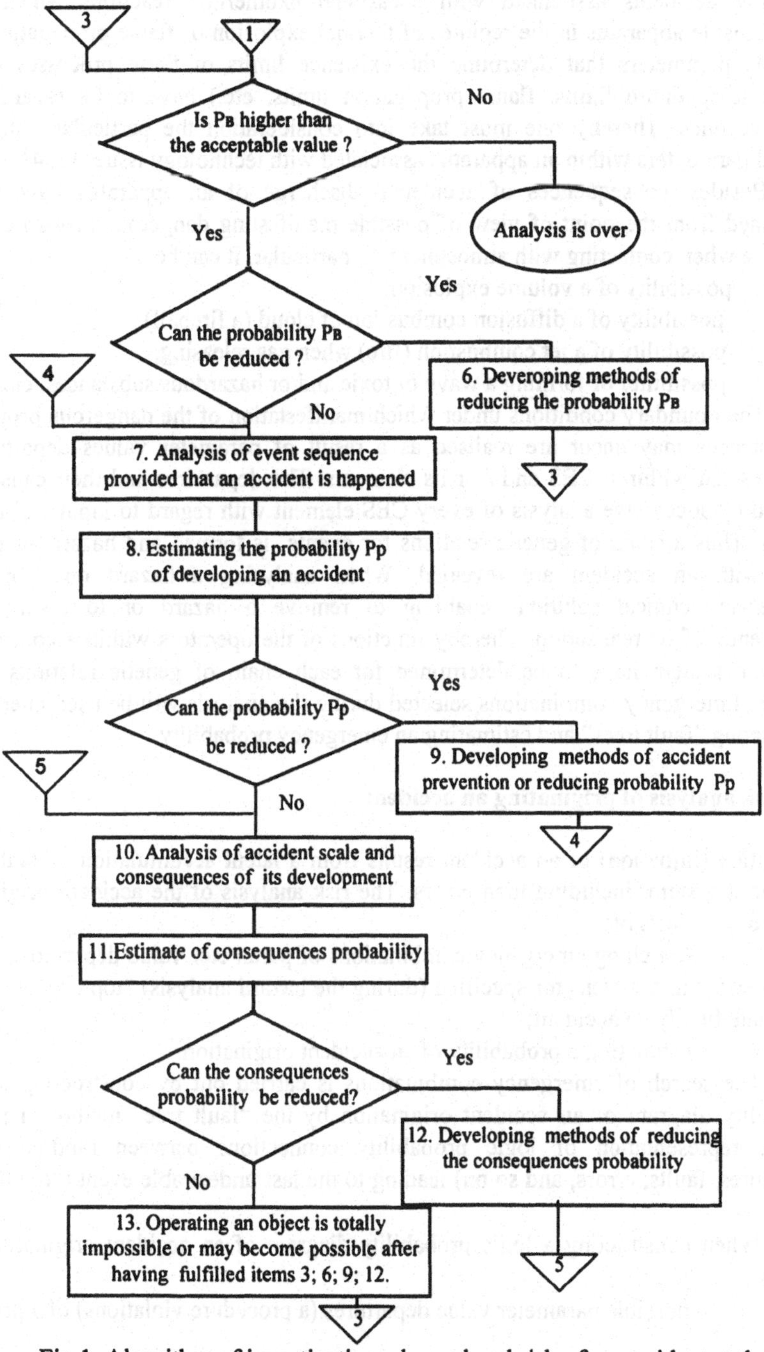

Fig.1. Algorithm of investigating a hazard and risk of an accident at the industrial objects

For accidents associated with occasional exothermic reactions which may proceed inside apparatus in the regimes of thermal explosion or flame propagation, the operation parameters that determine the existence limits of these processes (auto-ignition temperature limits, flame propagation limits, etc.) have to be regarded as restrictive ones. Thereby one must take into consideration the particular values of physical parameters within an apparatus associated with technology issues [3, 4].

Besides, consequences of accidental discharge of an apparatus have to be determined from the point of view of possible manifesting dangerous properties of a substance when contacting with atmosphere. In particular, it can be

- possibility of a volume explosion,
  possibility of a diffusion combustion of cloud (a fireball),
  possibility of a jet combustion (fire) when gas releasing,
  possibility of forming a wave of toxic and/or hazardous substances, etc.

The boundary conditions under which manifestation of the dangerous properties of substances may occur are realised as a result of parameter values departure of processes run within a CES and / or its elements. The departures and their causes are revealed by successive analysis of every CES element with regard to input and output streams. Thus a chain of genetic relations for events is formed and hazardous events ended with an accident are revealed. When analysing a hazard one considers engineering-technical solutions enabling to remove a hazard or to dismiss any contingency of its realisation. Thereby functions of the operators within a control and protection system have to be determined for each chain of genetic relations under analysis. Emergency combinations selected during the analysis will be used later when constructing "fault trees" and estimating an emergency probability.

3. Risk analysis of originating an accident

Origination (initiation) of an accident results from a latent accumulation of faults and errors in a system, including man errors. The risk analysis of the accident origination (initiation) consists of:

- searching emergency combinations of parameter value departures, faults, and errors in a system for specified (during the hazard analysis) "top events" which initiate finally an accident;
- estimating a probability of an accident origination.

The search of emergency combinations is carried out by constructing a logic probability diagram of an accident origination by the "fault tree" method, i.e. by a graphic representation of logic probability connections between random events (departures, faults, errors, and so on) leading to the last undesirable event ("top event") [5].

When constructing a logic probability diagram of an accident origination one regards:

- possible parameter value departures (a procedure violations) of a process;

– causes of these departures;

– mechanical break-downs and faults of equipment, pipeline and fitting elements;

– faults of instrumentation, signalling, automatic control and emergency protection systems;

– personnel errors.

To establish the genetic relations stipulated by the parameter departures (a procedure violations) of a process it will be used results of the hazard analysis.

When revealing causes of the departures there will be examined faults of the equipment and fittings, break-downs, as well possible technological causes associated with a violation of operation modes of functionally connected systems.

Having regarded the possible departures and their causes, an analysis of the instrumentation, signalling, automatic control and emergency protection systems enabled to monitor and protect against dangerous parameter departures is carried out for every process or operational procedure as analysed. The analysis is executed successfully for every departure.

If there is a stand-by facility it should be taken into consideration that after the main element fault a stand-by element can also be faulted during the main element restoration. It refers to equipment, fittings, instruments, control and protection means, and so on.

A very important element of analysis is the problem of interaction between man and system. When determining the possible man errors it is necessary to analyse his functions in designing, monitoring and controlling the working process, in particular:

– errors when developing and designing a system;

– errors when manufacturing, mounting and constructing;

– errors when repairing and redesigning;

– errors when operating.

When determining a probability it is necessary to take into account various causes of a man error, a possibility of a system state change and departures as result of subsequent erroneous decisions and actions of the personnel.

When analysing errors that can be made during repair, redesign and operation it's necessary to consider possibility of creeping the errors when making decisions and subsequent fulfilling decisions in accordance with official duties.

If the object is serviced by several operators an error probability must be determined with taking account of a duplication of actions. The analysis is continued until starting elementary events will be revealed whose probability can be determined on the basis of available statistical or reference data.

The data used for estimating a probability of starting elementary event (the departure) include:

– estimates of an operational reliability on the basis of statistical data on departures, faults and errors at the object as investigated;

– data on operational reliability of the element as regarded and on personnel errors received by Scientific Centre RIZIKON at the other plants (objects) similar to the regarded one (from the data base);

– data on reliability of the instrumentation, equipment, fittings and other elements referred within the certificates, instruction manuals of the products and other specifications;

– data referred in scientific and technical publications and the reference books, included data on operator errors in the man-machine system.

When constructing a "fault tree" and estimating a probability of an accident, the computer program "RELIABILITY" developed by SC RIZIKON is used. The program contains a data base of "emergency combinations" referred to the typical situations. In the case of non-typical situations the same program is used in the edit mode to create new logical probability combinations. The program enables to determine a maximum "emergency combination".

The quantitative estimation of an accident probability on the basis of the analysis of faults, errors and failures arisen during the object operation enables to make a decision on sufficiency (or insufficiency) of a system safety (see Fig. 2).

4. Simulation and prediction of accident consequences

If at least one of the emergency combinations analysed in the "fault tre " is realised it becomes impossible to prevent an accident by controlling the production process. A development of uncontrolled hazardous processes can lead to different accidents with various damage scales and various consequences. Accidents associated with an explosion, fire or dangerous and toxic substance ejection are most characteristic for the industrial objects. Each of these phenomena has specific injury factors, which gravity and range of action are determined by a power and conditions of an ejection.

When simulating an explosion one regards:

– explosions with fracturing a shell of equipment and pipelines as a result of pressure increase owing to uncontrolled physical or chemical processes within the equipment;

– explosions of condensed substances within the equipment or in the atmosphere after an ejection;

– volume explosions of gas or vapour-gas clouds after an ejection of compressed and liquefied gases or overheated liquids.

The conditions of originating the accident obtained in analysis of a hazard determine the explosion parameters. A phase state of operating medium components, temperature and pressure, fraction of condensed substance transformed into the gas phase are important when calculating a substance mass involved into the explosion. When computing the shock wave parameters one used the results of research works on explosions within unrestricted space based on the principle of a dynamical similarity [6 - 11].

Depending on situation it will be regarded ground, air or adit explosions. One calculates injury factors of an explosion which values define consequences:

– a pressure at shock wave front;

– a specific impulse;

– a duration of a compression phase;

- a maximum pressure within a ground at the given depth.

Consequences of people and structure constructions exposures to shock wave loads are estimated.

In the case of people it is determined how many people will at certain distances sustain:

- light injuries;
- serious injuries;
- grave injuries with a possible fatal outcome;

severe injuries with a frequent fatal outcome.

In the case of machines, equipment and industrial structures it can be determined light, average, grave and full destruction depending on their shock wave resistance according to experimental data [6, 8, 10, 12].

At an actual industrial site the shock wave acts on various obstacles (buildings, plants and vessels placed on an open space). In such a case a real pattern differs substantially from the spherical explosion. Thereby, digital simulation is carried out with the code developed by us together with Aslanov S.K. and Tsarenko A.P.

When simulating a fire one regards:

- combustion of free and diked spills of combustible and light-ignitable liquids [13-15];
- diffusion combustion of unmixed clouds after an ejection of pressurised liquefied gases or overheated liquids ("fireball") [2, 10, 16].

When estimating consequences one regards the following injury factors of a fire:

- intensity of thermal radiation;
- average surface density of the thermal radiation of a flame;
- burn-out rate;
- limit range at which an ignition of materials is possible from exposure to the thermal radiation.

When estimating the fire consequences one determines, besides what was burnt in flame, possible losses from exposure to the thermal radiation are determined. In the case of people the zones are determined where burns of the first, second and third extents are possible as well as the zone of pain threshold. In the case of materials a possibility of their inflammation and the fire propagation is determined.

When simulating an ejection of hazardous and toxic substances one considers weather conditions, a wind direction and velocity, ejection conditions and other parameters in accordance with the methods of computation.

To estimate possible consequences one determines:

- equivalent quantity of the substance in the primary and secondary clouds;
- zone areas of an actual contamination and possible contamination;
- width of a contamination zone;
- time of reaching an object.

On the basis of the calculation results one determines a possible amount of injured people, included cases with the lethal outcome.

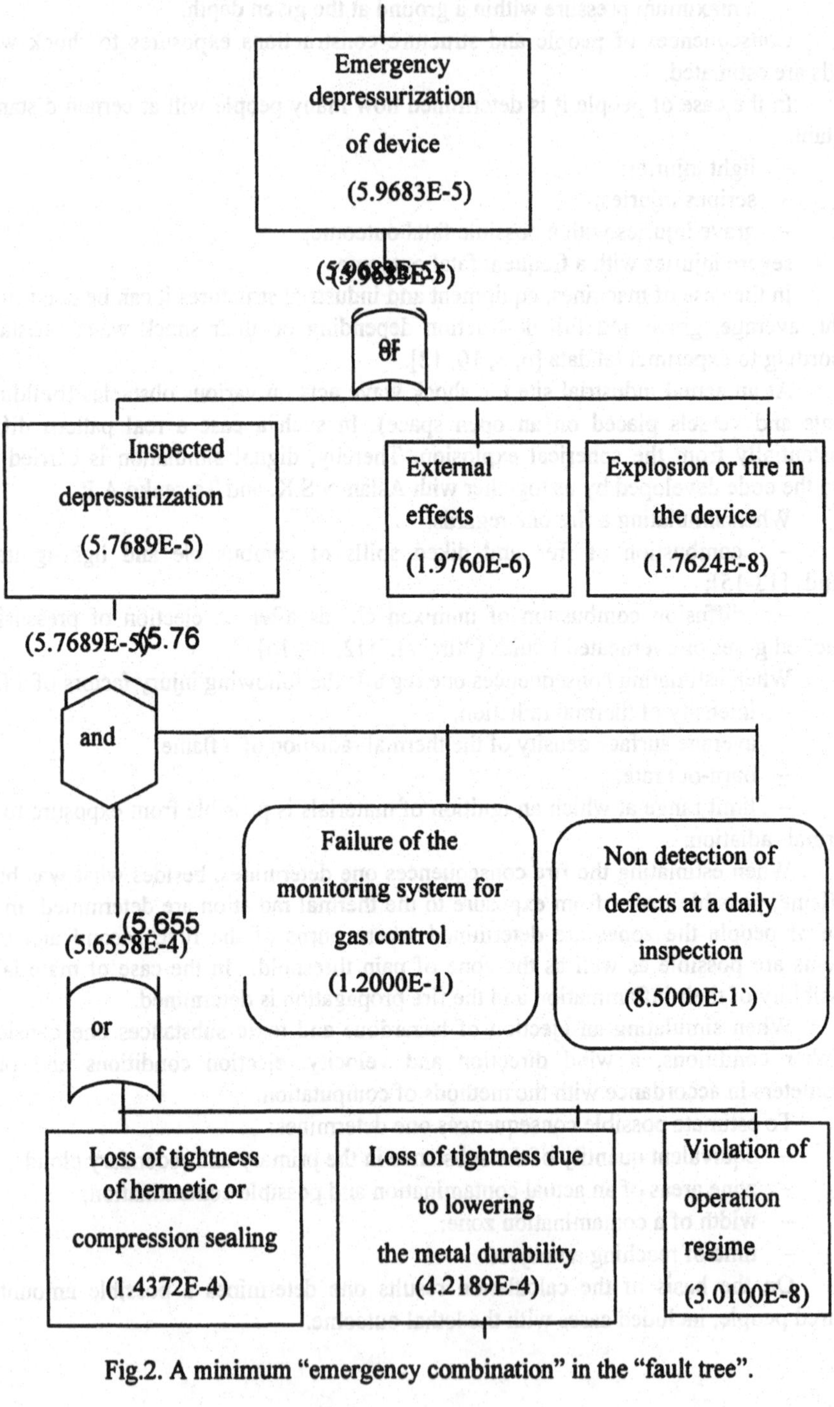

Fig.2. A minimum "emergency combination" in the "fault tree".

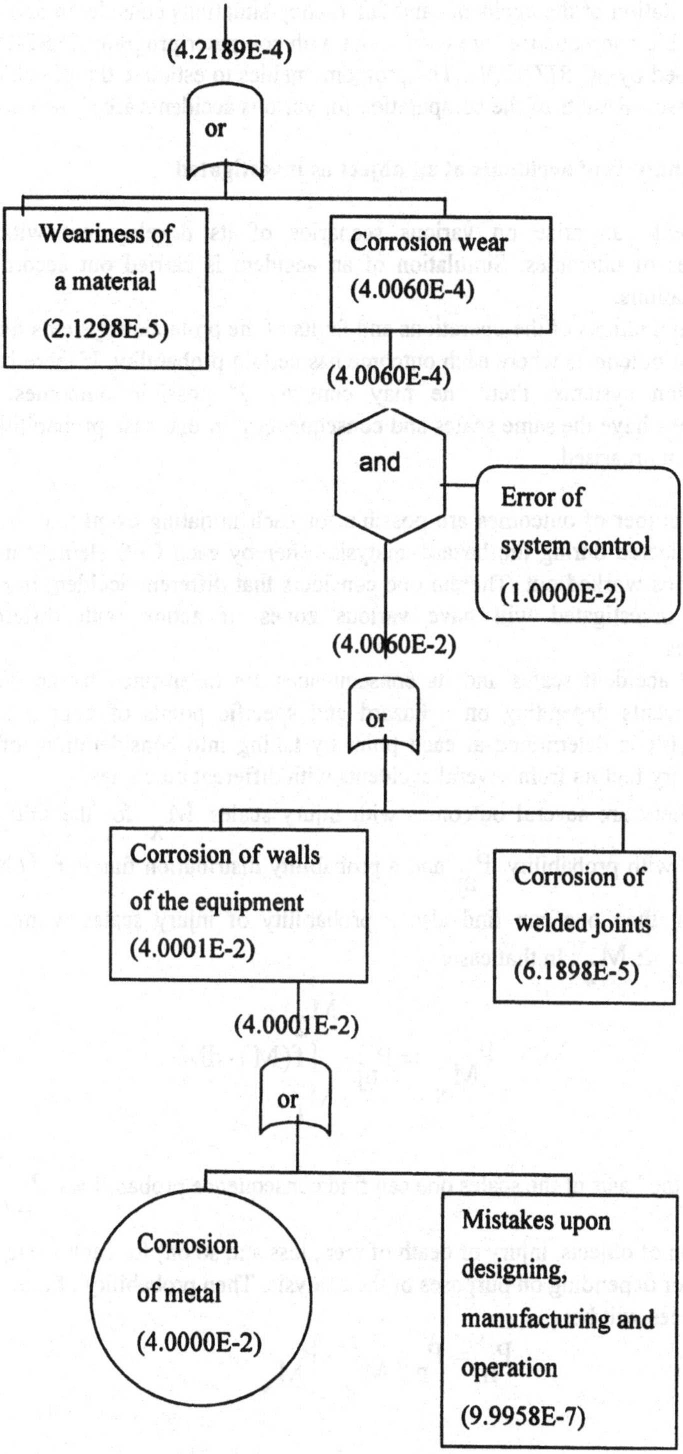

Fig.2. Continuation.

Simulation of the accidents and emergency situations considered and estimation of the possible consequences are carried out with computer program EXPERT, Version 1.3 developed by SC RIZIKON. The program enables to estimate the possible men and material losses. Results of the computation for various accidents are shown in Figs. 3-5.

5. Risk analysis of accidents at an object as investigated

The accident can arise on various scenarios of its development with different probabilities of outcomes. Simulation of an accident is carried out according to the typical situations.

Combinations of the operations and faults of the protection systems form a set of the accident outcomes where each outcome has certain probability. If there is number n of protection systems, then one may consider 2^n possible outcomes. Some of combinations have the same scales and consequences. In that case probabilities of such events are summarised.

A number of outcomes are possible for each initiating event that originates an accident selected during the hazard analysis. Thereby each CES element may have a few scenarios worked out. Therein one considers that different accident hazards at the object as investigated will have various zones of action with different injury probabilities.

The accident scales and its consequences are determined by an operation or initiating events depending on a hazard and specific points of depressurisation. A territorial risk is determined at each point by taking into consideration influences of various injury factors from several accidents with different outcomes.

If there are several outcomes with injury scales M_X for the initiating event considered with probability P_{Bj} and a probability distribution function $f(M)$ can be determined, then one can find also a probability of injury scales within the range $M_1 > M_X \geq M_2$. In that case:

$$P_{M_X} = P_{Bj} \cdot \int_{M_1}^{M_2} f(M) \cdot dM$$

On the basis of the scales one can find consequence probabilities P_{Π / M_X} (destruction of objects, injury or death of men, loss and so on) for each value of the injury factor depending on purposes of the analysis. Then probability of the consequences will be:

$$P_{\Pi} = P_{\Pi / M_X} \cdot P_{M_X}$$

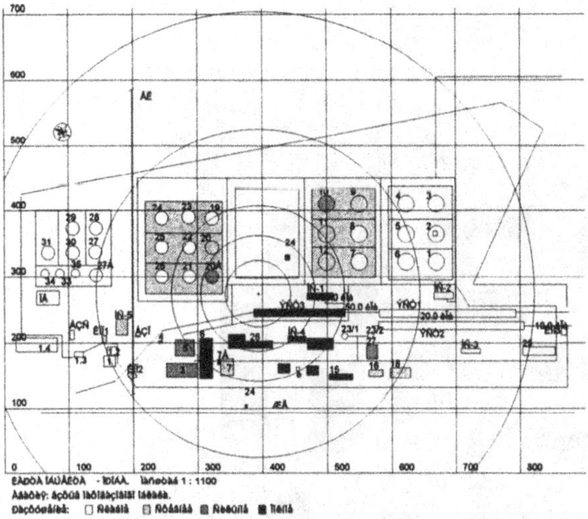

Fig.3. Results of simulating and estimating the explosion consequences of a propane-air mixture cloud at the raw-material and product storage

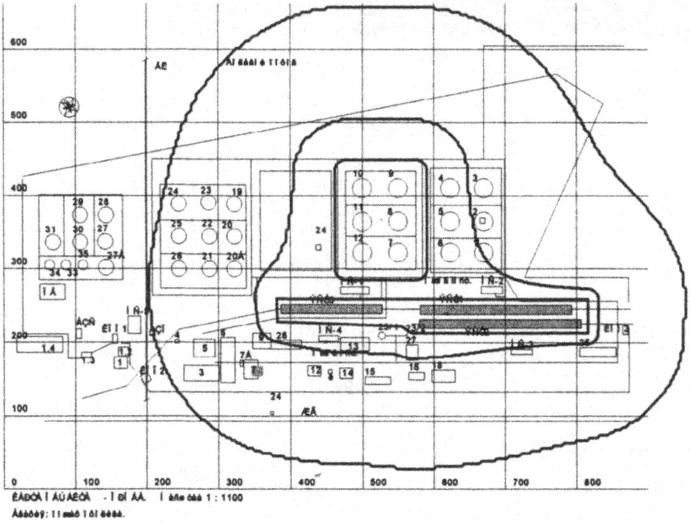

Fig.4. The simulation and estimation of fire consequences in case of spilling the petroleum products at the raw-material and product storage

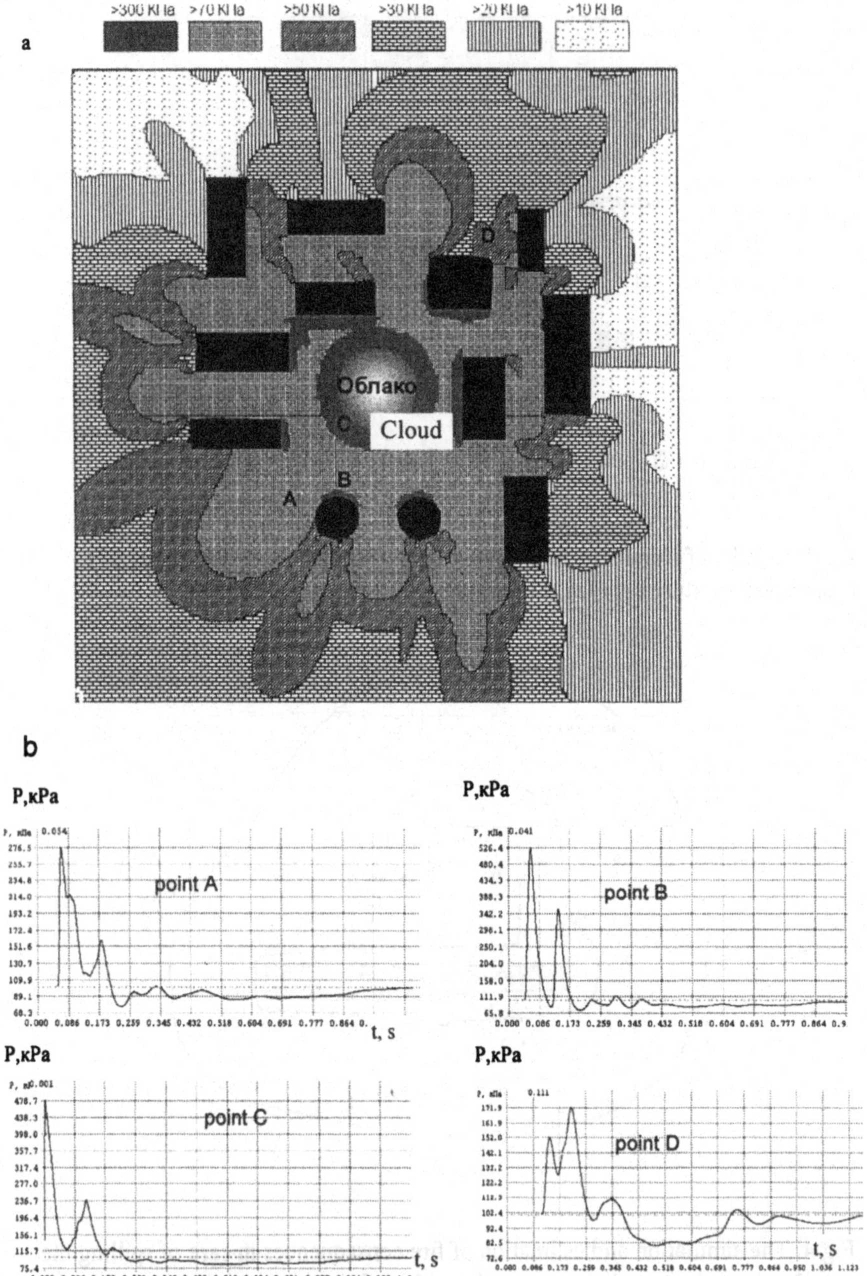

Fig.5. Numerical simulation of an explosion at the industrial site: a – instantaneous pressure field; b - P-t diagrams at the points A, B, C and D, accordingly.

To estimate a risk when simulating accidents and estimating a scale of their consequences one calculates probabilities of different outcomes and determines a total probability of a destruction and/or injury of people. At the territory analysed one determines a boundary of a given territorial risk level. If there is a zone where a risk is more than the acceptable value, then a possibility of its reduction is considered.

To estimate a territorial risk level one may use the program "Expert. Version 1.3". An example of estimating a territorial risk at one of oil refineries is shown in Fig. 6a.

a)

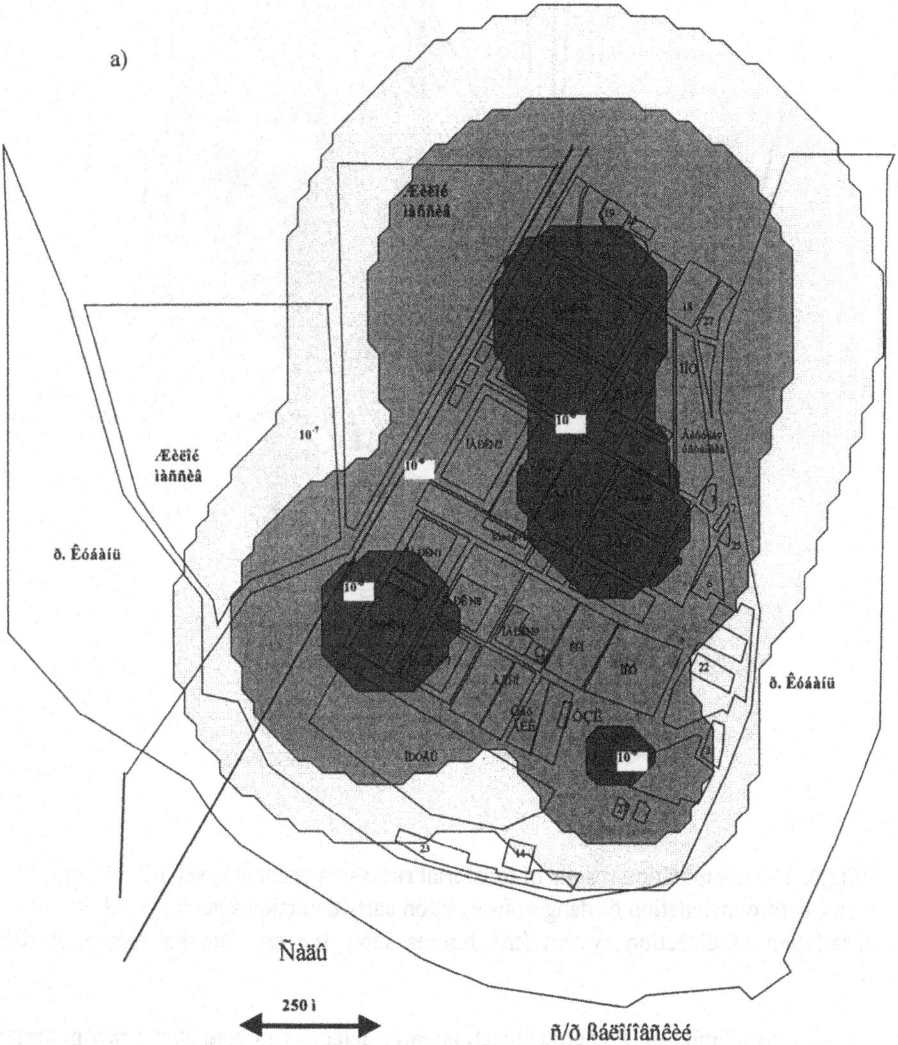

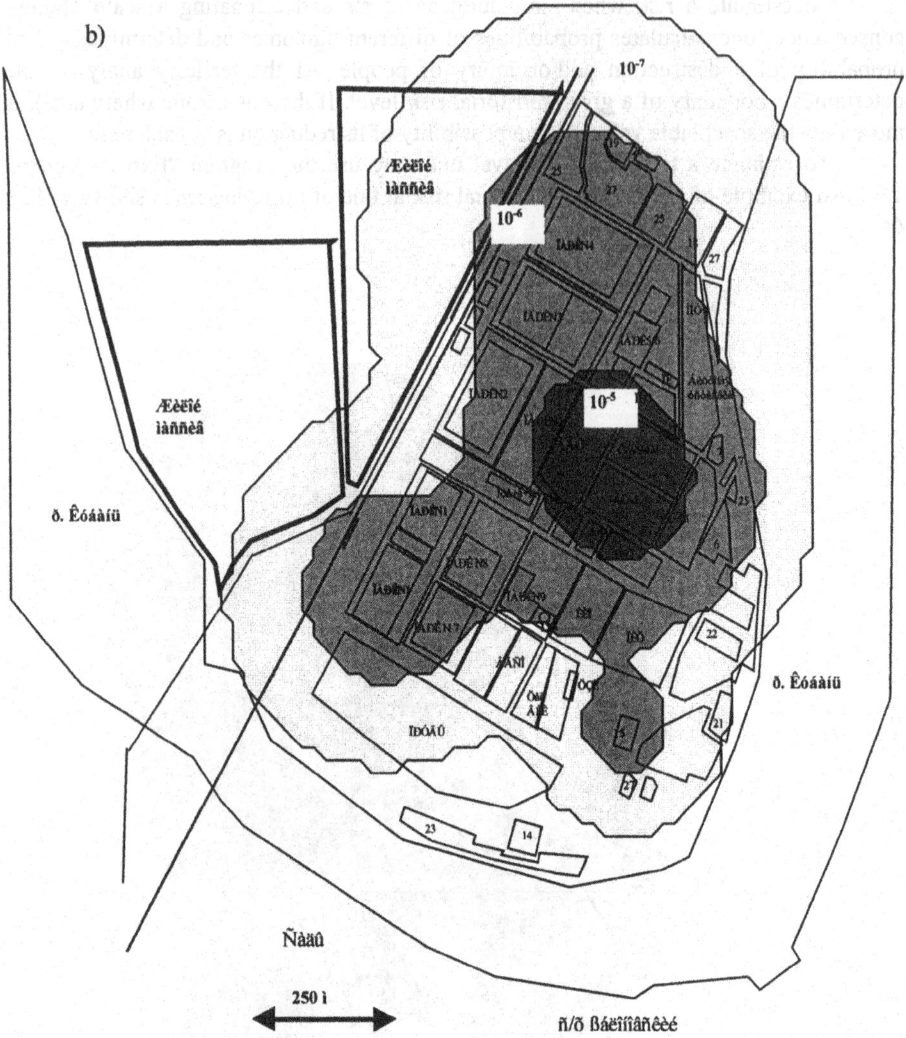

Fig.6. The computation results of territorial risk assessment at given oil refinery: **a** - before installation of dangerous ejection early detection system; **b** - after installation of detection system (the figures show the estimate for territorial risk level).

An installation of dangerous ejection early detection system at the raw-material and product storage enabled to reduce a risk level at residential districts of a city adjoined to the factory territory down to the acceptable value (see Fig. 6b).

6. References

1. Beschastnov, M.V. (1991) Industrial explosions. Estimation and prevention, Chimiya, Moscow.
2. Marshall, V. (1987) Major chemical hazards, Ellis Horwood Ltd., N.-Y.
3. Granovsky, E.A. (1980) Problems of applied combustion theory when developing explosion-safe production processes. - In: *Problems of explosion safety of production processes. Theses of reports on All-Union scientifically engineering conference.* Cherkassy, NIITEKHIM.
4. Granovsky, E.A. (1983) On a problem of explosion-safe implementation of the processes in chemical reactors, in: *Chemical reactors (theory, simulation, calculation). Proceedings of All-Union conference on chemical reactors*, **3**. Chimkent.
5. Kafarov, V.V., Meshalkin, V.P. et al. (1987) *Optimisation of reliability in chemical and oil-refinery industries,* Chimiya, Moscow.
6. Kogarko, S.M., Adushkin, V.V., Lyamin, A.G. (1965) Investigation of a spherical detonation of gas mixes, *Physics of Combustion and Explosion*, **2**.
7. *Physics of explosion* (1975), Ed. by Stanyukevitch K.P. 2nd edition revised, Nauka, Moscow.
8. Barstein, M.F., Borodachev, N.M., Blyumina, L.Kh. et al. (1981) *Dynamic design of structure response to specific impacts,* Ed. by Korenev B.G., and Rabinovitch I.M. Stroyizdat, Moscow.
9. Borisov, A.A., Gelfand, B.E. Tsyganov, S.A. (1985) On simulating shock waves originated upon detonation and combustion of gas mixes, *Physics of Combustion and Explosion*, **2**.
10. Baker, W.E., Cox, P.A., Westine, P.S., Kulesz, J.J., Strehlow, R.A. (1983) Explosion Hazards and Evaluation, Elsevier Scientific Publising Co., New York.
11. Aslanov, S.K., Golinsky, O.S. (1993) Integral theory of shock waves, Academy of Sciences, Kiev.
12. Demyanenko, G.P. et al. (1989) Protection of national production units against the weapons of mass destruction. The reference book. Vysshaya shkola, Kiev.
13. Shebeko, Yu.N., Shevchuk, A.P., Kolosov, V.A., Smolin, I.M., Brilev, D.R. (1995) Assessment of individual and social risks of accidents associated with fires and explosions for open-air production units, *Fire & Explosion Safety*, **1**.
14. Drisedell, D. (1990) Introduction to fire dynamics. Stroyizdat, Moscow (Translation from English).
15. Krishna, S. Mudan, (1984) Thermal radiation hazards from hydrocarbon pool fires, *Progress in Energy and Combustion Science*, **10**.
16. Zeldovitch, Ya. B., Rayzer, Yu.P. (1966) Physics of shock waves and high temperature hydrodynamic phenomena, Nauka, Moscow.

THE PROBLEMS OF POROUS FLAME - ARRESTERS

V.S.BABKIN
Institute of Chemical Kinetics and Combustion SB RAS
Novosibirsk 630090,
Russia

1. Introduction

It is known that the porous flame-arresters are based on the concept of critical diameter. They are used in the pipe lines with combustible gases when it is necessary to prevent accidental propagation of deflagration or detonation. In spite of apparent simplicity of the devices there are sometimes failure in their function due to as a rule inadequacy of operating conditions to technical requirement of the design. It is the consequence of some simplification of real processes of flame propagation and quenching in narrow channels and porous media. As a rule when designing the flame-arresters there are not taken into account some complicated factors like mass force action, gas flow properties (velocity, turbulence, multiphase), force interaction of the flow with a porous medium, chemical properties of the porous medium surface, and etc. Laminar burning velocity is the only used characteristic of reactivity of combustible gas. But it does not always adequacy represent the behaviour of various types of flames, e.g., cool, chain and thermal. It is usually not taken into account the boundary conditions of entering of combustion wave into porous bed of the flame-arrester and arising in this case of non-stationary and steady-state combustion regimes. These factors may cause failure in operation of porous flame-arresters and may lead to disaster.

In the present paper we discuss some principle problems of the porous flame-arresters. The question is not to revise the principles of flame-arresters but to anticipate arising of new complicated factors and to evaluate their influence on effectiveness of the flame-arresters. We start with discussion of possible combustion regimes that may arise in the flame-arrester bed.

2. Steady-state Gas Combustion Regimes in Inert Porous Media

The studies performed for the last two decades show that in inert porous media some different steady-state gas combustion regimes can be realised. They differ in determining parameters, the structure of combustion wave, the parametric dependence

V.E. Zarko et al. (eds.), Prevention of Hazardous Fires and Explosions, 199–213.

of propagation velocity, the mechanism of reaction transfer, etc. We report shortly some characteristic features of these regimes [1].

Low velocity regime (LVR) is realised, as a rule, in combustible gas flowing through the inert porous medium (PM). The flow (filtration) velocity is one of the important parameters of the process. Depending on the value of this velocity the thermal combustion wave propagates either up or downstream or is stationary in respect to PM. The wave propagation velocity is about 10^{-4} m/s. A strong inter-phase heat exchange and longitudinal heat transfer by PM are of importance for the mechanism of combustion wave propagation. Figure 1 shows the dependencies of combustion wave velocity u on the gas flow velocity in the bed of a porous material, V. The gas flow

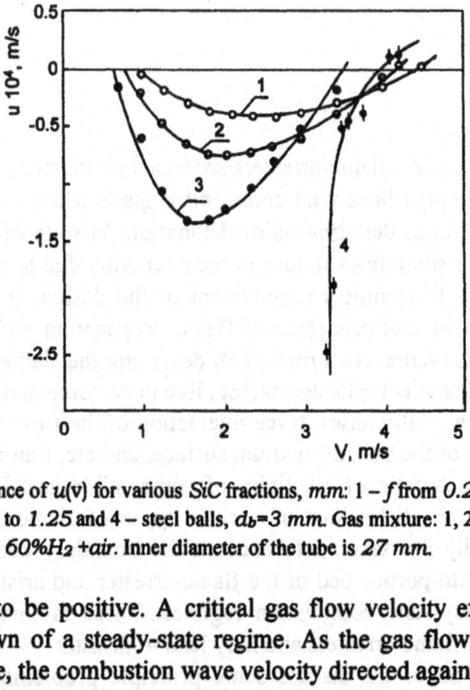

Figure 1. Dependence of *u*(v) for various *SiC* fractions, *mm*: 1 – *f* from *0.25* to *0.41*; 2 – *f* from *0.41* to *0.63*; 3 - *f* from *1* to *1.25* and 4 – steel balls, d_b=3 *mm*. Gas mixture: 1, 2, 3 – *65%* H_2*+air*, 4 – *60%*H_2 *+air*. Inner diameter of the tube is *27 mm*.

direction is assumed to be positive. A critical gas flow velocity exists, below which there is the break-down of a steady-state regime. As the gas flow velocity increases above the critical value, the combustion wave velocity directed against the gas flow first increases, passes the maximum, and then decreases to zero (curves 1-3). As the velocity increases further, the wave changes its direction. In the coarse-grained beds there is the critical flow velocity of v =3.1 m/s below which the combustion passes to the high velocity regime.

The high velocity regime (HVR) unlike low velocity one may be realised in PM with the initially quiescent combustible gas in pores. It is characterised by combustion wave velocities in the range 10^{-1} to 10 m/s and the inter-phase heat exchange intensity lower than that in LVR. Combustion is substantially subsonic, i.e. the process is isobaric in the open system. Compared to LVR, for HVR more pronounced is the aerodynamic rather than thermal interaction between PM and gas in the combustion wave. The gas flow, determined by external forces or by the combustion process itself, causes strong turbulence and, as a result, a high burning rate. The wide zone of chemical transformation and the influence of molecular gas properties on the propagation velocity, determined by the Lewis number, are typical of HVR.

Figure 2 shows the dependence of combustion wave velocity S in HVR and of the burning velocity of laminar flame S_u for air propane-and methane mixtures on the equivalence ratio ϕ. It is seen that, first, the turbulent burning velocities in HVR may exceed by order of magnitude the laminar ones. Second, the maximum values of a $(S/Su)_{max}$ ratio are shifted to the left for methane and to the right for propane flames with respect to the stoichiometric line ($\phi=1$).

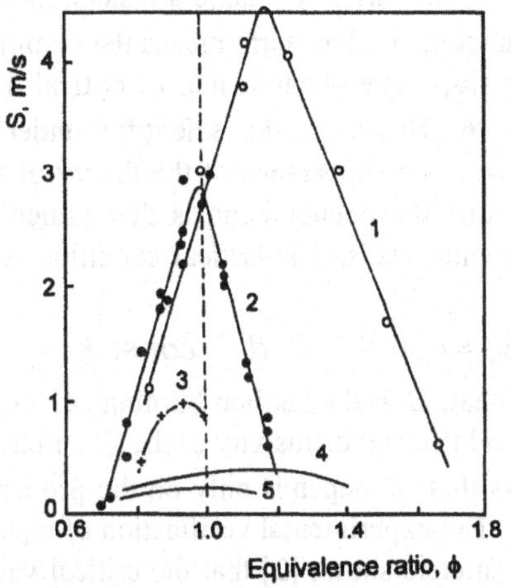

Figure 2. Dependencies of flame speed S (1-3) and burning velocity Su (4, 5) for air mixtures of propane and methane on equivalence ratio: 1 – propane, p_i=*0.1 MPa*, FPM d=*4.0 mm*; 2 – methane, p_i=*0.6 MPa*, SB d=*2.2 mm*; 3 – methane, p_i=*0.6 MPa*, SB d=*1.15 mm*; 4 – propane, p_i=*0.1 MPa*, 5 – methane, p_i=*0.6 MPa*.

Sonic velocity regime (SVR) may be realised with a change in system parameters that leads to an increase in flame velocity in HVR, e.g. increase in the initial pressure, laminar burning velocity, pore size, etc. In this case, the transition from HVR to SVR occurs, as a rule, by jumps. The flame velocities in SVR are about 100-700 m/s. A characteristic feature of SVR is the existence of a baric wave with a smoothly rising pressure in the front. In this case, the baric wave amplitude does not usually exceed the value of maximum pressure upon gas combustion under constant volume conditions. This regime is poorly explored.

Low velocity detonation (LVD) in inert PM exhibits the characteristic wave velocities of 500-1000 m/s. The intensity of a shock wave in this regime is insufficient for direct initiation of chemical reaction with a short delay. Gas is ignited at numerous hot spots resulting from the interaction of shock waves with each other and with porous medium elements.

Finally, *the normal detonation with heat and momentum losses (ND)* has characteristic velocities of 1500-2000 m/s and the usual mechanism of chemical reaction initiation, i.e. the gas self-ignition in the direct shock wave.

3. The Working Principle of Porous Flame-arresters

Flame-arresters are the devices used to localise (to quench) the flame. They are located on drainage and plume tubes, on the vessels with combustible liquids, on gas communications, etc. The flame-arresters protect equipment in which combustible gases or liquids are circulated, from flame penetration in emergency cases.

Although there is a great variety of flame-arresters, most of them are based on the principle of critical diameter, d^*. This term means the inability of flame to propagate in rather narrow tubes or slots. The phenomenon of critical diameter was discovered by Hamphry Davy in 1816. However, the scientific understanding of d^* has been developed much later with the appearance of the theory of laminar flame propagation. Zeldovich showed [2] that the phenomenon is determined by conductive heat losses from flame zone to channel walls. The critical conditions correspond to constancy of Peclet number

$$Pe^* = cS_u\rho_0 d/\lambda_0 \approx const \quad \text{or} \quad d^* = const\ k/S_u \qquad (1)$$

where c is the specific heat; S_u is the laminar burning velocity; ρ, λ, κ are the density, thermal conductivity, and thermal diffusivity of the gas mixture.

Equation (1) shows that d^* depends only on the properties of the gas mixture, its composition, and state. The experimental verification of equation (1) within wide range of the determining parameters shows [3] that the critical value of the Peclet number is $Pe^* \approx 65$. Spolding [4] has theoretically derived the limiting values of $Pe^* = 60.5$. Agreement between the theoretical and experimental data gave important arguments in favour of the thermal theory of flame, made background for the present methods of d^* determination, e.g. the method of flow cut (flame propagation direction from top to down) [5-7], and, finally, allowed one to create industrial flame-arresters [8].

The main element of attached flame-arresters is a semi-permeable porous barrier located in the protected passage. The bed of balls, sand, Rashig rings, metallic and ceramic filters, fibrous fillings, etc. are used as porous barriers. The sizes of the through PM channels must be below d^*.

A great body of information about d^* has appeared during a half of the century since the creation of the theory of critical diameter. As far as these data are often contradict to the theory it was necessary to discuss the problem of critical diameter [9]. As the d^* problems are closely related to the working efficiency of flame-arresters, these aspects will be considered in detail below.

Another circle of the problems, concerning flame-arresters, involves the latest achievements in the field of gas combustion in inert PM. Initially, it was indirectly assumed that the porous media, used in flame-arresters, simulates adequately narrow tubes with respect to laminar flame propagation and quenching. Thus, the data on d^*, based on laminar flame properties, are sufficient for designing effective flame-arresters.

It appears, however, that constructing flame-arresters, it is necessary to take into account not only the properties of laminar flames and the d^* value, but also the above mentioned wave combustion regimes. This is due to the fact that each steady-state regime has its own boundaries of existence with respect to mixture composition, pressure, temperature, and other parameters. The boundaries of existence mean the flame propagation limits beyond which the steady-state combustion is impossible (either flame quenching or blow-off) or the boundary of transition from one steady-state regime to another. The intrinsic mechanisms of limits and transitions have been studied for all regimes insufficiently. Therefore, one cannot a priori claim that at the limits of flame propagation in HVR, SVR and other regimes the same conditions will be satisfied as for laminar flames, i.e. $Pe^*_{LF} = Pe^*_{LVR} = Pe^*_{HVR} = \ldots Pe^*_{ND}$.

In addition, the problem of flame-arresters includes not only the question of the possibility of realising one or other regime in a porous bed of flame-arrester, but also the questions of non-stationary transition of combustion wave from free space in PM. The latter depends on many factors, including characteristics of the combustion wave in free space, combustion regime in PM, filtration velocity, v, and the co-ordinate of combustion wave with respect to the boundaries of PM, x/L (L being the flame-arrester length).

4. The Influence of Combustible Mixture Motion on Flame Quenching in Narrow Channels and Porous Media

As follows from equation (1), the concept of critical diameter is equivalent to concept of flammability limit in one parameter, i.e. the tube diameter, and corresponds to particular case of the general problem of combustion break-down.

In the theory of flame propagation limits [2] it is implicitly assumed that the motion of gas is non- essential factor. This is exhibited in the fact that the Nusselt number Nu, determining heat losses from flame zone to tube walls, is assumed constant. As a result, Eq. (1) contains no $d^*(v)$ dependence in which υ is the gas velocity.

On the other hand, the certain influence of gas motion is quite possible. First, as v increases the heat exchange in tubes increases, and, thus, the flame propagation limits must be narrower. Moreover, with increasing velocity, and transition to turbulent flow regime (increasing turbulence intensity u'), one may enter the area of hydrodynamic flame quenching [10]. Finally, the gas flow may be a heat-carrier in the case where the flame can heat tube walls up to high temperatures (see below).

It is shown [9] that taking into account of gas motion leads to the critical condition, containing additional determining parameter, i.e. gas velocity

$$Re = c_1 Z^{-4/3} Pe^{8/3}, \quad d^* = (c_2 \upsilon^{3/5} Z^{4/5} k)/S_u^{8/5} \tag{2}$$

upon convective quenching, and

$$Re \approx Pe^2, \quad d^* = (c_3 \upsilon k)/S_u^2 \tag{3}$$

upon hydrodynamic quenching of the flame.

Here $Re = \mathrm{v}d/\nu$, $Pe = S_u d/k$, $Z = E(T_b - T_0)/RT_b^2$.

The data for the stoichiometric propane-air mixtures show (Fig. 3) that in experiment [11] the exponent in the $Re \approx Pe^n$ dependence is in the range of $n=1.9\text{-}2.1$ whereas the values calculated by Eqs. (2) and (3) are $n=2.7$ and $n=2$, respectively. Thus, we have strong arguments to reckon that the critical diameter depends on the filtration velocity of the combustible mixture.

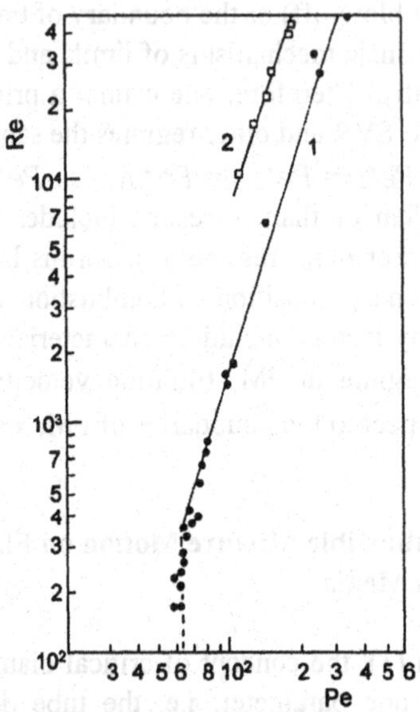

Figure 3. Dependence *Re(Pe)* for propane-air mixtures. 1 – flame propagation in inert porous media under various initial pressures [12], 2 - flame propagation limits in the tube with inner diameter of *3/8* inches under p_i=*0.1 MPa* [11].

In the case of porous media, in the HVR, the gas burns [12] in pore channels by the turbulence mechanism. The velocities of fresh and burnt gases in pore channels are non-uniform due to the pulsation character of flowing. The flame propagates by the relay mechanism, i.e. through the quickest pulses which, however, are limited from above by gas thermal relaxation and flame quenching in pore channels

$$Re = c_4 Z^{-3/2} Pe^3, \tag{4}$$

where $Re = vd/\nu$, v is the critical gas velocity, and d is the equivalent diameter of the pore channel in expression.

Thus, compared to single channels, in HVR far from the flammability limits the convective critical condition (4) in the channels is realised not simultaneously, i.e. the flame extinguishes only locally. In this case, the process of combustion as a whole does not stop, and flame quenching in the quickest pulses plays only the stabilising role.

5. The Lewis Number Effects in Narrow Channels and Porous Media

When the flame propagates in the channels near the limits, the heat losses cause non-uniform distribution of temperature over the flame front, which leads to a local change in the burning velocity and to flame front distortion. Front distortion results in the effects of preferable diffusion, front stretch, an increase in the total flame area, etc. This, however, is neglected in [2] and in a number of later papers that use the one-dimensional approximation and assume the equality of Lewis number to unity.

On the other hand, the dynamics of distorted flames has recently been studied very actively. The focus of attention is the topological properties of turbulent and cellular flames, stretch-effects, Lewis number effects, etc. To take into account the curvature, front stretch, preferable diffusion effects a great number of equations have been proposed, beginning with the pioneering papers [13-15], taking into account a corresponding correction for the value of laminar burning velocity of the flat flame front. The expression below was proposed in [16]

$$S_{u0} - S_u = L\alpha, \qquad (5)$$

where S_{uo} and S_u are the burning velocities of the flat and distorted flames, respectively; L is the Markstein length; α is the flame stretch rate. It is assumed that α takes into account the effects of flame stretch and curvature.

The behaviour of distorted flames in the case of strong heat losses has been studied in less detail. The propagation of distorted flames in a rectangular tube near the limits with $Le=1$ has been numerically studied in [17]. It is shown that with heat losses the chemical reaction rate near the wall is negligibly small. The flame zone is widened. As the value of heat losses increases, the flame front curvature increases. Similarly, the problem of flame propagation and quenching between two parallel plates and in cylindrical tubes has been studied in [18,19]. It is demonstrated that in the channels with diameter close to the critical one, the flame has the convex (towards fresh mixture) flame front. A decrease in the Lewis number causes a substantial change in the flame structure, form and propagation velocity.

For the channels of critical sizes, of major importance are the diffusion and thermal processes. In this case, using the hypothesis of preferable diffusion [20] it is assumed that in the flame, convex towards fresh mixture, the light, most mobile reagent enriches the zone of chemical reaction in the vicinity of the leading point due to the difference in the diffusion coefficients of fuel and oxidiser. This results in a local increase in the burning velocity in the case where the reagent is limiting and in a decrease in the burning velocity when there is excess of the reagent in the fresh mixture. An increase in the burning velocity with respect to S_{uo} and the low values of the critical Pe number, based on S_{uo}, are expected at the lean limits of hydrogen and methane flames and the flames of heavy hydrocarbons with either air or oxygen at the rich limit. This hypothesis was substantiated in [21]. The authors, studying the flame propagation limits in inert PM in the high velocity regime, show that the critical Pe number does not remain constant with varying mixture composition, as follows from Eq. (1), but increases with the enrichment of the mixture with fuel in hydrogen and methane flames and decreases in the rich hydrocarbons flames (Fig. 4, full circles; Table 1). Eq. (1) may be corrected with respect to the Lewis number using Eq. (5) [21]. To this end, Eq. (5) is written in the non-dimension form

$$Pe = Pe_0(1 - Ma \cdot Ka), \tag{6}$$

where Pe and Pe_0 are the Peclet numbers, constructed by S_u and S_{u0}, respectively; $Ma=L/\delta_0$ is the Markstein number; $Ka=\delta_0 \alpha / S_{u0}$ is the Karlovitz number; $\delta_0=\kappa/S_{u0}$ is the flame front thickness. The experimental data on the limiting pressures of the hydrogen-air flames in PM from polyethylene granules with $d=0.1$ cm (PG) and from polyurethane foam with $d=0.28$ cm (PF), the Markstein lengths L from [22], the values of burning velocities S_{u0} from [22] without any corrections for the dependence of S_{u0} on pressure, were used to calculate the values of Ma, δ_0 and Pe_0*. The results obtained for Pe from Eq. (6) are given in Fig. 4 by the empty circles. It is seen that relationship (6) is in better agreement with the condition of constancy at the limit than Eq. (1). The estimate obtained does not pretend to give a high accuracy due to its approximate character but demonstrates a correct approach to the solution of the problem discussed.

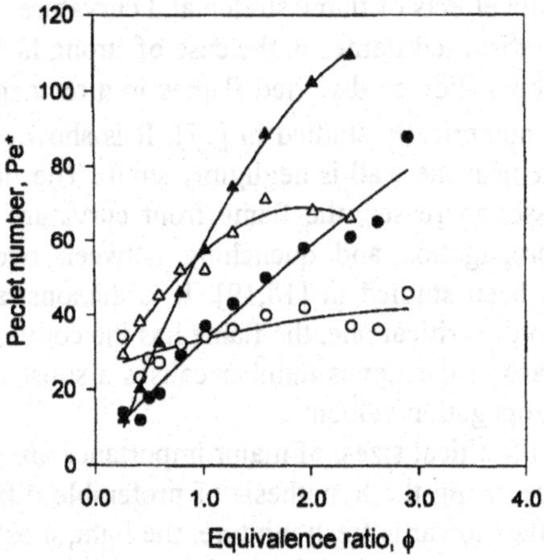

Figure 4. Dependence of critical Peclet number on H_2 content in hydrogen-air mixture for various porous media [21]. Triangle points correspond to polyurethane foam, $d=2.8$ mm, p_l from 0.042 to 0.097 MPa. Circle points correspond to polyethylene grains, $d=1.0$ mm, p_l from 0.074 to 0.267 MPa.

TABLE 1. Limit conditions of hydrocarbon-air flames in porous media.

Fuel content, %	Limit pressure, P_l, *Mpa*	Pe^*	Flame speed at the limit, S^*, *m/s*	S^*/S_u
Propane-air flames. Polyurethane foam, d=*0.28 cm*				
3.50	0.095	65	1.0	2.9
3.75	0.080	63	1.2	3.0
4.00	0.070	60	1.5	3.5
4.50	0.057	50	1.6	3.7
5.00	0.056	43	1.5	3.9
5.50	0.058	31	1.1	4.0
6.00	0.063	23	0.7	3.9
6.50	0.086	20	0.4	3.4
Methane-air flame. Steel balls, d=*0.12 cm*				
7.15	0.70	44	-	-
7.25	0.70	46	0.20	1.8
7.50	0.60	48	-	-
8.00	0.52	53	0.39	2.2
8.50	0.45	57	0.50	2.3
9.50	0.43	67	0.50	1.9

6. Propagation Limits for Combustion Waves

As term the "limit" we understand here the parametric boundary of existence of given steady-state combustion regime at which flame quenching occur. From the point of view of the discussed problem the next three problems are of interest, namely, a) limit conditions of steady-state regimes, b) the most dangerous of them i.e. ones having the lowest limit parameters of p_o, T_o, d etc; c) the conditions of correct measuring of the limits.

Limit conditions of the high velocity regime has been investigated in [21,23] where it has been shown that:

- the region of HVR is more narrow than that of the unconfined laminar flames;
- the flame velocities near the HVR limits are of order of magnitude of laminar burning velocities, more accurately, 2-6-fold as the ones, corresponding to the given experimental conditions (Table 1);
- the extreme points of pressure p^*_{min} and the maximum velocities at limits $(S^*/S_u)_{max}$ are shifted relative to the $S_{u\,max}$ position towards the region of rich propane and lean methane and hydrogen air mixtures (Fig. 5, Table 1) [23];
- the critical Peclet numbers are within the range of 20-90 and have a tendency to a decrease in propane mixtures upon enrichment and in methane and hydrogen mixtures upon depletion with fuel (Fig. 4, Table 1).

Thus, the critical Peclet numbers within a 70% spread are in agreement with the universal value of $Pe=65$ [3]. A similar conclusion has been drawn for HVR in [24]. Some excess of S over S_u may be due to weak turbulence, flame front curvature typical for narrow channels, the effects of Lewis number, etc. Taking into account the fact that far from the limits the values of an S/S_u ratio may amount to 20 and more, it is assumed that as p_i decreases and approaches the limit, the turbulence intensity decreases and flame velocity tends to the laminar burning one. This explains the fact why the limiting conditions for HVR and for laminar flames are the same, $Pe^*_{HVR} \approx Pe^*_{LF}$.

With increasing values of parameters such as p_i, S_u, d, according to (4), the velocity of steady-state flames increases and at some critical values of these parameters HVR transits by jumps into SVR. This transition in the hydrogen-air mixtures has been studied in [23]. The authors determined not only the transition boundaries but also the velocities of flame propagation in HVR over the entire concentration region for various initial pressures. It has been shown that within the entire range of the equivalence ratio $Pe^*_{HVR} \approx Pe^*_{LF}$ and $Pe^*_{SVR} \geq Pe^*_{HVR}$.

The Pe^*_{SVR} and Pe^*_{HVR} values are most close in highly reactive mixtures (near $\phi=1$) and differ substantially in poor mixtures (far from $\phi=1$, Fig. 6).
The latter conclusion is confirmed by the data of [25, 26] according to which in the oxygen mixtures at the limit of steady-state regimes with $S=525$-$840\ m/s$

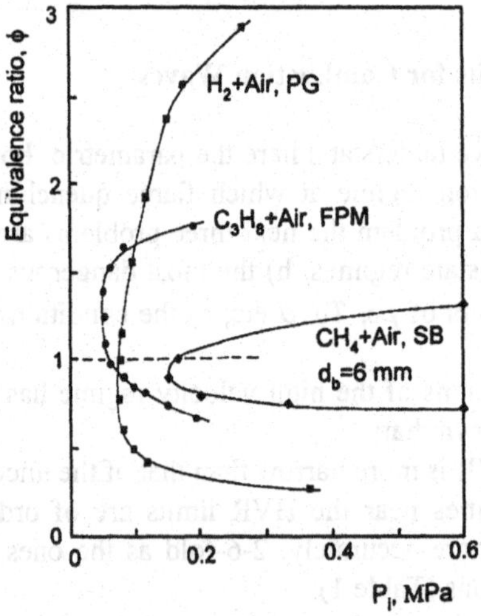

Figure 5. **Pressure limits of flame propagation. Hydrogen-air flames in the polyethylene grains PM (PG); propane-air flames in PM made of foil (FPM); methane-air flames in the steel balls PM, d_b=6 mm, (SB)**

$Pe^*_{SVR} \approx Pe^*_{LF}$ while in the air mixtures with $S=500$-$600m/s$ $Pe^*_{SVR} \geq Pe^*_{LF}$. A paradoxical result of the identity of limiting conditions $Pe^*_{HVR} \approx Pe^*_{LF}$ in combustion waves with different structures, velocities and the mechanisms of propagation, may be explained on a basis of Fig. 6. In fact, there is a very narrow area of HVR existence which is difficult to be resolved experimentally. Moreover, to satisfy the equality $Pe^*_{SVR} \approx Pe^*_{LF}$ the steady-state HVR does not need to exist.

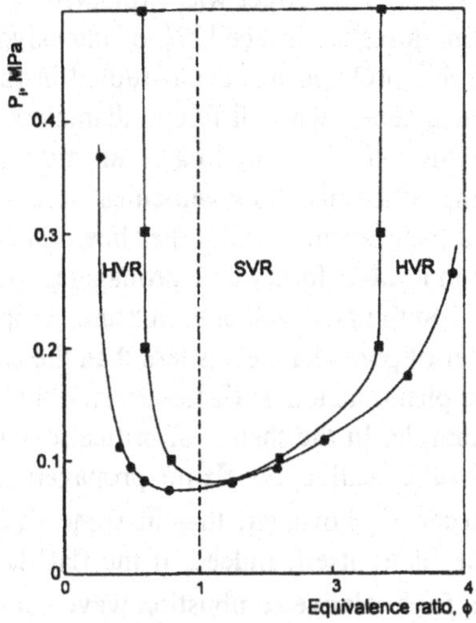

Figure 6. Flame propagation limits and regime transition boundary of hydrogen-air flames in the polyethylene grains PM [23].

Some remarks, however, should be made for the critical conditions for normal detonation with heat and momentum losses and of the low velocity detonation in PM.

It is thought that detonation can be initiated in a more narrow range of compositions than deflagration [3, p.331]. It was concluded [26] that in inert fillings the steady-state detonation velocity decreases and the minimum velocity (near the limit) depends weakly on mixture composition and PM particle size and amounts to 400-800 m/s. Thus, as the wave velocity decreases, there is transition from ND to LVD and further to SVR. Rozlovsky [3, p.330], studying the possibility of detonation passage through inert PM, has concluded that the "flame-arresters ... quenching deflagration, are also effective in the case of detonation. ... The penetration of flame through the channels of flame-arresters is independent on the mechanism of combustion propagation." In his opinion "...deflagration rather than detonation is quenched in narrow channels, detonation, in fact, ceases earlier." We agree with this viewpoint although it is beyond reason to assume that in all cases, including PM, the area of deflagration existence is always wider than that of detonation.

Finally, we note that the question of the correctness of the determination of limits mentioned above, is still open to argument. This is the problem of the non-stationary transition from free and semi-confined space to porous medium and of the formation one of the steady-state regimes. The non-steady state combustion processes in PM are actually unstudied either theoretically or experimentally. Researchers, however, know the difficulties, connected with the excitation of steady-state regimes in PM, arising in some cases.

7. Burning-through of Flame-arresters

Using flame-arresters, it was observed that in many cases the flame penetrates through a porous bed of a flame-arrester with an equivalent diameter of pore channels smaller than the critical one [3, 8, 27]. This effect was explained by various concepts, e.g., of flame stabilisation on flame-arrester surface [27], of intensive chemiluminescence [28], etc. Being highly actual, this problem has been studied in [29, 30]. Experiments were carried out in open quartz tubes with different diameters filled with grained inert material through which the premixed hydrogen-air mixtures were passed at room temperature. The gas composition and flow velocities were varied within a wide range. The flame was initiated at the open tube end. It has been established that in the PM bed the steady-state combustion wave forms and propagates with a bright luminous flat front. Its velocity depends on the flow velocity, mixture composition, tube diameter and other factors. The diameter of pore channels is less than the critical diameter (Fig. 7).

Let us make clear the phenomenon of flame-arrester burning-through using a single narrow channel as an example. In the theory of critical diameter it is usually assumed that the temperature of tube wall upon flame propagation and quenching remains constant, $T_s=T_0$. It is expected, however, that in some cases the tube walls may be strongly heated up by the flame itself. Indeed, if the flat flame propagates in the tube with the moving mixture, for a relative combustion wave velocity u/S_u, we get

$$U/Su = (v/Su) - 1 \qquad (7)$$

where v is the gas velocity. Obviously, with $v/S_u=1$, i.e. in a slow front motion ($u \approx 0$) the possibility of tube heating appears.

Upon tube heating the heat from the combustion product zone can be transferred through the wall to the zone of fresh gas (Fig. 7). In this case gas enthalpy, temperature and chemical reaction rate in the flame are increased. Thus, the effective velocity of gas combustion increases. Three characteristic states are possible in the wave behaviour (Fig. 7), namely, counter-flow (a), co-flow (b), and stationary wave (c). The thermal conditions in the wave depend on the direction of wave propagation. An important consequence of wall heating and heat recuperation, mentioned above, is the possibility of combustion front propagation in the channels with diameter smaller than the critical one. This combustion regime, called the low velocity regime (LVR), is possible in inert porous media [31] and sometimes in single narrow tube [32]. In PM the conditions for LVR are more favourable due to smaller heat losses.

At the present time, LVR is known quite well. The determining regime parameters and the parametric dependence of wave velocity have been established; the thermal structure and flame front stability have been studied; the origin and limits and the mechanisms of wave propagation have been determined; the LVR theory has been developed. The results have been generalised in survey [1, 33, 34].

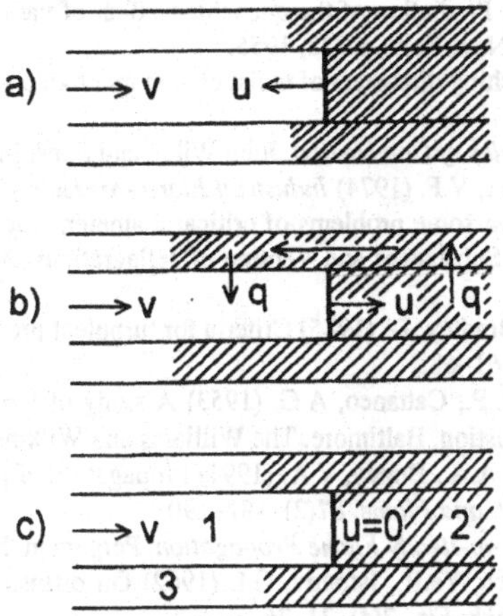

Figure 7. Scheme of movement of combustion waves in the tube with heated walls (a) – counter-flow movement, (b) – co-flow movement, (c) – standing wave, 1 – fresh mixture, 2 – combustion products, 3 – tube walls. Heated walls are shown with dashed lines, q – heat flux.

Thus, the phenomenon of burning-through is not due to the overheating of flame-arresters by the stabilised flame or due to other facts, but is caused by the initiation and propagation of combustion wave in LVR over flame-arresters, i.e. by the initiation and propagation of the thermal wave with its own unusual characteristics. Taking this into account, it is necessary to develop a novel approach to the choice of the porous material for flame-arresters, to the methods of their testing and of their protection against burning-through. The theory of combustion wave penetration through a flame-arrester is not then limited to the phenomenon of flame flashback in HVR, SVR, LVD, ND and should take into account the LVR mechanism. Besides, the theory should also include the problems of flame stabilisation and blow-off at the exit from flame-arrester and of flame localisation in porous medium near the surface [35-37].

Note that the problem discussed are also important for some other applications, e.g., for the Botha-Spalding burners used to study flat flames and their structure, to measure the burning velocity and other characteristics [38]; for porous radiant burners used as the sources of heat fluxes (IR-emitters) in industrial technologies.

This work was supported by the grant from the European Union (INTAS-96-1173) and by the grant of Russian Basic Research Foundation (No 98-03-32308).

References

1. Babkin, V.S. (1993) Filtrational combustion of gases. Present state of affairs and prospects, *Pure and Applied Chemistry*, **65(2)**, 335–344.
2. Zeldovich, Ya.B. (1941) Theory of limit of quiet flame propagation, *Zh. Prikl. Mekh. Tekh. Fiz.*, **11(1)**, 159-169.
3. Rozlovskii, A.I. (1980) *Basic principles of explosion safety under operation with combustible gases and vapours*, "Khimiya", Moscow.
4. Spalding, D.B. (1957) A theory of flammability limits and flame quenching, Proc.Roy. Soc., L. **A240**, No. 1220, 83–100.

5. Berlad, A.L., Potter, A.E. Prediction of the quenching effect of various surfacegeometries, 5-th Symp. (Intern.) on Combustion, New-York, p.728, 1955.
6. Grove, J.R. (1966) Quenching diameters of moist carbon monoxide – air mixtures, *Combust. and Flame*, **10(3)**, p. 388.
7. Kuo, K.K. (1986) *Principles of Combustion*, John Wiley and Sons, Inc, p.326.
8. Strizhevskii, I.I., Zakaznov, V.F. (1974) *Industrial Flame-Arresters*, 'Khimiya", Moscow.
9. Babkin, V.S. (in press) On some problems of critical diameter, Proc. of the 2-nd Intern. Seminar, Fire and Explosion Hazard of Substances and Venting ofDeflagrations. All – Russian Research Institute for Fire Protection, Moscow.
10. Abdel-Gayed, R.G. and Bradley, D. (1985) Criteria for turbulent propagation limits of premixed flames, *Combust. and Flame*, **62**, 61–68.
11. Starkman, E.S., Haxby, L.P., Cattaneo, A.G. (1953) A study of free flames in turbulent streams, 4-th Symp. (Intern.) on Combustion, Baltimore. The Williams and Wilkins Co., 670–673.
12. Babkin, V.S., Korzhavin, A.A., Bunev, V.A. (1991) Propagation of premixed gaseous explosion flames in porous media, *Combust. and Flame*, **87(2)**, 182–190.
13. Markstein, G.N. (1964) *Non-steady Flame Propagation*. Pergamon, New-York.
14. Barenblatt, G.I., Zeldovich, Ya.B., Istratov, G.I. (1962) On diffusion - thermal stability of a laminar flame, *Zh. Prikl. Mekh. Tekh. Fiz.*, **2(4)**, 21–26.
15. Joulin, G., Mitani, T. (1981) Linear stability analysis of two – reactant flames, *Combust and Flame*, **40(3)**, 235–246.
16. Clavin, P. (1985) Dynamic behavior of premixed flame fronts in laminar and turbulent flows,Prog. Ener. Combust. Sci., **11(1)**, 1-59.
17. Benkhaldoun, F., Larroututou, B., Denet, B. (1989) Numerical investigation of the extinction limit of curved flames, *Combust. Sci. and Tech*, **64**, 187–198.
18. Hackert, C.L., Ellzey, J.L., Ezekoye, O.A. (1998) Effects of thermal boundary conditions on flame shape and quenching in ducts, *Combust. and Flame*, **112(1/2)**, 73–84.
19. Zulinyan, G.A., Makhviladze, G.M., Melihov, V.I. (1991) Numerical investigation of laminar flame form and structure, Preprint, Moscow, the Institute of Mechanical Problems AN SSSR, **499**, 50 p.
20. Lewis, B., Elbe, G. (1961) *Combustion Flames and Explosion of Gases*, Academic Press Inc. N.Y. and London.
21. Korzhavin, A.A., Bunev, V.A., Potytnyakov, S.I., Babkin, V.S., Bradley, D., Lawes, M. Submitted to *Combust. and Flame*.
22. Dowdy, D.R., Smith, D.B., Taylor, S.C., Williams, A. (1990) The use of expanding flames to determine burning velocities and stretch effects in hydrogen/air mixtures, 23-d Symp. (Intern.) on Combustion. The Combustion Institute, Pittsburgh, pp. 325–332.
23. Korzhavin, A.A., Bunev, V.A., Babkin, V.S., Lawes, M., Bradley, D. (1993) Regimes of gas combustion in porous media and conditions for their existence,Proc. of the Russian – Japanese Seminar on Combustion. The Russian Section of the Combustion Institute, Chernogolovka, pp. 97–99.
24. Pinaev, A.V. (1994) Combustion modes and flame propagation criteria for an encumbered space, *Combust. Explos. and Shock Waves*, **30(4)**, 448-461.
25. Lyamin, G.A., Pinaev, A.V.(1985) Supersonic (detonation) combustion of gases in inert porous media, *Doklady Acad. Nauk SSSR*, **283(6)**, 1351–1354.
26. Pinaev, A.V., Lyamin, G.A. (1989) Fundamental laws governing subsonic and detonating gas combustion in inert porous media, *Combust. Explos. and Shock Waves*, **25(4)**, 448-458.
27. Glikin, M.A., Bityutskii, V.K., Kroshina, O.G., Savitskaya, L.M. (1981) The design princeples of industrial flame-arresters, *Chemical Industry*, 7, 428–430, (in Russian).
28. Mal'tseva, A.S., Steblev, A.V., Frolov, Yu.E., Rozlovskii, A.I. (1981) Direct radiative heating with intensive chemi-luminescence, *Doklady Acad. Nauk SSSR*, **258(2)**, 406–410, (in Russian).
29. Babkin, V.S., Drobyshevich, V.I., Laevsky, Yu.M., Potytnyakov, S.I. (1982) On the mechanism of combustion wave propagation in a porous medium during the gas filtration,*Doklady Acad. Nauk SSSR*, **265(5)**, 1157–1161.
30. Babkin, V.S., Potytnyakov, S.I., Laevsky, Yu.M., Drobyshevich, V.I. (1982) Fire resistance of flame-arresters, in *Fire prevention*, VNIIPO,.Moscow, pp. 111–114.
31. Babkin, V.S., Drobyshevich, V.I., Laevsky, Yu.M., Potytnyakov, S.I. (1983) Filtration combustion of gases, *Combust. Explos. and Shock Waves*, **19(2)**, 147–155.

32. Zamaschikov, V.V. (1997) Experimental investigation of gas combustion regimes in narrow tubes, *Combust. and Flame*, **108(3)**, 357–359.
33. Babkin, V.S., Laevsky, Yu.M. (1987) Seepage gas combustion, *Combust. Explos. and Shock Waves*. **5**, 531–547.
34. Laevsky, Yu.M., Babkin, V.S. (1988) Filtrational gas combustion, in *Heat Waves Propagation in Heterogeneous Media*. Ed. Matros Yu.Sh. 'Nauka" Novosibirsk, pp. 108–145.
35. Takeno, T., Sato, K., Hase, K. (1981) A theoretical study on an excess enthalpy flame, 18-th Symp. (Intern.) on Combustion. The Combustion Institute, Pittsburgh, pp. 465–472.
36. Buckmaster, J., Takeno, T. (1981) Blow-off and flashback of an excess enthalpy flame, *Combust. Sci. and Tech*, **25**, 153 – 158.
37. Eng, J.A., Zhu, D.L., Law, C.K. (1995) On the structure, stabilization, and dual response of flat-burner flames, *Combust. and Falme*, **100**, 645–652.
38. Botha, J.P., Spalding, D.B. (1954) The laminar flame speed of propane/air mixtures with heat extraction, Proc. Roy. Soc., **A 225**, 71–96.
39. Mital, R., Gore, J.P., Viskanta, R. (1997) A study of the structure of submerged reaction zone in porous ceramic radiant burners, *Combust. and Flame*, **111(3)**, 175–184.

IGNITION AND EXTINCTION OF SOLID PROPELLANTS BY THERMAL RADIATION

L. DE LUCA* and L. GALFETTI†
Facoltà di Ingegneria, Politecnico di Milano
32 Piazza Leonardo da Vinci, 20133 Milan, MI, Italy

Abstract

A survey of some peculiar ignition features of solid propellants is conducted in an attempt to identify possible civil applications of this specialized technological area. The relevance of the associated phenomena of dynamic extinction and static extinction, or pressure deflagration limit, emerges. It is shown that strong enough dynamic disturbances of different nature can permanently quench a solid propellant flame if the final operating pressure is below the deflagration limit (and no radiant energy is provided). This suggests concrete means of dealing with fire accidents in civil applications.

Nomenclature

$\tilde{a},\tilde{b}$ = dimensional coefficients used in Eq. 26, 27

a_b = factor of ballistic steady burning rate law, $(cm/s)/(atm)^n/(cal/cm^2 s)^{n_q}$,used in Eq. 15

$\tilde{A}_s$ = multiplicative factor of Arrhenius pyrolysis, $(cm/s)/(atm)^{n_s}/(cal/cm^2 s)^{n_{sq}}$,in Eq. 14

c = specific heat, *cal/g K*

d_{cr} = critical diameter, *cm*

$\tilde{E}_{(...)}$ = activation energy, *cal/mole*

$\tilde{f}(x), \tilde{f}_m(x) \equiv$ functions describing spatial distribution of impinging radiation

*Corresponding author. Professor, Dipartimento di Energetica, Ph.:(39-2)2399-3912, Fax: (39-2)2399-3940,
e-mail: DeLuca@icil64.cilea.it

†Professor, Dipartimento di Energetica

V.E. Zarko et al. (eds.), Prevention of Hazardous Fires and Explosions, 215–250.

I_s = external radiant flux intensity, *cal/cm²s*
k = thermal conductivity, *cal/cm s K*
m = mass burning rate, *g/cm²s*
$n \equiv [\partial \ln \overline{m} / \partial \ln \overline{p}]_{T_1,\bar{q}=const'}$ pressure sensitivity of steady burning rate, nondim.
$n_q \equiv [\partial \ln \overline{m} / \partial \ln \overline{I}_s]_{T_1,\bar{p}=const'}$, radiation sensitivity of steady burning rate, nondim.
n_s = pressure exponent of pyrolysis law, nondim., defined by Eq. 14 (Arrhenius)
n_{sq} = radiation exponent of pyrolysis law, nondim., defined by Eq. 14 (Arrhenius)
$n_{Ts} \equiv [\partial \ln \overline{T}_s / \partial \ln \overline{p}_s]_{T_1,\bar{q}=const'}$ pressure sensitivity of steady surface temperature, nondim.
$n_{Tsq} \equiv [\partial \ln \overline{T} / \partial \ln \overline{I}_s]_{T_1,\bar{p}=const'}$ radiation sensitivity of steady surface temperature, nondim.
N_s = radiation fraction absorbed below reacting surface layer, defined in Eq. 6
p = pressure, *atm*
p_{ref}, T_{ref} = reference pressure (68 *atm*), reference temperature (298 *K*)
$\tilde{q}$ = energy flux intensity, *cal/cm²*
$q_{ref} = \rho_c c_{ref} r_{b,ref} (T_{s,ref} - T_{ref})$ reference energy flux, *cal/cm² s*
Q= heat release, cal/g (positive if exothermic)
r_b= burning rate, cm/s
$r_{b,ref}, T_{s,ref}$= reference burning rate $r_b(p_{ref})$, reference surface temperature $T_s(p_{ref})$
=
r_λ = average reflectivity of burning surface
$\Re$ = universal gas constant; 1.987 *cal/mole K*
t = time coordinate, *s*
T= temperature, *K*
T_I= initial propellant temperature, *K*
x= space coordinate, *cm*

Greek Symbols
α = thermal diffusivity, *cm²s*
γ = flame elongation parameter for distributed flames, introduced in Eq. 18 and Eq. 20
$\Gamma = \sqrt{k_c \rho_c c_c}$ = condensed phase thermal responsivity, *cal/cm²* $\sqrt{s}K$
ε = nondim. chemical reaction rate
$\zeta = mc_g/k_g x$, nondim. group
λ = wavelength
ρ = density, *g/cm³*
$\sigma_p \equiv [\partial \ln \overline{m} / \partial T_1]_{\bar{p},\bar{q}=const'}$ initial temperature sensitivity of steady burning rate, *1/K*

Subscripts
c = condensed phase
c,s = surface from the condensed-phase side
f = flame; also: final
g,s = surface from the gas-phase side
g = gas
i = initial
ref = reference
s = burning surface

Superscripts
$\overline{(\ldots)}$ = steady-state value
$\widetilde{(\ldots)}$ = dimensional value
$\langle\ldots\rangle$ = average value
$(\ldots)_0$ = value at t = 0

Abbreviations
AP = Ammonium Perchlorate
CTPB, *HTPB* = Carboxyl-Terminated Polybutadiene, Hydroxyl -Terminated Polybutadiene
DBP = double-base propellants
FM = Flame Modeling
HMX =cyclotetramethylene tetranitramine
KTSS = Krier-T'ien-Sirignano-Summerfield
NC, *NG* = Nitrocellulose, Nitroglycerin
PBAA = Polybutadiene Acrylic Acid
PDL = Pressure Deflagration Limit
PU = Polyurethane
QSHOD = Quasi-Steady gas phase, Homogeneous condensed phase, One-Dimensional
Sa = salicylic acid
ZN = Zeldovich-Novozhilov

1. Introduction

Motivations, objectives, and the plan of presentation of this survey are discussed in this section.

1.1. MOTIVATIONS

Solid propellants allow cheaper and prompt propulsion missions but reduced performances and little control, as compared to other kinds of fuels. Composite or heterogeneous solid propellants are typically employed for space propulsion, while double-base or homogeneous propellants are commonly used for tactical rockets or gun propulsion; a suitable mixing between the two classes of solid propellants is often found in high energy propulsive missions. In all cases, the importance of a proper ignition transient is obvious and this explains the large amount of theoretical and experimental work dedicated to solid propellant ignition.

In addition, burning near pressure deflagration limit (PDL) and dynamic extinction phenomena are sometimes of strong interest in rocket propulsion. Ignition and controlled extinction for high altitude operations of rocket motors, pulsed operations of micro-trusters for satellite attitude control, transient processes in general are some of the most relevant applications. Moreover, PDL is important for handling hazards, storage safety, and ageing of energetic materials. Also it is recalled that low pressure burning in general, i.e. close to PDL, allows easier diagnostics of combustion wave structure and properties by magnifying time and space scales.

In most cases, PDL and dynamic extinction phenomena are accompanied by oscillatory combustion. PDL, dynamic extinction, and oscillatory combustion phenomena are different manifestations of intrinsic burning instability of energetic materials. In general, they overlap creating confusing effects in experimental investigations, as denoted by poor reproducibility and large scattering of the collected data. For an appropriate set of operating conditions, the three phenomena will just merge in one.

The above concepts, mainly developed in the framework of rocket propulsion, can find useful civil applications. This fallback is the main theme of the present paper.

1.2. OBJECTIVES

The purpose of this paper is to:

- offer a quick survey of ignition properties of solid propellants;
- offer a survey of PDL and dynamic extinction effects;
- propose a comprehensive physical and mathematical transient model including these effects;
- suggest useful civil applications of these peculiar effects.

1.2. PLAN OF PRESENTATION

The plan of presentation is as follows. First, a general mathematical model will be proposed in Sec. 2 retaining the classical QSHOD assumptions; the condensed phase is assumed chemically inert while surface gasification is assumed of a generalized Arrhenius type. The governing set of equations has to be integrated numerically; a reasonable agreement will be found among physical observations and numerical trends. Then, a survey of peculiar phenomena of interest for civil applications is conducted: ignition, PDL or static extinction, and dynamic extinction will separately be reviewed respectively in Sec. 3, 4, 5. For each of these phenomena, a literature survey will be conducted, the main experimental techniques summarized, typical results recalled, and the effects of a possible external radiation illustrated. Finally, some techniques of fire extinction based on fundamental combustion knowledge are proposed in Sec. 6.

2. QSHOD Modeling

Following the pioneering work by Zeldovich [1] in 1942, two main approaches, known as Zeldovich-Novozhilov (ZN) method and Flame Modeling (FM) method, have emerged to study unsteady combustion of solid propellants. Both share the basic assumptions of Quasi-Steady gas phase, Homogeneous condensed phase, and One-Dimensional burning processes (QSHOD framework). Within this framework and for pressure perturbations only, linear stability analyses were first presented by Denison and Baum [2] in 1961 for premixed flames by the FM method and Novozhilov (e.g.: [3] [4] [5]) in 1965 by the ZN method. Both papers, in the *linear* approximation of the problem, relaxed the assumption of constant surface temperature until then accepted. Systematic investigations were carried out by Culick for a variety of flame configurations (e.g.: [6] [7]) and Krier et al. (KTSS, [8]) for diffusive flames. This paper deals with ignition processes in the framework of the FM method.

2.1. FORMULATING THE MASTER QSHOD PROBLEM

A one-dimensional strand of homogeneous material is assumed burning with a quasi-steady gas phase subjected to changes in time of pressure and/or external radiation. In general, thermophysical properties are allowed to be pressure and/or temperature dependent. The strand is burning, with no velocity coupling, in a vessel at uniform pressure; radiation, if any, consists of a radiant flux originated *exclusively* from a continuous external source of thermal nature and interacting *exclusively* with the condensed phase. Overall, assume no radiation scattering, no photochemistry, no condensed phase chemical reactions, and no external forces. Define a Cartesian axis with its origin anchored at the burning surface and positive in the gas-phase direction. When needed, nondimensional quantities are obtained by taking as reference those values, maybe only nominal, associated with the conductive thermal wave in the condensed phase under adiabatic burning at 68 atm.

Let the condensed phase $(x<0)$ be a semi-infinite slab of uniform and isotropic composition, and T_1 a fixed value. The energy equation and associated conditions [9] [10] [11] are

$$\rho_c c_c(T)\left[\frac{\partial T}{\partial t}+r_b(T_s)\frac{\partial T}{\partial x}\right]=\frac{\partial}{\partial x}\left[k_c(T)\frac{\partial T}{\partial x}\right]+N_s\tilde{f}(x)\left(1-\overline{\overline{r}}_\lambda\right)I_s(t) \quad (1)$$

$$T(x,t=0)=T_0(x) \quad (2)$$

$$T(x\to-\infty,t)=T_1 \quad (3)$$

$$\left(k_c\frac{\partial T}{\partial x}\right)_{c,s}=\left(k_g\frac{\partial T}{\partial x}\right)_{g,s}+Q_s\rho_c r_b(T_s)+(1-N_s)\left(1-\overline{\overline{r}}_\lambda\right)I_s(t) \quad (4)$$

where the function $T_0(x)$ is in general a known initial temperature profile required for ignition or extinction studies. Notice that at $x = 0$

$$\tilde{q}_{c,s}\equiv\left(k_c\frac{\partial T}{\partial x}\right)_{c,s}\qquad \tilde{q}_{g,s}\equiv\left(k_g\frac{\partial T}{\partial x}\right)_{g,s} \quad (5)$$

while

$$N_s=\exp\left(-\overline{\overline{a}}_\lambda\frac{\alpha_c}{r_b}\frac{\Re\overline{T}_s}{\overline{E}_s}\right) \quad (6)$$

has been estimated, for the standard Michelson profile and extending its validity to transient burning, as suggested in [12] and [13] based on [14].

For $x<0$

$$\tilde{f}(x)\geq 0 \quad and \quad \int_{-\infty}^{0^-}\tilde{f}(x)dx=1 \quad (7)$$

making sure that the externally impinging radiation is totally absorbed, no matter how weakly, within the condensed phase. Typically, for a monochromatic radiation source find

$$\tilde{f}(x)=\overline{\overline{a}}_\lambda\exp\left(\overline{\overline{a}}_\lambda x\right) \quad (8)$$

At least in principle, the $\tilde{f}(x)$ evaluation can easily be extended to more complex radiative configurations. For example, let us consider a polychromatic radiant flux impinging on the burning surface with net intensity $I_s(t)$. In such a case, it is convenient (see Appendix B in [15]) to define the following polychromatic function

$$\tilde{f}(x) = \int_0^\infty \tilde{f}_m(x,\lambda)\, d\lambda \tag{9}$$

being $\tilde{f}_m(x,\lambda)$ in turn a monochromatic function to be assigned depending on the optical properties of the propellant condensed-phase and nature of the radiation source impinging on the burning surface. Typically, for an absorbing but nonscattering condensed-phase with average thickness $1/\overset{=}{a}_\lambda$ of the optical absorption layer, one finds (Kottler, 1964)

$$\tilde{f}_m(x,\lambda) = \overset{=}{a}_\lambda \left(\frac{I_\lambda}{I}\right)_s \exp\left(\overset{=}{a}_\lambda x\right) \tag{10}$$

and thus

$$\tilde{f}(x) = \int_0^\infty \overset{=}{a}_\lambda \left(\frac{I_\lambda}{I}\right)_s \exp\left(\overset{=}{a}_\lambda x\right) d\lambda$$

being $(I_\lambda / I)_s\, d\lambda$ the spectral fraction of the total external radiant energy impinging on the burning surface.

The (net) heat release H_s of gasification reactions concentrated at the burning surface is

$$Q_s(p,T_s) = \overline{Q}_s(\overline{p}) + c_g\left(\overline{T}_s - T_s\right) - \int_{T_s}^{\overline{T}_s} c_c(T)\, dT \tag{11}$$

where the dependence $\overline{Q}_s(\overline{p})$ under steady-state operations is assumed known.

Under these circumstances, for thermophysical properties depending on pressure only, the nontrivial time-invariant steady-state solution is found in general as

$$\overline{T}(x) = T_l + \left(\overline{T}_s - T_l\right)\exp\left(x\overline{r}_b/\alpha_c\right) + \frac{N_s\left(1-\overset{=}{r}_\lambda\right)\overline{I}_s}{\rho_c c_c \alpha_c\left(\overline{r}_b/\alpha_c - \overset{=}{a}_\lambda\right)}\left[\exp\left(\overset{=}{a}_\lambda x\right) - \exp\left(x\overline{r}_b/\alpha_c\right)\right] \tag{12}$$

or in particular, if $\overline{r}_b/\alpha_c = \overset{=}{a}_\lambda$, as

$$\overline{T}(x) = T_l + \left[\overline{T}_s - T_l - \frac{N_s\left(1-\overset{=}{r}_\lambda\right)\overline{I}_s}{\rho_c c_c \alpha_c} x\right] \exp\left(x\overline{r}_b/\alpha_c\right) \tag{13}$$

The time-invariant steady-state solution, for temperature dependent thermophysical properties, is discussed in [9] pp. 546-547.

2.2. THE AQSHOD PROBLEM

The above general set of equations is enough to conduct intrinsic stability analyses of the steady-state solution. But specific submodels for the surface pyrolysis and gas-phase transient flame need to be implemented to assess quantitative trends and evaluate time histories of burning processes.
The test case is a model propellant whose properties are listed in Table 1 (modified from [16] p. 811 and [17] pp. 73-78). Solutions of the governing set of equations are found by numerical integration according to the finite difference procedures and scale transformations discussed in [18].

2.2.1. *Surface Pyrolysis*

Let the burning surface be an infinitesimally thin planar surface subjected to one-step, irreversible gasification processes. A general form of the Arrhenius exponential law, commonly used for concentrated pyrolysis, is

$$r_b(p,I_s)=\tilde{A}_s p^{n_s} I_s^{n_{sq}} \exp\left(-\frac{\tilde{E}_s}{\Re T_s}\right) \tag{14}$$

For the explicit pressure dependence, a power law with finite n_s is commonly accepted; likewise, for the explicit radiation dependence, a power law with finite n_{sq} is proposed. In principle, the powers n_s and n_{sq} should be evaluated experimentally. But it is shown in [19] that for discrete pyrolysis functions

$$\tilde{A}_s=\frac{a_b \bar{p}^{n} \bar{I}_s^{n_q} \exp\left[\sigma_p\left(T_1-T_{ref}\right)\right]}{\bar{p}^{n_s} \bar{I}_s^{n_{sq}} \exp\left[-\frac{\tilde{E}_s/\Re}{\bar{T}_s}\right]} \tag{15}$$

$$n_s=n-n_{T_s}\frac{\tilde{E}_s/\Re}{\bar{T}_s} \tag{16}$$

$$n_{sq}=n_q-n_{T_{sq}}\frac{\tilde{E}_s/\Re}{\bar{T}_s} \tag{17}$$

being n and n_q evaluated by steady-state burning rate measurements while n_{T_s} and $n_{T_{sq}}$ by steady-state surface temperature measurements; for details, see [19]. Due to the practical importance of the Arrhenius kind of pyrolysis, this particular problem is conveniently called the AQSHOD (Arrhenius surface pyrolysis, Quasi-Steady gas phase, Homogeneous condensed phase, and One-Dimensional burning processes) problem and will specifically be addressed in this paper. Notice, however, that the study of ignition transients require in principle the fully nonsteady set of equations. This concentrated surface gasification, enforced by most investigators, is a pure phenomenological approach. No attempt is made in this paper to go beyond these phenomenological limits; for details, see [20].

2.2.2. *Transient Flame Model*

Let the gas-phase mixture be a semi-infinite column of thermally perfect components with no interactions with radiation. The flow is assumed to be one-phase, laminar, nonviscous, low-subsonic, and with Lewis number =1. This aspect of the problem, described in detail elsewhere [9] [16] [11], is only quickly summarized here. Based on experimental information, a mechanistic transient model was developed by this research group for distributed or spacewise thick flames. Results from all quasi-steady transient flames models accepted in the literature can be recovered as special cases. For distributed flames of thickness x_f with the maximum heat release rate site approaching the burning surface, the nonlinear heat feedback [9] is

$$\tilde{q}_{g,s}(p,m) = mQ_g \frac{\gamma+1}{\zeta_f} F(\gamma;\zeta_f) \tag{18}$$

where

$$F(\gamma) \equiv 1 - \frac{(-1)^{\gamma}\gamma!}{(\zeta_f)^{\gamma}} exp(-\zeta_f) + \sum_{i=1}^{\gamma} \frac{(-1)^{i}\gamma!}{(\gamma-i)!(\zeta_f)^{i}} \tag{19}$$

being $\zeta_f = mc_g / k_g x_f = m^2\langle t_g^+ \rangle$ a nondimensional quantity evaluated at the flame front and $\langle t_g^+ \rangle$ an average characteristic time parameter of the gas phase to be defined for each flame model. The additional assumption of flames *linearized* in the exponential term, i.e., ζ_f large, can be invoked to analytically compute frequency response functions but not for ignition or extinction studies. The parameter γ is used to describe the space elongation of the flame; for details, see [9] p. 537. Notice anyway that $\gamma = 0$ recovers the classical KTSS distributed flame, while the case $\gamma > 0$ deals with distributed flames of larger thickness.

In this work, analytical solutions are provided for $\gamma = 0$ or $\gamma = 1$, yielding

- $F(\gamma = 0) \equiv 1 - exp(-\varsigma_f)$ i.e., the standard KTSS nonlinear flame
- $F(\gamma = 1) \equiv 1 - 1/\zeta_f + \dfrac{exp(-\varsigma_f)}{\varsigma_f}$.

Thus, the associated nonlinear heat feedback is

$$\tilde{q}_{g,s}(p,m) = mQ_g \frac{\gamma+1}{\zeta_f}\left[1 - \frac{\gamma}{\varsigma_f} - \left(-\frac{1}{\varsigma_f}\right)^{\gamma} \exp\left(-\varsigma_f\right)\right]. \tag{20}$$

In the spirit of quasi-steady gas phase, the energy release $Q_g(p)$ in the gas phase is defined once for all by an integral energy balance carried out, under steady-state standard operating conditions (adiabatic and 300 K initial temperature), from the burning surface to the flame end

$$\overline{Q}_g(\overline{p}) + \overline{Q}_s(\overline{p}) \equiv \overline{Q}(\overline{p}) = \int_{T_{ref}}^{\overline{T}_s} c_c(T)\,dT + c_g\left(\overline{T}_f - \overline{T}_s\right) \tag{21}$$

The transient flame temperature is provided by a gas-phase energy balance for any quasi-steady flame as

$$T_f(t) = T_s(t) + \frac{Q_g(p) - \tilde{q}_{g,s}(p,m)/m(t)}{c_g}. \tag{22}$$

The (average) characteristic gas-phase time parameter is written as

$$< t_g^{+}(p;m) > \equiv f(p)g(m) \tag{23}$$

where the function $f(p)$ is evaluated under steady operations but in the spirit of gas-phase quasi-steadiness is assumed valid under transient conditions as well. The function $g(m)$ depends primarily on temperature and requires a specific submodel; notice that for convenience, in the sense of gas-phase quasi-steadiness, reference is made to the variable m, the instantaneous mass burning rate. In all developments of this paper, only $g(m)=1$ is considered; results for $g(m) \neq 1$ were reported in [21], for example.

2.2.3. *Forcing Functions*
Finally, any forcing term acting on the burning propellant from outside has to be explicitly assigned in time: typically

$$p(t) = \textit{externally assigned} \tag{24}$$

$$I_s(t) = \textit{externally assigned} \tag{25}$$

according to the wanted pressure and/or radiation driven transient burning.

2.2.4. *Thermophysical Properties*

In general, linear expressions are implemented but other expressions are allowed as well. For example, both AP-based composite propellants and DBP feature a condensed-phase specific heat linearly increasing with temperature

$$c_c(T) = c_{ref} \cdot \left[1 + \tilde{a}\left(T - T_{ref}\right)\right] \tag{26}$$

being $\tilde{a} > 0$ usually. Condensed-phase thermal conductivity may also be seen as linearly changing with temperature

$$k_c(T) = k_{ref} \cdot \left[1 + \tilde{b}\left(T - T_{ref}\right)\right] \tag{27}$$

being typically $\tilde{b} \geq 0$ for DBP but $\tilde{b} \leq 0$ for AP-based composite propellants; in this case, thermal diffusivity is deduced from measured or assumed thermal conductivity. An alternative approach is to deduce thermal conductivity from measured thermal diffusivity (e.g., see [22] [23])

$$k_c(T) = \alpha_c(T) \cdot \rho \cdot c_c(T) \tag{28}$$

In either case, both for pure AP and AP-based mixtures, a sharp decrease of thermal conductivity and thermal diffusivity seems to be observed when AP crystals transform from orthorhombic to cubic at some transition temperature (513 K for slow heating rates); but this interpretation is polluted to an unknown extent by condensed phase chemical reactions. In addition, a slight decrease of thermal conductivity and thermal diffusivity is observed for increasing temperature when AP is orthorhombic. For details, see [24].

3 Ignition

3.1. BACKGROUND

Comprehensive reviews were offered by Summerfield et al. in 1963 [25], Price et al. in 1966 [26], Barrère in 1968 [27], Merzhanov and Averson in 1971 [28], Hermance in 1984 [29], Vilyunov and Zarko in 1989 [30], and Lengellé in 1991 [31].

"Ignition of solid propellant is the process occurring between the initial application of some energetic stimulus to a quiescent chunk of propellant and the full-scale combustion of the propellant. The term 'ignition' is used commonly in two ways: 1) the process of achieving full-scale combustion, and 2) the point in time at which sufficiently full-scale combustion is deemed to occur. Ignition of the propellant is thus a transient process with a definite point of initiation but an endpoint that depends completely upon the definition of that endpoint" [29] pp. 241-242. The energetic stimulus can be chemical, thermal, or even photochemical. Most recent work in solid propellant ignition has used radiant energy by a power laser as the stimulus for ignition. This is due to the fact that experimenters can select the heat flux to be applied to the propellant sample independently on all other environmental parameters, such as pressure, initial temperature, and chemical composition of the pressurizing gas. On the other hand, radiative ignition experiments are peculiar because, in addition to surface heating, in-depth heating usually occurs and the tested propellant sample is not submerged in the hot reactive gases environment.

Theoretically, the critical or first ignition site triggering a thermal runaway process was assumed to be in the condensed-phase (exothermic, pressure independent condensed-phase reactions) or in the gas-phase (fuel and oxidizer vapors, produced by decomposition of the condensed-phase, mix by counter-diffusion and react exothermically in the gas-phase allowing pressure sensitivity) or at the solid/gas interface (exothermic reactions started hypergolically by chemical attack of the propellant surface by vigorous oxidizing gases such as fluorine, again allowing pressure sensitivity). Both the gas-phase theory and hypergolic (also known as heterogeneous) theory show a marked decrease of the ignition delay for increasing oxidizer concentration and/or increasing pressure. This question is still unsolved; probably the controlling mechanism is different for different propellants and different operating conditions. However, practical means exist to predict ignition delays under a variety of conditions: in this paper numerical integration of the governing FM set of equations is used. A recent approximate but closed form solution [31] consists in saying that ignition occurs when the surface temperature rise is larger to a detectable degree (say 15%) than the corresponding inert material temperature rise; this allows to deduce convenient analytical solutions of the ignition delay (seen as an irreversible thermal degradation) in terms of energy flux. This ad-hoc, continuous heating, condensed-phase criterion was shown to yield accurate enough results for a variety of solid propellants. Under operating conditions typical of rocket motors, ignition delay was found to be triggered by the thermally least stable ingredient; to be insensitive to the heating mode, pressure, and nature of the environment gas; and to be quite larger than the time needed to establish a full equilibrium flame.

3.2. SOME TYPICAL IGNITION TRANSIENTS

A general radiant ignition map is illustrated in Fig. 1 [15]. A preliminary important remark is that first gasification (sometimes called first ignition) and self-sustained ignition are separate events. Detailed experimental analyses showed that a succession of phenomena including gas evolution from the heated surface, faint IR (infrared) emission from the heated surface and gas-phase, and substantial flame development takes place serially during continuous radiation. On the other hand, go/no-go testing reveals that the boundary of self-sustained ignition and possibly of dynamic extinction are crossed later. This physical interpretation is confirmed by several high-speed shadowgraph movies [32] [15] [33] [34]. Notice, however, that the details of this general trend can strongly be affected by the specific nature of the tested material.

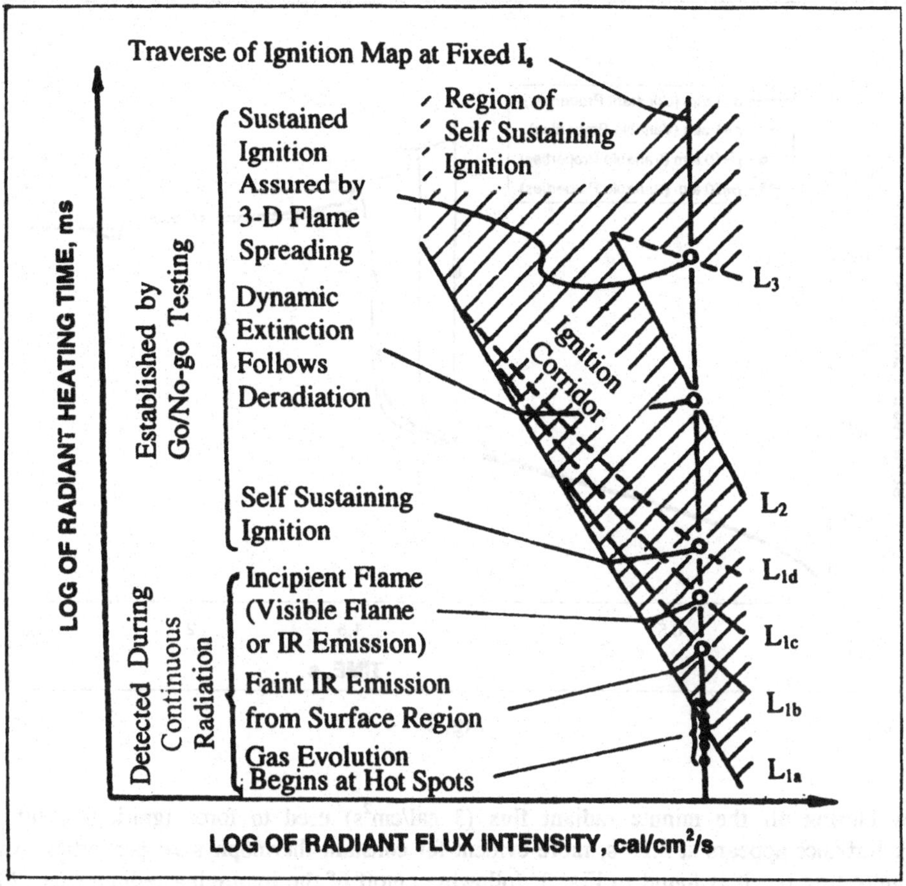

Fig. 1

By numerical solution of the basic set of equations illustrated in Sec. 2, one can study trends and bounds. For the assumed test case, several ignition transients were computed under radiant energy stimulus; for a matter of space, plots in terms of surface temperature vs time and mostly for fully opaque condensed-phase are reported. The effect of constant or variable condensed-phase thermophysical properties is illustrated in Fig. 2, at 1 atm, for a minute external radiant flux (3 cal/cm^{2}s). It is seen that allowing variable thermophysical properties strongly damps out the transient burning induced by the thermal runaway associated with the ignition process. Increasing pressure from 1 to 10 or 50 atm makes the thermal runaway sharper and sharper. Notice in particular that

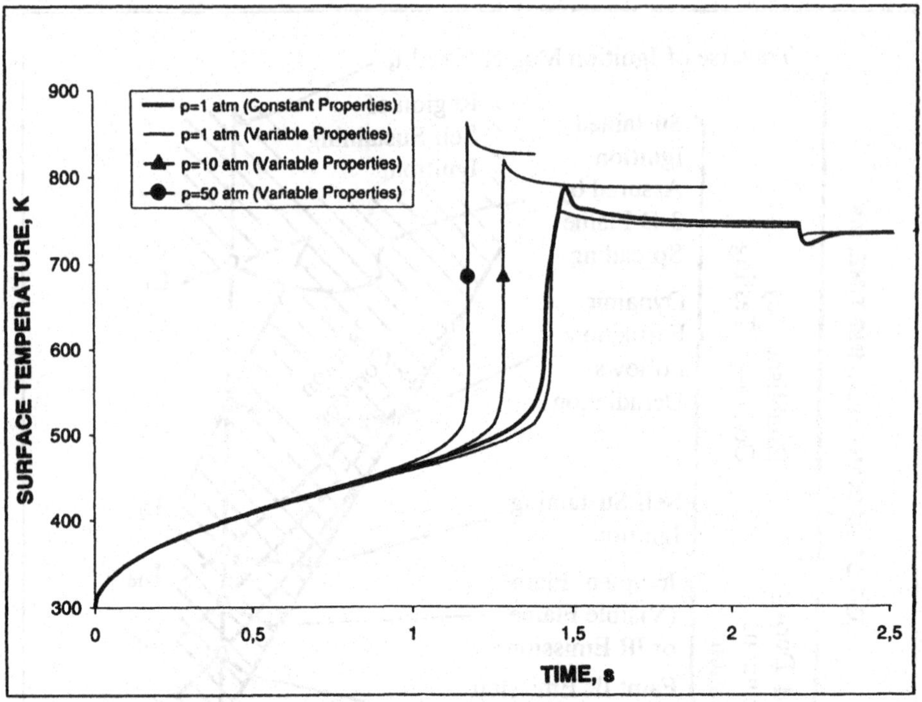

Fig. 2

by cutting off the minute radiant flux (3 cal/cm^{2}s) used to force ignition, a little disturbance appears at 2.25 s, more evident for constant thermophysical properties. An interesting result is found in Fig. 3: following cutoff of the impinging radiant flux (70 cal/cm^{2}s) at 0.03 s, the previous perceivable disturbance yields now a clear extinction process at 1 atm if a faster radiation cutoff (b_r = 100 instead of 5) is enforced. This effect is understood to be due to dynamic extinction following fast deradiation, and it is a consequence of the combined fast radiation cutoff and low operating pressure.

These representative results give an idea about the variety of possible solutions and their sensitivity to the computational details. However, for the purposes of this paper, the main interest lies in the dynamic burning behavior induced by the radiation cutoff; due to the usefulness of this potential application, some peculiar effects of dynamic burning are reviewed in next sections.

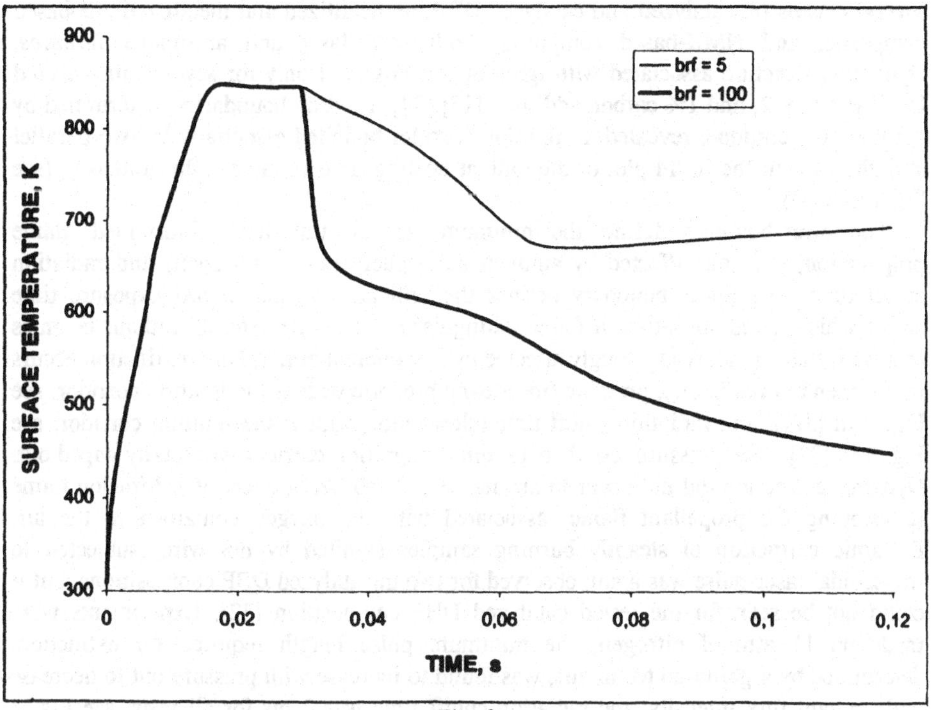

Fig. 3

3.3. FAST DERADIATION

Initial studies on dynamic extinction by fast deradiation overlap with those on radiative ignition. In 1970 Ohlemiller, at Princeton University, first observed dynamic extinction by fast deradiation while constructing an ignition map of a noncatalyzed DBP (a NC composition) by a CO_2 laser source, providing up to 120 cal/cm^{2}s at the target, in the range of 3 to 4 atm of nitrogen. It seems that, independently, the discovery was made in Russia by Mikheev in his Candidate's Dissertation published at about the same time (1970) but not easily available. From successive work published in the open literature, one would guess that Mikheev's data concerned uncatalyzed DBP tested in a furnace at low radiant flux (few cal/cm^{2}s) at 1 atm of air or nitrogen. Supporting evidence for the dynamic character of this phenomenon and further experimental data were successively offered mainly at Princeton University ([15][32][34]) and in Russia ([35][36][37][38] and [39] pp. 173-178). The radiative pulse experiment on steadily burning propellants, reported in [32], was suggested by Yu.A. Gostintsev during a stay at Princeton University.

Systematic experimental results were collected at Princeton University using a 100 W continuous wave, multimode CO_2 laser emitting at 10.6μm in the far infrared. Two mechanical shutters, of the iris leaf type, provided a trapezoidal radiation pulse; the action time of the shutters could be regulated at 1-10 ms. Ignition tests were performed, mainly in the range of 5 to 21 atm of nitrogen or air, for 12 propellants representative of several classes (uncatalyzed and catalyzed DBP, unmetallized and metallized AB-based composite, and HMX-based composite) both with laser and arc-image furnaces. Dynamic extinction associated with ignition was observed only for several uncatalyzed DBP with 0, 0.2, and 1% carbon additive [15] [34]. Ignition boundaries, determined by a go/no-go technique, revealed an ignition corridor bounded essentially by two parallel straight lines in the ln /ln plot of the radiant heating time vs radiant flux intensity (see Fig. 6 in [34]).

The lower boundary defines the minimum exposure time for self-sustained flame propagation; it is not affected by ambient atmosphere (air or nitrogen) and radiation cutoff time. The upper boundary defines the critical (i.e., maximum) exposure time above which a radiation-driven flame extinguishes when the external radiant beam is removed; this boundary is strongly affected by the ambient atmosphere (extinction occurs in nitrogen but not in air), pressure (increasing pressure widens the ignition corridor, see Fig. 2 in [32]), and radiation cutoff time (slow cutoff widens the ignition corridor, see Fig. 3 in [32]). For pressure less than 11 atm the ignition corridor was totally wiped out. Dynamic extinction did not occur in air due to the stabilizing effect of a diffusion flame (enveloping the propellant flame) associated with the oxygen contained in the air. Dynamic extinction of steadily burning samples (ignited by hot wire) subjected to trapezoidal laser pulse was again observed for two uncatalyzed DBP compositions, but it could not be seen for the tested catalyzed DBP composition [32]. Experiments were made up 11 atm of nitrogen; the minimum pulse length required for extinction, determined by a go/no-go technique, was found to increase with pressure but to decrease with radiant flux intensity; the radiation cutoff time was 2 ms for all tests; see Fig. 4 (adapted from Fig. 4 in [32]). Experiments in air were not conducted; composite propellants were not tried.

All this was confirmed by high-speed movies (see Fig. 1 in [32]) and photodetectors. However, the same propellants extinguishable at the laser apparatus could not be extinguished when tested, in similar operating conditions, at an arc-image apparatus. This used as source a 2.5 kW high-pressure xenon arc lamp with spectral emissions in the visible (similar to sunlight, peak near 0.55 μm), except for some nonequilibrium high-intensity bands in the near infrared. Tests at the arc-image apparatus were conducted only for the ignition runs.

The different results obtained with uncatalyzed DBP and with two radiative sources were mainly attributed to different optical properties (volumetric absorption and scattering are strongly wavelength dependent; uncatalyzed DBP are much more opaque in the far infrared than in the visible) and larger radiant flux cutoff time with the arc-image apparatus (about twice with respect to the laser apparatus). Other minor differences were also operative: spatial structure of the radiant beams (less sharp with the arc-image source), sample positioning (protruding from a metallic holder at the laser apparatus but flush at the arc-image; this may cause heat sink effects), different beam geometry (collimated and perpendicular for the laser but converging and oblique for the arc-image). The different results between uncatalyzed DBP vs catalyzed DBP and AP-based composite, observed at both laser and arc-image apparatus, were attributed to the different flame structure and different radiation interaction with the combustion products. Indeed, both catalyzed DBP and AP-based composite, give a relatively higher ratio of gas-phase heat feedback to the burning surface heat release; this is known to be stabilizing. Moreover, radiation is gently terminated and effectively decreased by carbonaceous filaments and particulates generated by the burning samples; this is stabilizing and occurs especially for catalyzed DBP. However, it is expected, in principle, that all propellants should manifest dynamic extinction by fast deradiation if tested in the appropriate range of operating conditions.

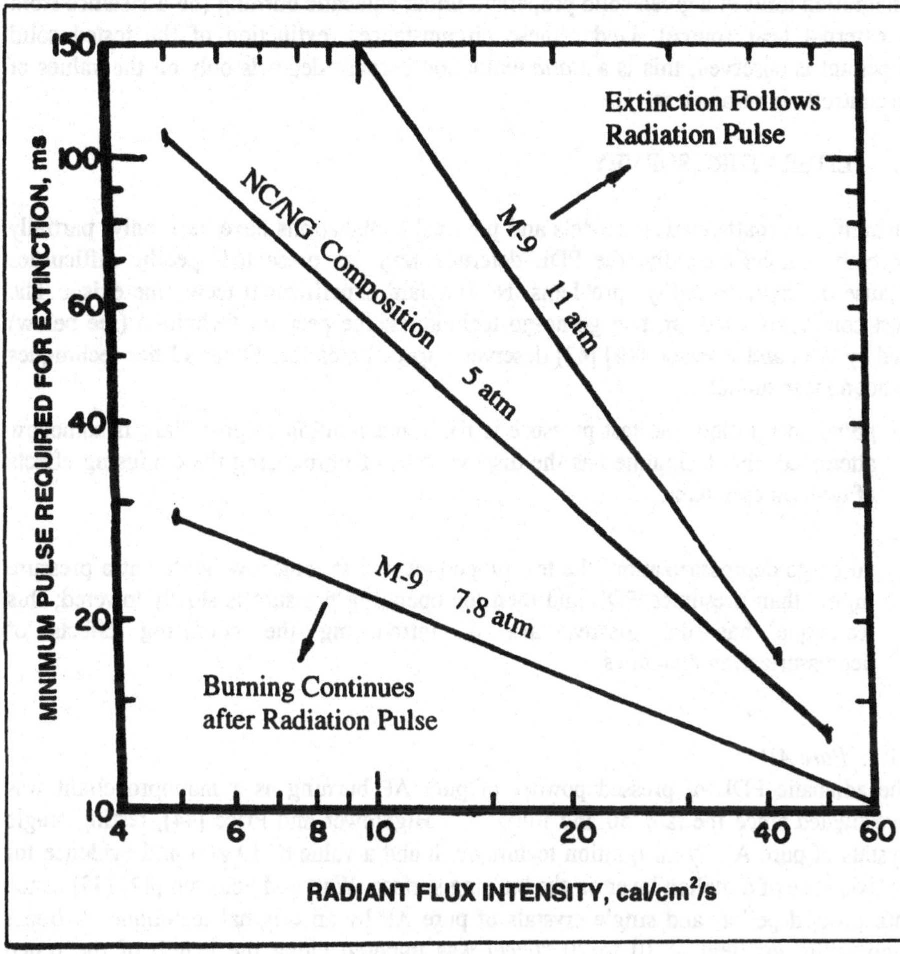

Fig. 4

4. Pressure Deflagration Limit

PDL is defined as that minimum pressure under which no self-sustained steady burning can occur; this is the view usually taken in the Western literature. In the Russian literature, the corresponding view of critical diameter d_{cr} is taken; this is defined as that minimum sample size under which no self-sustained steady burning can occur due to excessively large heat losses [40] [41]. The two concepts of PDL and d_{cr} are related. They help in understanding ignition phenomena details and also have a practical influence on how to run experimental tests in general. Whether in terms of pressure or size, deflagration limits are a property of energetic materials but depend on the testing as well as external operating conditions (initial temperature, external radiation, etc.). In most cases reported in the literature, experimental and theoretical analyses concern the deflagration limit of a given solid propellant under adiabatic burning (no assistance from an external heat source). Under these circumstances, extinction of the tested solid propellant is observed; this is a *static* extinction because depends only on the values of the controlling parameters.

4.1. LITERATURE SURVEY

Theoretically, mathematical models and physical explanations have been only partially successful. Experimentally, the PDL determination has presented specific difficulties because of "reproducibility" problems. No standard experimental technique exists. The most commonly used are two go/no-go techniques; the peculiar technique (see below) used by Watt and Petersen [42] [43] deserves a special mention. Other ad-hoc techniques do not appear suitable.

- go/no-go ignition: the test pressure is fixed and ignition of propellant is somehow attempted; this technique has the disadvantage of introducing the confusing effects of ignition dynamics.

- go/no-go depressurization: the test propellant is first somehow ignited at a pressure higher than presumed PDL and then the operating pressure is slowly lowered; this technique has the disadvantage of introducing the confusing effects of depressurization dynamics.

4.1.1. *Pure AP*

The adiabatic PDL of pressed powder of pure AP burning as a monopropellant was investigated since the late '50 and initial '60. Hightower and Price [44], testing single crystals of pure AP by an ignition technique, found a value of 19 atm and evidence for the existence of a molten layer on the burning surface. Watt and Petersen [42] [43] tested both pressed pellets and single crystals of pure AP by an original technique. A linear temperature gradient of 10 to 20 *°C/cm* was imposed along the length of the tested sample and the deflagration wave, triggered at the warmer end, propagated up to the point where the sample temperature was too low to support steady burning. Since the conductive thermal wave thickness of the samples was much less than the externally imposed thermal gradient, the deflagration wave was essentially steady. The interesting result of a unique curve, for pressed pellets and single crystals of pure AP, correlating PDL with the sample temperature was obtained. PDL decreases linearly from 26.2 to 15.3 atm when the sample temperature increases from -40 *°C* to 50 *°C*, in particular the value of 19 atm is recovered at ambient temperature. The results were found to be independent on the nature of the inert pressurizing gas.

Radiative heat losses from the burning surface were the first cause invoked for the existence of a lower limit pressure for self-sustained steady deflagration. However, Johnson and Nachbar [45] theoretically and Horton and Price [46] experimentally proved that heat losses, in particular of radiative nature from the burning surface, are far too small to explain the observed PDL values (at least of AP steady deflagration).

A different approach was taken by Guirao and Williams [47] in 1971 in modeling AP steady deflagration between 20 to 100 atm. They assumed exothermic condensed-phase reactions in a liquid layer covering the burning surface and deflagration completion in the adjacent gas-phase of the AP unreacted in the liquid-phase (about 30%). The thickness of the liquid-layer was estimated 2 μm at most, and decreasing with decreasing pressure. When the surface temperature falls below the AP melting point, 833 ± 20 K, the model predicts that the condensed-phase reactions rate strongly decrease and the liquid layer thickness shrinks to zero. This was considered to cause the (lower) pressure deflagration limit. The addition of heat losses was verified to have no significant impact to the model.

Cohen Nir [48] [49], by testing a large number of pure AP pressed pellets by an ignition technique, obtained the following experimental findings:

- the adiabatic PDL decreases linearly for increasing strand density.
- PDL decreases asymptotically for increasing burning surface area. In particular, a minimum strand size exists below which no steady deflagration can be observed; this critical dimension corresponds to a value of the ratio (cross section area)/(strand perimeter) of the order of 0.5.
- PDL decreases less than linearly with increasing initial temperature.
- the following criterion was heuristically adopted to interpret experimental results: near PDL, the burning rate is not zero but assumes a constant finite value of about 0.18 cm/s.

4.1.2. *AP-Based Composite Propellants*

A careful review of previous work was presented by Steinz and Summerfield [50][51][52] by testing a large number of propellant formulations by a depressurization technique. First, ignition was given in nitrogen above 0.3 atm, and then pressure was reduced "as slowly as possible". The adiabatic PDL value "was taken as the lowest value (out of a number of tests) at which the propellant could be made to burn (p. 128 of [51]). Reproducibility was estimated within 10%, but the possible presence of side burning inhibitors caused PDL to increase (e.g., from 0.045 to 0.060 atm). The results obtained can be summarized as follows:

- in the subatmospheric pressure range, burning rates on a $ln(\bar{r}_b)$ vs $ln(\bar{p})$ plot in general follow a straight line whose slope increases from 0.4-0.5 to 0.7 making burning more pressure dependent.

- The adiabatic PDL is not less than 0.05 for most tested propellants, but falls down to 0.005 when a protective ash layer retaining heat near the burning surface is formed.

Petersen et al [53] studied low pressure burning with the intent to increase the PDL value for enhanced safety of solid propellant handling and rocket motor operation. By properly combining several chemical and physical factors, the authors were able to raise PDL above 3 atm for the current AP-based composite propellants with only a slight loss in performances. In particular, PDL could be increased by: replacing hydrocarbon binders (PBAA) with oxygenated binder (PU or polyesters); increasing binder content (demonstrated only for PU binder); using small AP particles.

4.1.3. *Double-Base Propellants*

A good amount of experimental data is provided by Kubota in his Ph.D. thesis [54] pp. 27-28 and Fig. 20; the results obtained (probably by an ignition technique) can be summarized as follows:

- below 1.5 atm, burning rates on a $ln(\bar{r}_b)$ vs $ln(\bar{p})$ plot in general follow a straight line whose slope falls from 0.7 to 0.2 making burning little dependent on pressure.

- The adiabatic PDL is around 0.4 atm for most tested propellants, but it falls down to 0.1 atm for compositions containing 1% *CuSa* (salicylic acid) as burning rate catalyst.

4.2. CURRENT UNDERSTANDING

Cookson and Fenn [55], by testing AP-based composite propellants by slow depressurization of the combustion chamber, concluded that some finite value of pressure exists for which burning rates vanish even under adiabatic combustion. The important remark was made that PDL is sensibly dependent on the strand size: there is a minimum strand size below which steady combustion is not allowed no matter how large the pressure is; viceversa, there is a minimum pressure below which steady combustion is not allowed no matter how large the strand size is.

Likewise, the dependence of the adiabatic PDL on the geometric details of the tested sample was observed in many experimental investigations [42] [43] [48] [49] [51] [55] [56]. The *shape factor*, defined as the ratio (area of the cross section)/(perimeter of the cross section) of the propellant sample, is the controlling parameter. With reference to Fig. 5 (adapted from [56]), the adiabatic PDL decreases when the shape factor increases until a minimum pressure is reached below which self-sustained steady deflagration cannot be achieved no matter how large the shape factor is. A different geometry of the samples (square or rectangular cross sections) does not affect the results as long as the shape factor is conserved unmodified [56] [57].

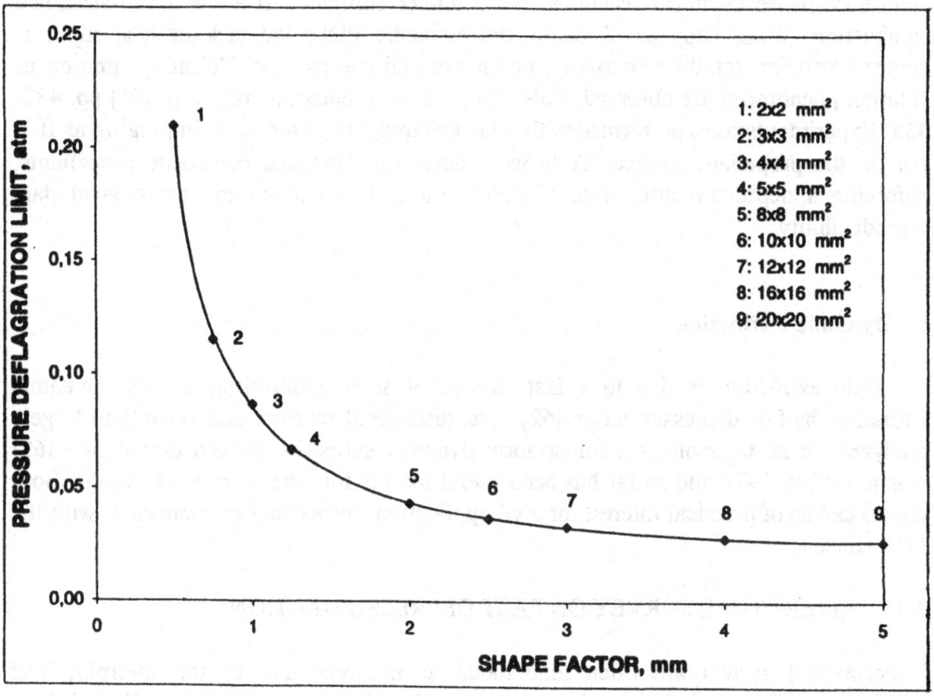

Fig. 5

The horizontal asymptote is expected to correspond to the adiabatic PDL of a strand with infinite cross section. This trend can be explained by observing that the cooling effects of the gaseous medium surrounding the propellant for small shape factors is comparable with the flame heat feedback to the burning surface. When the strand size is large enough, heat losses to the surroundings become negligible. This implies that, for large values of the shape factor, the adiabatic PDL only depends on the burning propellant nature. Thus, the adiabatic PDL is an intrinsic property of the tested material. Typical values are in the range 0.01 atm (AP-based composite propellants) to 0.1 atm (DBP); PDL is 19 atm for pure AP, but much larger for pure AN [58].

Deflagration limits, in particular PDL, can be seen as a manifestation of intrinsic instability [59] [60] .

4.3. EFFECTS OF EXTERNAL RADIATION

By definition, steady burning of solid propellants is impossible below PDL. The total heat input, provided by gas-phase heat feedback as well as surface and perhaps condensed-phase chemical reactions, is no longer sufficient to allow a self-sustained combustion. What happens if, below the adiabatic PDL, the lack of heat input is compensated for, totally or partially, by an external energy flux? Pulsating ignition or ablation phenomena are observed. This aspect of the problem is treated in [61] pp. 432-435. Experiments were performed with a far infrared CO_2 laser beam impinging at 10.6 μm on the propellant surface. Tests were made on AP-based composite propellants, enforcing a depressurization rate of 0.001 atm/s, low enough to ensure good data reproducibility.

5. Dynamic Extinction

Dynamic extinction is due to a fast *change* of some controlling factors. Dynamic extinction by fast depressurization [62] was discovered in 1961 and since then largely analyzed for its technological importance; dynamic extinction by fast deradiation [63] was proved in 1972 and so far has been useful mainly for laboratory applications. Both effects can be of practical interest for civil applications, especially in connection with the PDL concept.

5.1. LITERATURE SURVEY ON FAST DEPRESSURIZATION

Experimental results are often ambiguous to interpret due to the interplay and overlapping of several factors. Attention is preferably focused on data collected from laboratory burners (ranging from depressurization strand burners to simulated rocket combustion chambers) rather than actual rocket motors where fluid dynamic effects may be dominant.

In 1961 Ciepluch [62][63][64] conducted the first systematic experimental study of depressurization transients in a laboratory combustion chamber closely simulating an actual motor. Fast depressurization was obtained by suddenly opening a chamber vent hole. Initial chamber pressure in the range of 34-82 atm and ambient pressures down to 3.5 mm Hg were explored. The burning transient was followed by measuring simultaneously combustion luminosity (primarily in the visible range) and chamber pressure. Several AP-based composite propellants were tested; few data on DBP were also reported. Although no quantitative extinction criterion was formulated, the following conclusions were reached:

- A critical depressurization rate exists below which burning continues and above which extinction occurs.
- The critical depressurization rate increases linearly as the chamber pressure prior to venting increases.

- The critical depressurization rate is substantially affected by the propellant composition.

- Reignition may follow extinction if the depressurization is not too fast and/or nozzle back pressure is not too low.

- The critical boundary between (permanent) extinction and continued burning was found to be a straight line in the *dp/dt* vs p_i (initial pressure) linear plot by a go/no-go technique.

An important critical review was given by Merkle et al. [65][66] in 1969. Since the pressure decay in Ciepluch's experiments was exponential, the maximum depressurization rate occurred at the very beginning of the transient burning. This was erroneously interpreted to suggest a tight dependence of dynamic extinction on the initial depressurization rate. Merkle et al. [65][66] recognized that dynamic extinction depends on the entire *p(t)* curve and furnished a new quasi-steady flame model somewhat accounting for the propellant heterogeneity. However, they did not formulate a quantitative extinction criterion. For numerical simulation purposes only, a critical value of surface temperature (600 K) was empirically picked up, below which chemical reactions are considered too weak to sustain deflagration waves. Merkle et al. [65] [66] also produced a rather complete set of data by systematically testing several AP-based composite propellants and one catalyzed DBP. A special laboratory combustor was designed to minimize erosive burning effects and to cause unidimensional cigarette burning of the sample. Fast pressure decay was obtained by rupturing a double-diaphragm system. Pressure and light emission (as seen by a photomultiplier in the visible range) were simultaneously recorded. Extinction was considered to occur when zero light emission from the flame could be observed. Venting was always to atmospheric pressure; because of this, reignition was nearly always observed. The following conclusions were reached:

- The critical boundary between extinction (permanent or temporary) and continued burning was found by a go/no-go technique to be always straight in the *dp/dt* vs p_i linear plot.

- The tested catalyzed DBP (N-5 composition) is much easier to extinguish than any of the tested AP-based composite propellants.

Some of the above conclusions may be in conflict with other experimental studies in which permanent extinction was recorded. Merkle et al. [65][66] adopted the permanent extinction criterion only for tests using a catalyzed DBP, since no visible radiation can be detected from DBP burning at low pressures.

A series of depressurization tests was conducted in a strand burner connected to a much larger pressurized dump chamber through an exhaust orifice in [67]. This arrangement allowed venting to two different final pressures: 1 and 3.25 atm. Before depressurization, the exhaust orifice was kept close by a double diaphragm system. The tested propellant was *AP.CTPB.Al₂O₃/83.16.1*. For each test the steady burning rate before depressurization was measured by a fuse wire system. The depressurization tests were of exponential nature, so that the maximum depressurization rate occurs at the very beginning of the depressurization transient. The go/no-go extinction boundary is again a straight line the $(dp/dt)_{max}$ vs p_i plot, if a sufficiently large pressure excursion is enforced. In addition, increasing final pressure increases the go/no-go extinction boundary. The maximum depressurization rate required for extinction at 3.25 atm final pressure is larger by one order of magnitude than 1 atm final pressure. A novel experimental result is the behavior of the go/no-go extinction boundary when the initial pressure approaches the final pressure (see Fig. 11 in [67]). The go/no-go extinction boundary, for decreasing initial pressure p_i but fixed final pressure p_f, decreases linearly down to a minimum and then turns upward. Obviously, no dynamic extinction can occur for $p_i = p_f$ and, therefore, a vertical asymptote is expected for the go/no-go extinction boundary.

5.2. CURRENT UNDERSTANDING

A paper by T'ien [68] in 1974 was the first explicitly aimed at establishing an extinction criterion for fast depressurization. T'ien claims that for depressurization transients, if the instantaneous burning rate drops below the unstable burning rate solution at the final pressure, extinction will occur. The dynamic stability boundary for fast deradiation was shown to coincide with that for fast depressurization in [69].

6. Extinction Applications

The capability of the model proposed in Sec. 2 to reproduce experimental results was already verified several times in the past. For example, see Fig. 6 and Fig. 7 (respectively adapted from Fig. 9 and Fig. 8 in [18]); see also Fig. 10 and Fig. 15 in [18]. In Fig. 6 a satisfactory comparison between numerical and experimental results concerning laser ignition of an AP-based composite solid propellant (with and without 1% carbon addition) is shown. Radiant flux intensity ranges from about 10 to 100 cal/cm²s, while no pressure effect is noticeable in the interval from 5 to 21 atm. Condensed-phase thermophysical properties were independently calibrated on the steady-state thermal profiles. The effect of the condensed-phase optical transparency is illustrated in Fig. 7, comparing two limiting cases for $\overline{\overline{\alpha}}_{\lambda} = 250\ cm^{-1}$ and $500\ cm^{-1}$.

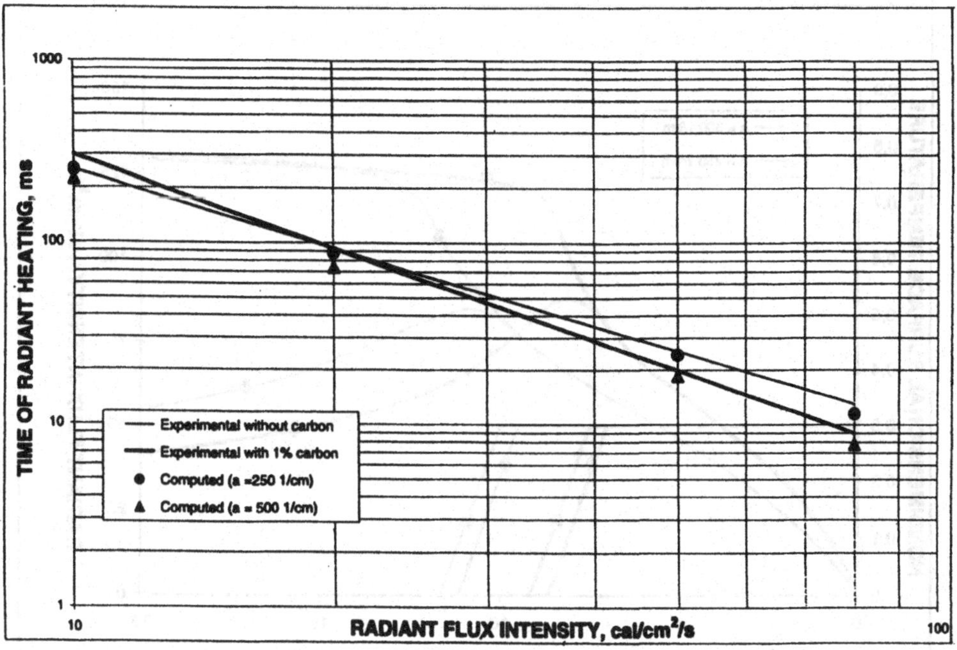

Fig. 6

The go/no-go ignition transients were computed under a radiant flux of 50 cal/cm^2s with b_r = 30. Larger optical transparency $\left(\overline{\overline{\alpha}}_\lambda = 250\ cm^{-1}\right)$ implies a delayed ignition runaway, since the radiant energy deposition is spread over a thicker portion of the condensed-phase.

Dynamic extinction can effectively be used for safety applications in terms of both fast deradiation and fast depressurization; see Figs. 8-9. For example, in Fig. 8 (variable thermophysical properties) a fast pressure drop from 2 to 0.1 atm with b_p =100 brings about dynamic extinction, while combustion continues for a radiation drop with b_r=1000 from 3 to 0 cal/cm^2s. It is, however, important that final pressure be less than PDL of the tested material to make sure that reignition will not occur.

The two dynamic effects are combined in Fig. 9 (variable thermophysical properties). A fast pressure drop from 2 to 0.1 atm with b_p =100 brings about continued combustion; likewise, a fast radiation drop with b_r =1000 from 3 to 0 cal/cm^2s again brings about continued combustion. But dynamic extinction occurs if the two dynamic disturbances are simultaneously enforced.

Forcing a fast depressurization below the pressure deflagration limit is the safest way to impose permanent extinction. Pressure pulses can also be effective, but cannot

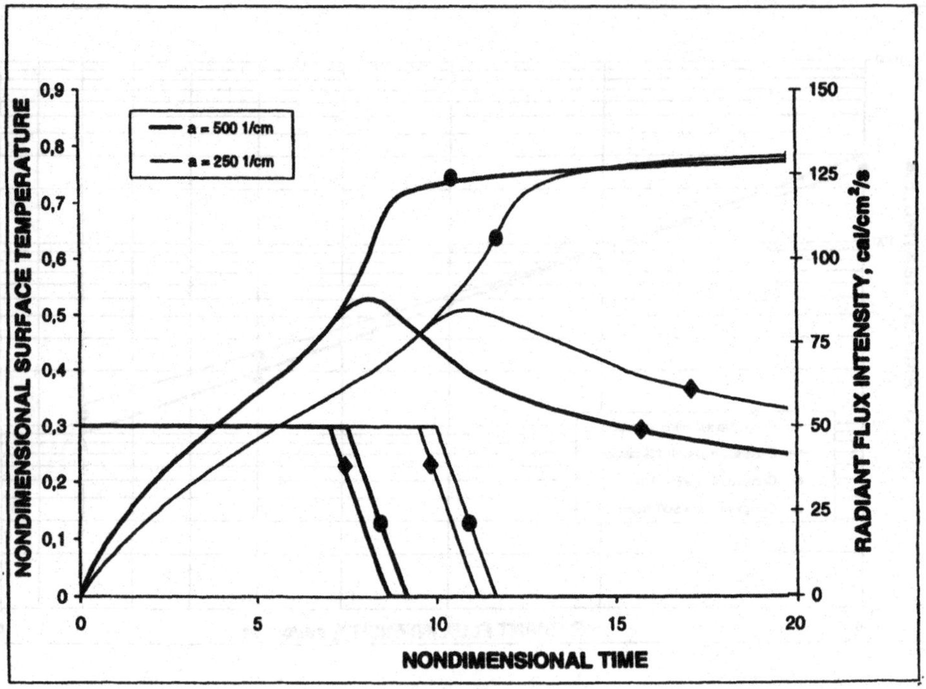
a = 500 1/cm
a = 250 1/cm
0,9
0,8
0,7
0,6
0,5
0,4
0,3
0,2
0,1
0
NONDIMENSIONAL SURFACE TEMPERATURE
150
125
100
75
50
25
0
RADIANT FLUX INTENSITY, cal/cm²/s
0
5
10
15
20
NONDIMENSIONAL TIME

Fig. 7

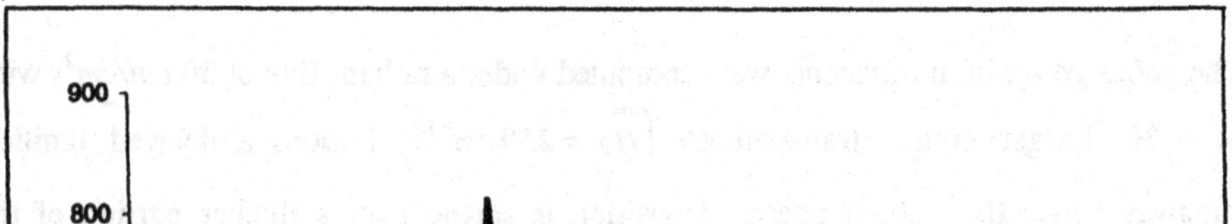
900
800
700
600
500
400
300
SURFACE TEMPERATURE, K
0
0,5
1
1,5
2
2,5
3
3,5
4
TIME, s

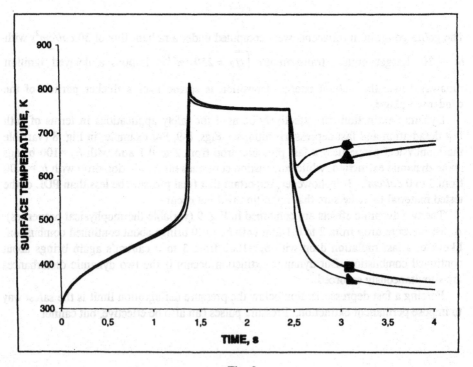

Fig. 8

assure permanent extinction. Fast deradiation and radiation pulses yield permanent extinction only if the final radiant flux is zero and simultaneously pressure is below the pressure deflagration limit. To assure the maximum effect with radiation disturbances, it is necessary to concentrate radiation absorption at the burning surface rather than in-depth; in turn this requires choosing the most suitable radiation wavelength for each burning material (ideally, one would wish a fully transparent gas-phase but a totally opaque condensed-phase). On the other hand, strongly luminous flames are little prone to dynamic extinction since the natural radiation from the flame is capable to promote ablation and/or full reignition even for very low operating pressures.

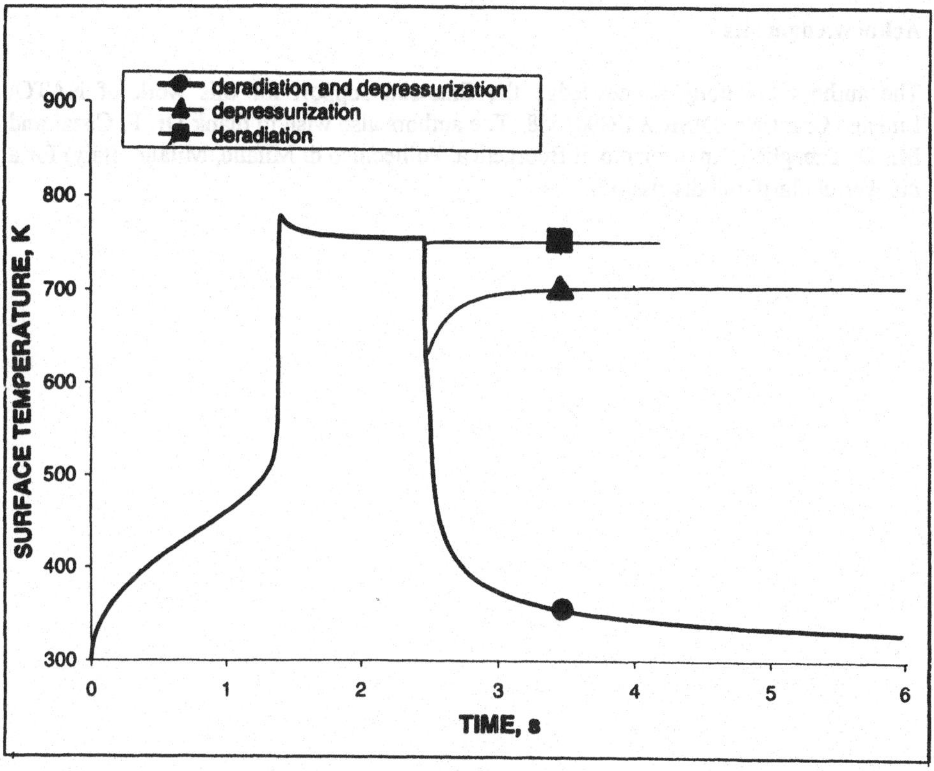

Fig. 9

7. Conclusion

Ignition studies of solid propellants allow not only to better understand and predict transients of accidental fires in general, but also to devise techniques to avoid or extinguish fires based on fundamental knowledge. In particular, deflagration limits can effectively be used to prevent accidental fires while dynamic effects can be exploited for extinction. Within the explored range of relevant parameters, the dynamic extinction effects were found to be strongly affected by the flame nature (thickness, thermokinetic parameters, etc.) but little affected by temperature dependent condensed-phase thermophysical properties.

Acknowledgments

The authors gratefully acknowledge the financial support for this work of NATO\ Linkage Grant No. DISRM.LG961388. The authors also wish to thank Dr. F. Cozzi and Mr. D. Tabaglio (Dipartimento di Energetica, Politecnico di Milano, Milano, Italy) for a number of clarifying discussions.

TABLE 1

Properties of Model Solid Propellant Used as Test Case

Assumed or Measured Properties

initial sample temperature,	$T_1 = 300$ K
condensed phase density,	$\rho_c = 1.609$ g/cm^3
condensed phase specific heat,	$c_c(T) = 0.3\,[1 + 4.775E-04\,(T-300)]$ cal/g K
condensed phase thermal diffusivity,	$\alpha_{ref} = 2.020E-03$ cm^2/s
condensed phase absorption coefficient,	$\bar{\bar{a}}_\lambda = 500$ cm^{-1}
surface activation energy,	$\tilde{E}_s = 1.600E+04$ cal/mole
gas phase activation energy,	$\tilde{E}_g = 2.000E+04$ cal/mole
gas phase specific heat,	$c_g = 0.45$ cal/g K
gas phase thermal conductivity,	$k_g = 0.100E-03$ cal/cm s K
average molecular mass of gas products,	$\bar{\bar{M}} = 28.4$ g/mole
flame model parameters:	$\alpha(p) = 0,\ \beta(p) = 1,\ \gamma(p) = 0$

Computed Properties

condensed phase thermal conductivity,	$k_{ref} = 0.975E-03$ cal/cm s K
condensed phase thermal responsivity,	$\Gamma_{ref} = \sqrt{k_{ref}\rho_c c_{ref}} = 2.170E-02$ cal/cm^2 $\sqrt{s}$ K

Standard (adiabatic and 300 K initial temperature) Ballistic Properties

steady burning rate, $\bar{r}_b = 0.123(\bar{p})^{0.473}$

steady burning surface temperature, $\bar{T}_s = 734.6(\bar{p})^{0.0297}$

steady adiabatic flame temperature, $\bar{T}_f = 2684(\bar{p})^{0.01}$

steady surface heat release (positive if exothermic), $\bar{Q}_s = 96.5(\bar{p})^{0.049}$

pressure exponent of pyrolysis law, $n_s = n - n_{T_s}\dfrac{\tilde{E}_s}{\Re \bar{T}_s} = 0.473 - 0.0297\dfrac{1.600E+04}{1.987 * 734.6(\bar{p})^{0.029}}$

radiation exponent of pyrolysis law, $n_{sq} = 0$

Reference Standard (adiabatic and 300 K initial temperature) Properties at 68 atm

pressure,, $p_{ref} = 68$ atm

temperature, $T_{ref} = 300$ K

burning rate $r_{b,ref} = 0.905$ cm/s

surface temperature, $T_{s,ref} = 832.7$ K

adiabatic flame temperature, $T_{f,ref} = 2799.7$ K

surface heat release (positive if exothermic) $Q_{s,ref} = 118.7$ cal/g

condensed phase characteristic length, $\alpha_{ref}/r_{b,ref} = 22.3\ \mu\text{m}$

condensed phase characteristic time, $\alpha_{ref}/r_{b,ref}^2 = 2.466$ ms

condensed phase sink, $Q_{ref} = c_{ref} \cdot (T_{s,ref} - T_{ref}) = 159.81$ cal/g

gas phase heat release, $Q_{g,ref} = Q_{ref} + c_g \cdot (T_{f,ref} - T_{s,ref}) - Q_{s,ref} = 926.26$ cal/g

flame thickness $(\gamma = 0)$, $x_{f,ref} = \dfrac{k_g}{c_g} \dfrac{1}{\rho_c r_{b,ref}} \dfrac{Q_{g,ref}}{c_{ref} \cdot (T_{s,ref} - T_{ref}) - Q_{s,ref}} = 34.4\ \mu\text{m}$

thermal flux, $\tilde{q}_{ref} = \rho_c r_{b,ref} Q_{ref} = 232.71\ \text{cal/cm}^2\text{s}$

List of Figure Captions

- **Fig. 1 - A general radiative ignition map (adapted from [33]).**

- **Fig. 2 - Surface temperature vs time during a radiative ignition transient, at different pressures, for constant and variable thermophysical properties (under 3 cal/cm^{2}s with $b_r = 500$).**

- **Fig. 3 - Surface temperature vs time during a radiative ignition transient, for variable thermophysical properties, at 1 atm and under 70 cal/cm^{2}s showing dynamic extinction for $b_r = 100$ but continued burning for $b_r = 5$.**

- **Fig. 4 - Dynamic extinction of steadily burning noncatalyzed DBP following a laser pulse (adapted from [32]).**

- **Fig. 5 - Evaluating PDL for AP.HTPB/86.14 (adapted from [56]).**

- **Fig. 6 - Validation of the ignition model by comparing experimental and numerical results of a composite propellant under laser radiation (adapted from [18]).**

- **Fig. 7 - Effect of condensed-phase optical transparency for a go/no-go ignition transient at 5 atm under 50 cal/cm^{2}s with $b_r = 30$ (adapted from [18]).**

- **Fig. 8 - Surface temperature vs time during a radiative ignition transient, for constant vs variable thermophysical properties, showing continued combustion for fast depressurization from 2 to 0.1 atm with $b_p = 100$, whereas dynamic extinction occurs if fast deradiation too is enforced from 3 cal/cm^{2}s with $b_r = 1000$.**

- **Fig. 9 - Surface temperature vs time during an ignition transient, for variable thermophysical properties, showing continued combustion for fast depressurization from 2 to 0.1 atm or for fast deradiation from 3 cal/cm^{2}s, whereas dynamic extinction occurs if the two dynamic disturbances are simultaneously enforced.**

References

1. Zeldovich, Ya. (1942) On the combustion theory of powder and explosives, *Journal of Experimental and Theoretical Physics*, **12**, 498-510.
2. Denison, M.R. and Baum, E. (1961) A Simplified Model of Unstable Burning in Solid Propellants, *ARS Journal*, **31**, 1112-1122.
3. Novozhilov, B.V. (1965) Stability criterion for steady-state burning of powders, *Journal of Applied Mechanics and Technical Physics*, **6(4)**,157-160.
4. Novozhilov, B.V. (1973) *Nonstationary Combustion of Solid Rocket Fuel,* Nauka, Moscow, Russia. Translation AFSC FTD-MT-24-317-74.
5. Novozhilov, B.V. (1992) Theory of Nonsteady Burning and Combustion Stability of Solid Propellants by the ZN Method, in L. DeLuca, E.W. Price, and M. Summerfield (eds.), *Nonsteady Burning and Combustion Stability of Solid Propellants,* AIAA Progress in Astronautics and Aeronautics, AIAA, Washington, DC, USA, volume **143**, chapter 15, pp. 601-641.
6. Culick, F.E.C. (1968) A review of calculations for unsteady burning of solid propellant, *AIAA Journal*, **6(12)**, 2241-2255.
7. Culick, F.E.C. (1969) Some problems in the unsteady burning of solid propellants, Technical report NWC TP-4668, Naval Weapons Center, China Lake, CA, USA.
8. Krier, H., T'ien, J.S., Sirignano, W.A., and Summerfield, M. (1968) Nonsteady burning phenomena of solid propellants: Theory and experiments, *AIAA Journal*, **6(2)**, 278-288.
9. DeLuca, L.T. (1992) Theory of Nonsteady Burning and Combustion Stability of Solid Propellants by Flame Models, in L. DeLuca, E.W. Price, and M. Summerfield (eds.), *Nonsteady Burning and Combustion Stability of Solid Propellants,* AIAA Progress in Astronautics and Aeronautics, AIAA, Washington, DC, USA, volume 143, chapter 14, pp. 519-600.
10. DeLuca, L.T., Mazza, P., and Verri, M. (1995) Intrinsic stability of solid rocket propellants burning under thermal radiation, in *Proceedings of CNES-CNRS-ONERA Conference, Bordeaux, France, 11-15 Sep* 1995, volume 1, pp. 77-90.
11. DeLuca, L.T., Verri, M., and Jalongo, A. (1997) Intrinsic Stability of Energetic Solids Burning under Thermal Radiation, in W.A. Sirignano, A.G. Merzhanov, and L.T. DeLuca (eds.), *Progress in Combustion Sciences - in Honor of Ya.B. Zel'dovich,* AIAA Progress in Astronautics and Aeronautics, AIAA, Reston, VA, USA, volume 173.
12. Son, S.F., and Brewster, M.Q. (1993) Linear burning rate dynamics of solids subjected to pressure or external radiant flux oscillations, *Journal of Propulsion and Power*, **9(2)**, 222-232.
13. Brewster, M.Q., and Son, S.F. (1995) Quasi-steady combustion modeling of homogeneous solid propellants, *Combustion and Flame,* **103 (1-2)**, 11-26.
14. Ibiricu, M.M., and Williams, F.A. (1975) Influence of externally applied thermal radiation on the burning rates of homogeneous solid propellants, *Combustion and Flame*, **24(2)**, 185-198.

15. DeLuca L.T. (1976) Solid Propellant Ignition and Other Unsteady Combustion Phenomena Induced by Radiation, PhD. thesis, Aerospace and Mechanical Sciences Department, Princeton University, Princeton, NJ, USA.
16. DeLuca, L.T., DiSilvestro, R. and Cozzi, F. (1997) Intrinsic combustion instability of solid energetic materials, *Journal of Propulsion and Power*, **11(4)**, 804-815, 1995. See also Comments, Journal of Propulsion and Power, **13(3)**, 454-456.
17. Derghi A. and Tabaglio, D. (1997) Influenza delle proprietà termofisiche nella combustione di propellenti solidi per endoreattori, Master's thesis, Dipartimento di Energetica, Ingegneria Aeronautica, Politecnico di Milano, Milan, Italy.
18. Galfetti, L., Riva, G., and Bruno, C. (1992) Numerical Computations of Solid Propellant Nonsteady Burning in Open or Confined Volumes, in L. DeLuca, E.W. Price, and M. Summerfield (eds.), *Nonsteady Burning and Combustion Stability of Solid Propellants,* AIAA Progress in Astronautics and Aeronautics, AIAA, Washington, DC, USA, volume 143, chapter 16, pp. 643-687.
19. DeLuca, L.T. (1997) Radiation-driven steady pyrolysis of energetic solid materials, Technical Report CI97-3, California Institute of Technology, Karman Laboratory of Fluid Mechanics and Jet Propulsion, Pasadena, CA, USA.
20. DeLuca, L.T., Verri, M., Cozzi, F., and Colombo G. (1996) Revising the pyrolysis jacobian, in *NATO Advanced Research Workshop (NATO-ARW) on Peaceful Utilization of Energetic Materials*, Fourth International Conference on Combustion (ICOC-96), Saint Petersburg, Russia, 3-6 Jun 1996.
21. Cozzi, F., Cristiani, R., and DeLuca L.T. (1993) Stability of radiation driven burning of solid rocket propellants, IAF Paper No. 93-S.2.467.
22. Stark, J.A., and Taylor, R.E. (1985) Determination of thermal transport properties in ammonium perchlorate, *Journal of Propulsion and Power,* **1(5)**, 409-410.
23. Shoemaker, R.L., Stark, J.A., and Taylor, R.E. (1985) Thermophysical properties of propellants, *High Temperature – High Pressures,* **17**, 429-435.
24. Zanotti, C., Volpi, A., Bianchessi, M. and DeLuca L.T. (1992) Measuring Thermodynamic Proprieties of Burning Propellants, in L. DeLuca, E.W. Price, and M. Summerfield (eds.), *Nonsteady Burning and Combustion Stability of Solid Propellants,* AIAA Progress in Astronautics and Aeronautics, AIAA, Washington, DC, USA, volume 143, chapter 5, pp. 145-196.
25. Summerfield, M., Shinnar, R., Hermance, C.E., and Wenograd, J. (1963) A critical review of recent research on the mechanism of ignition of solid propellants, Technical Report Aeronautical Engineering Laboratory Report No. 661 also AF-AFOSR-92-63, Princeton University, Princeton, NJ, USA.
26. Price, E.W., Bradley, H.H.Jr., Dehority, G.L., and Ibiricu, M.M. (1966) Theory of ignition of solid propellants, *AIAA Journal*, **4(7)**, 1153-1181.
27. Barrère, M. (1968) Solid propellant ignition: General considerations, *La Recherche Aérospatiale*, **123**, 15-28.
28. Merzhanov, A.G., and Averson, A.E. (1971) The present status of the thermal ignition theory: An invited review, *Combustion and Flame*, **16**, 89-124.
29. Hermance, C.E. (1984) Solid-Propellant Ignition Theories and Experiments, in K.K. Kuo, and M. Summerfield (eds.), *Fundamentals of Solid-Propellant*

Combustion, AIAA Progress in Astronautics and Aeronautics, AIAA, New York, N.Y., USA, volume 90, chapter 5, pp. 239-304.

30. Vilyunov, V.N., and Zarko, V.E. (1989) *Ignition of Solids*, Elsevier, Amsterdam and New York.
31. Lengellé, G., Bizot, A., Duterque, J., and Amiot, J.C. (1991) Ignition of solid propellants, *La Recherche Aérospatiale*, pp. 1-20.
32. Ohlemiller, T.J., Caveny, L.H., DeLuca, L.T., and Summerfield, M. (1972) Dynamic effects of ignitability limits of solid propellants subjected to radiative heating, in *14th Symposium (International) on Combustion*, The Combustion Institute, Pittsburgh, PA, USA, pp. 1297-1307.
33. DeLuca, L.T., Caveny, L.H., Ohlemiller, T.J., and Summerfield, M. (1976) Radiative ignition of double base propellants, i: Some formulation effects, *AIAA Journal*, **14(8)**, 940-946.
34. DeLuca, L.T., Ohlemiller, T.J., Caveny, L.H., and Summerfield, M. (1976) Radiative ignition of double base propellants, ii: Pre-ignition events and source effects, *AIAA Journal*, **14(8)**, 1111-1117.
35. Mikheev, V.F., and Levashov, Yu.V. (1973) Experimental study of critical conditions during the ignition and combustion of powders, *Combustion Explosion and Shock Waves*, **9(4)**, 438-441.
36. Klevnoi, S.S. (1971) Quenching of explosives following interruption of the ignition light flux, *Combustion Explosion and Shock Waves*, **7(2)**, 147-155.
37. Assovskii, I.G. (1973) Transient combustion of powder subjected to intense light, *Combustion Explosion and Shock Waves*, **9(6)**, 765-773.
38. Gostintsev, Yu.A. (1974) Extinction of a steady state burning powder by a pulse of thermal radiation, *Combustion Explosion and Shock Waves*, **10**, 686-689.
39. Boggs, T.L., and Zinn, B.T. (1978) *Experimental Diagnostics in Combustion of Solids*, volume 63 of *AIAA Progress in Astronautics and Aeronautics*, AIAA, Washington, DC, USA.
40. Annikov, V.E., and Kondrikov, B.N. (1967) *Influence of Pressure on Burning Ability of High Explosives*, Higher School Publishing House, Moscow, Russia, in Russian, pp. 300-306.
41. Kondrikov, B.N., and Novozhilov, B.V. (1974) Critical combustion diameter of condensed substances, *Combustion Explosion and Shock Waves*, **10(5)**, 580-587, 1974.
42. Watt, D.M., and Petersen, E.E. (1969) Relationship between the limiting pressure and the solid temperature for the deflagration of ammonium perchlorate, *J. Chem. Phys.*, **50(5)**, 2196-2198.
43. Watt, D.M., and Petersen, E.E. (1970) The deflagration of single crystals of ammonium perchlorate, *Combustion and Flame*, **14**, 297-302.
44. Hightower, J.T. and Price, E.W. (1967) Combustion of ammonium perchlorate. *11th Symposium (International) on Combustion*, pp. 463-472.
45. Johnson, W.E., and Nachbar, W. (1962) Deflagration limits in the steady linear burning of a monopropellant with application to ammonium perchlorate, in *8th*

Symposium (International) on Combustion, Williams and Wilkins, Baltimore, MD, USA, pp. 678-689.
46. Horton, M.D. and Price, E.W. (1962) Deflagration of pressed ammonium perchlorate, *ARS Journal*, **32**, 1745.
47. Guirao, C., and Williams, F.A. (1971) A model for ammonium perchlorate deflagration between 20 and 100 atm, *AIAA Journal*, **9**(7), 1345-1356.
48. Cohen Nir., E. (1972) Effet de la temperature initiale sur la vitesse de combustion et sur la pression limite de deflagration du perchlorate d'ammonium, *La Recherche Aerospatiale*, **2**, 75-84.
49. Cohen Nir., E. (1973) An experimental study of the low pressure limit for steady deflagration of ammonium perchlorate, *Combustion and Flame*, **20**, 419-435.
50. Steinz, J.A., Stang, P.L., and Summerfield, M. (1968) The burning mechanism of ammonium perchlorate-based composite solid propellants, AIAA Preprint 68-658.
51. Steinz, J.A., and Summerfield, M., (1969) Low pressure burning of composite solid propellants, Technical Report 88.
52. Steinz, J.A., Stang, P.L., and Summerfield, M. (1969) The burning mechanism of ammonium perchlorate based composite solid propellants, Technical Report AMS-830, Princeton University, Princeton, NJ, USA, PhD. Thesis, Aerospace and Mechanical Sciences Department.
53. Petersen, J.A., Reed Jr., R., and McDonald, A.J. (1967) Control of pressure deflagration limits of composite solid propellants, *AIAA Journal*, **5**, pp. 764-770.
54. Kubota, N., Ohlemiller, T.J., Caveny, L.H., and Summerfield, M. (1973) The mechanism of super–rate burning of catalyzed double base propellants, PhD. Thesis, Department of Aerospace and Mechanical Sciences, AMS-1087, Princeton University, Princeton, NJ, USA.
55. Cookson, R.E., and Fenn, J.B. (1970) Strand size and low temperature deflagration limit in a composite propellant, *AIAA Journal*, **8**, pp. 864-866.
56. Zanotti, C., Colombo, G., Grimaldi, C., Carretta, U., Riva, G., and DeLuca L.T. (1986) Oscillatory burning of composite propellants near pressure deflagration limit, in *Proceedings 15th International Symposium on Space Technology and Science*, Tokyo, Japan, pp. 243-254.
57. Zanotti, C., and Giuliani, P. (1994) Pressure deflagration limit of solid rocket propellants: Experimental results, *Combustion and Flame*, **98**, 35-45.
58. Glaskova, A.P. (1967) Effect des catalyseurs sur la deflagration du nitrate d'ammonium et de ses melanges, *Explosifs*, **1**(**1**), 5-13.
59. Bruno, C., Riva, G., Zanotti, C., Dondè, R., Grimaldi, C. and DeLuca, L.T. (1985) Experimental and theoretical burning of solid rocket propellants near the pressure deflagration limit, *Acta Astronautica*, **12**(**5**), 351-360.
60. DeLuca, L.T., Riva, G., Zanotti, C., Dondè, R., Lapegna, V., and Oleari, A. (1984) The pressure deflagration limit of solid rocket propellants, in AGARD PEP 63rd(A) Specialists *Meeting on Hazard Studies for Solid Propellants Rocket Motors*, Paper 22, AGARD, Paris, France.
61. Zanotti, C., Carretta, U., Grimaldi, C., and Colombo, G. (1992) *Self-Sustained Oscillatory Burning of Solid Propellants: Experimental Results*, in L.T. DeLuca,

E.W. Price, and M. Summerfield (eds.), *Nonsteady Burning and Combustion Stability of Solid Propellants,* AIAA Progress in Astronautics and Aeronautics, AIAA, Washington, DC, USA, volume **143**, chapter 11, pp. 399-439.

62. Çiepluch, C.C. (1961) Effects of Rapid Pressure Decay on Solid Propellant Combustion, *ARS Journal*, **31**, pp.1584-1586.
63. Ciepluch, C.C. (1962) Effect of Composition on Combustion of Solid Propellants During a Rapid Pressure Decrease, Technical Report NASA TN D-1559.
64. Ciepluch, C.C. (1964) Spontaneous Reignition of Previously Extinguished Solid Propellants, Technical Report NASA TN D-2167.
65. Merkle, C.L., Turk, S.L. and Summerfield, M. (1969) Extinguishment of Solid Propellants by Rapid Depressurization, PhD. Thesis, Aerospace and Mechanical Sciences Department, AMS-880, Princeton University, Princeton, NJ, USA.
66. Merkle, C.L., Turk, S.L. and Summerfield, M. (1969) Extinguishment of Solid Propellants by Depressurization: Effects of Propellant Parameters, *AIAA Paper 69-176*, AIAA Washington, DC, USA.
67. Dondè, R., Riva, G., and DeLuca, L.T. (1984) Experimental and theoretical extinction of solid rocket propellants by fast depressurization, *Acta Astronautica*, **11(9)**, 569-576.
68. T'ien, J.S. (1974) A Theoretical Criterion for Dynamic Extinction of Solid Propellants by Fast Depressurization, *Combustion Science and Technology*, **9**, pp. 37-39.
69. DeLuca, L.T. (1975) Nonlinear stability analysis of solid propellant combustion, in *2nd International Symposium on Dynamics of Chemical Reactions*, CNR, Roma, Italia, pp. 245-256.

AN ASSESSMENT OF IGNITION HAZARD FOR SHIELDED ENERGETIC MATERIALS AND ITS RELATION TO FLAMMABLE CHEMICALS

A.G.KNYAZEVA[1] and V.E.ZARKO[2]

[1] Tomsk State University, Physico-Technical Faculty, Tomsk 634050, Russia; [2] Institute of Chemical Kinetics and Combustion, Russian Academy of Sciences, Novosibirsk 630090, Russia

1. Introduction

Experimental and theoretical study of ignition characteristics for solid substances under conditions of complex heat exchange is of importance for different applications. Particularly, it concerns ignition of shielded chemically active materials. There are numerous examples of investigations on reaction initiation in composite heterogeneous systems, studies of combustion of macro heterogeneous systems with large size inclusions, studies of fire and explosion safety.

In mathematical simulation of such problems great number of factors should be taken into consideration. Experimental verification and substantiation of proposed theoretical models are very important. At the same time the experimental data can be used to obtain global parameters characterizing ignited system.

The present paper deals with simple formulation of the ignition problem for shielded reactive solid materials and with some particular cases of ignition of materials through evaporated and decomposed shields.

2. Problem formulation

Let us start with simple formulation of the problem.

Figure 1 shows surface of semi-infinite layer of condensed substance in which exothermic chemical reaction can be excited. The reactive substance is shielded with an inert material of thickness l. The axis x is directed into the bulk of substance. If the temperature of free surface of shield is constant and equals T_s, and thermal contact between two substances is ideal ($\Delta = 0$, Fig.1), the temperature distribution in the system can be described on the basis of solution of conjugate thermal conductivity problem (1)-(6).

V.E. Zarko et al. (eds.), Prevention of Hazardous Fires and Explosions, 251–264.

$$c_1\rho_1\frac{\partial T_1}{\partial t}=\frac{\partial}{\partial x}\lambda_1\frac{\partial T_1}{\partial x},\quad 0<x<l, \tag{1}$$

$$c_2\rho_2\frac{\partial T_2}{\partial t}=\frac{\partial}{\partial x}\lambda_2\frac{\partial T_2}{\partial x}+Q_2\varphi(T_2),\quad x>l, \tag{2}$$

$$T_1\big|_{x=l}=T_2\big|_{x=l},\quad \lambda_1\frac{\partial T_1}{\partial x}\bigg|_{x=l}=\lambda_2\frac{\partial T_2}{\partial x}\bigg|_{x=l} \tag{3}$$

$$T_1=T_s \qquad \text{for } x=0 \tag{4}$$

$$\lambda_2\frac{\partial T_2}{\partial x}=0,\quad x\to\infty. \tag{5}$$

$$T_1=T_2=T_0,\quad t=0. \tag{6}$$

The goal is to determine the ignition characteristics, i.e. the temperature and the instant of time when runaway of exothermic reaction starts.

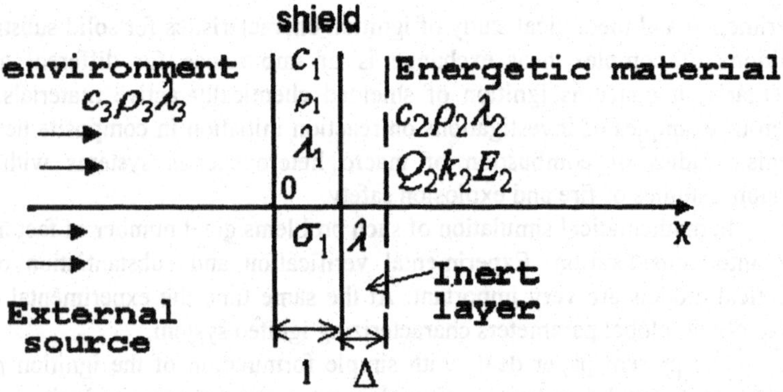

Figure 1. Schematic of problem formulation. Gap ($\Delta \neq 0$) between shield and energetic material corresponds to non-ideal thermal contact.

The following designations are used: T is the temperature; x is the space coordinate; t is the time; c_k, ρ_k, λ_k are the specific heat, density and thermal conductivity, respectively. Index $k=1$ relates to the shield and $k=2$ relates to the reagent. The heat release function is taken in the form

$$\varphi(T_2)=k_2\,exp\left(-\frac{E_2}{RT_2}\right) \tag{7}$$

where E_2 is the activation energy of chemical reaction.

We assume that the thermal properties of substances are constant and are taken at the temperature that is characteristic of ignition process.

The boundary condition corresponding to the action of a constant heat flux takes the form:

$$-\lambda_1 \frac{\partial T_1}{\partial x} = q. \tag{8}$$

If the shield has high thermal conductivity, its heating proceeds very fast, and one may ignore the temperature distribution in that layer. Integrating Eq. (1) within the limits from 0 to l and taking into account the condition (8) we obtain the boundary condition (9) after shifting the origin of coordinates to the reactive substance surface

$$-\lambda_2 \frac{\partial T_2}{\partial x}\bigg|_{x=0} = q - c_1\rho_1 l \frac{\partial T_1}{\partial t}\bigg|_{x=0}, \quad T_2 = T_1. \tag{9}$$

Based on above simple formulation of the problem, several mathematical problems can be solved that correspond to the ignition by radiant and convective fluxes and by hot body. Ignition of energetic materials through chemically reacting shield with taking into consideration failure of thermal contact between substances or heat loss into environment can also be examined.

The approximate analytical methods are convenient to solve the above mentioned problems [1-8]. The methods imply separation of the ignition period into two parts: the inert heating of substance and the reaction excitation in the heated layer. To find solution of the inert problem (without chemical source term in thermal conductivity equation) one may use known methods of thermal conductivity theory. We used the approximate method [9-11] based on the combination of Laplace integral transform method and asymptotic expansion of this solution into series about small or large physical parameters in space.

Two dimensionless parameters γ and ε appear in the formulation of the inert problem that allow to simplify essentially the problem solution in particular cases. Parameter γ is called the thermal thickness of the shield.

$$\gamma = \delta\sqrt{K_c/K_\lambda} \equiv l/\sqrt{\kappa_1 t_*}, \quad \kappa_1 = \lambda_1/(c_1\rho_1), \quad K_c = \frac{c_1\rho_1}{c_2\rho_2}, \quad K_\lambda = \frac{\lambda_1}{\lambda_2}$$

It is the ratio of its geometric thickness to the width of preheat layer which is formed in the shield during time t_*: The time t_* is called adiabatic induction time and T_* is the characteristic temperature (should be defined specifically in particular cases). Two limiting cases $\gamma \ll 1$ and $\gamma \gg 1$ are of interest for qualitative analysis of problem.

Dimensionless parameter ε is characteristic for all conjugate problems of thermal conductivity theory.

$$\varepsilon = (1 - K_\varepsilon)/(1 + K_\varepsilon), \qquad |\varepsilon| < 1.$$

The solution to the problem can be presented in the form of series about ε. Included in ε, the parameter K_ε is known as coefficient of thermal activity of one substance in relation to other. K_ε is calculated by

$$K_\varepsilon = \sqrt{K_c K_\lambda} = \sqrt{(\lambda_1 c_1 \rho_1)/(\lambda_2 c_2, \rho_2)}.$$

3. Ignition criteria

The instant of exothermic reaction excitation may be determined by use of the known the Zeldovich quasistationary criterion [3] and the Vilyunov adiabatic criterion [6], which have clear physical interpretation.

According to the **adiabatic criterion**, the rate of temperature rise due to external heating at the instant of ignition is equal to the rate of temperature rise due to chemical heat release on the surface of reactive material (or at the boundary between substances for conjugate problem). Equation (10) gives definition of the criterion:

$$\left(\frac{\partial T}{\partial t}\right)_{\substack{inert\\heating}} = \left(\frac{\partial T}{\partial t}\right)_{\substack{chemical\\reaction}} \tag{10}$$

The criterion works well if the surface temperature changes monotonously during whole heating time. The adiabatic time of ignition (or time of *triggering* the chemical reaction), $t = t_a$ and adiabatic ignition temperature, $T = T_a$ are obtained based on this criterion. These characteristics relate to the point of inflection on the curve "surface temperature - time" that can be recognized simply on the experimental records.

According to the **quasistationary criterion,** at the moment of ignition time the heat flux due to external heating is equal to the heat flux from reaction zone. The last one can be determined from solution of corresponding quasistationary problem. Equation (11) gives definition of the quasistationary criterion:

$$\left(\frac{\partial T}{\partial x}\right)_{\substack{x=0,\\external\\heating}} = \left(\frac{\partial T}{\partial x}\right)_{\substack{x=0,\\from\\reaction\ zone}} \tag{11}$$

This criterion can be properly used if the surface temperature changes slowly.

In the case corresponding to the adiabatic criterion, the maximum of temperature locates on the reagent surface while in the quasistationary case it may locate within the reaction zone. Below we consider some examples of ignition problem (without mathematical details).

4. Examples of problem solution

4.1. IGNITION BY HOT PLATE

The temperature time dependent behavior in the problem with boundary condition of first kind ($T_s = const$) for various thickness shield was studied numerically [12]. The heating of a thermally thin shield up to the temperature $T_* = T_0 + RT_s^2/E_2$, when exothermic reaction triggers, occurs quickly; the most part of ignition time is spent on the formation of thermal boundary layer in energetic materials of sufficient thickness. Thermally thin shield corresponds to inequality (12)

$$\sqrt{\kappa_1 t_i^o} >> l \qquad (12)$$

Here t_i^o is the ignition time for l equal zero.

If the shield is thermally thick, the inequality (13) is hold

$$\sqrt{\kappa_1 t_i^0} << l \qquad (16)$$

In this case the thermal boundary layer forms simultaneously with heating of shield to the temperature $T_* = T_0 + RT_s^2/E_2$. The features of formation of temperature field allow use for estimations for thermally thin shield the quasistationary criterion. In opposite case, when surface temperature changes monotonously, both criteria give identical results.

The formulae for calculation of the ignition time may be presented in the form of Eqs.(14) and (15) for thin and thick shields, correspondingly:

$$\frac{t_q}{t_q^0} = 1 + \frac{l}{\sqrt{\kappa_2 t_q^0}} \frac{\theta_0}{\sqrt{\pi}} \frac{\lambda_2}{\lambda_1}, \quad \sqrt{\kappa_2 t_q^0} >> l \qquad (14)$$

$$\frac{t_q}{t_q^0} = \frac{1+2m^2}{m^2} \cdot \frac{l^2}{\kappa_1 t_q^0}, \quad m = \frac{1-\varepsilon}{\sqrt{\pi}}, \qquad (15)$$

It follows from Eq.14 that calculated data can be treated in coordinates $ln[(t_q-t_q^0)/t_q^0]$, $ln(\lambda_2/\lambda_1 \sqrt{\kappa_2 t_i^0})$. It was revealed that the results of numerical calculations are described very reasonably with straight line in the above coordinates. This confirms feasibility of proposed analytical solution. The experimental results for ignition of nitrocellulose shielded with different thickness layers of teflon or polyethelene terephtalate were also well fitted in these coordinates (Fig.2) [12, 13] This indicates that simple mathematical model is quite adequately to describes the phenomenon under study.

4.2. IGNITION BY CONSTANT HEAT FLUX

Similar study was carried out for the problem of ignition of shielded energetic material by constant heat flux [11, 13], see Eq.8. It was found that both quasistationary and adiabatic criteria can be applied in this case. For example, if shield is thermally thin, the formula for ignition time has a form

$$\frac{t_a}{t_a^0} = 1 + \sqrt{2\pi}\,\frac{\gamma K_\varepsilon}{\sqrt{\theta_0}}\sqrt{\frac{t_a^0}{t_a}}$$

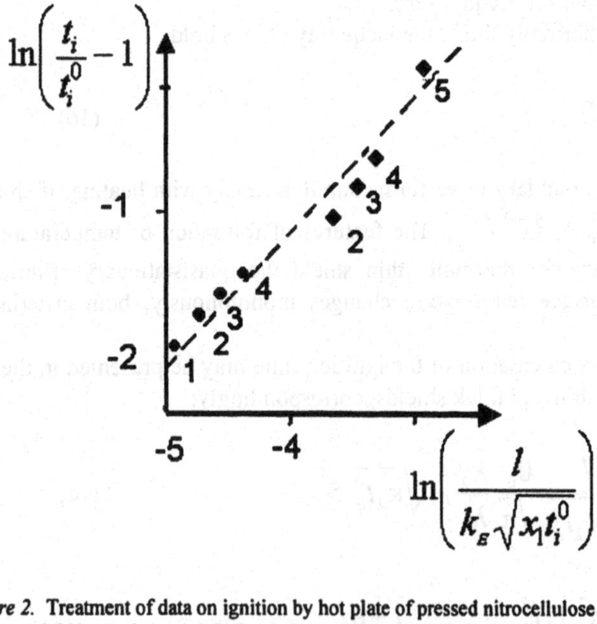

Figure 2. Treatment of data on ignition by hot plate of pressed nitrocellulose shielded with teflon layer (l = • - 0.002 cm, l = ♦ - 0.006 cm); Ts = 500°C (1), 510°C 9(2), 520°C (3), 530°C (4), 540°C (5).

Here $\theta_0 = \frac{T_a - T_0}{RT_a^2} E_2$ is the dimensionless temperature drop, and temperature $T_* = T_a^0$ can be determined from equation:

$$T_a^0 = T_0 + \frac{2}{\pi} \frac{q_0^2}{\lambda_2 \rho_2} \frac{exp(E_2/RT_2)}{Q_2 k_2}$$

If shield is translucent and external heating is performed by radiant flux, the analysis of problem is analogous to previous ones [14]. In this case, mathematical formulation of problem involves new parameter - an absorption index σ_1. The equation of thermal conductivity for shield and the condition for heat fluxes on the interface between substances is written as follows:

$$c_1 \rho_1 \frac{\partial T_1}{\partial t} = \frac{\partial}{\partial x} \lambda_1 \frac{\partial T_1}{\partial x} + q\, exp(-\sigma_1 x), \quad 0<x<l,$$

$$\lambda_1 \left.\frac{\partial T_1}{\partial x}\right|_{x=l} - \lambda_2 \left.\frac{\partial T_2}{\partial x}\right|_{x=l} = q\, exp(-\sigma_1 l)$$

There is one distinctive feature of the ignition behavior under study: the dependence of ignition time on absorption index is non monotonic (Fig.2). This is consequence of concurrence of two different ways of heat transfer: thermal conductivity and irradiation. However, the maximum of ratio of ignition times for translucent and opaque materials amounts only 1.1-1.2, therefore this effect may hardly be discovered in experimental studying of the ignition process.

Experimental investigation of ignition process for the system *"nitrocellulose-polyimide shield"* with various shield thickness was carried out on the basis of Xenon lamp set up. The ignition time was determined with indication by thermocouple inserted between substances.

The experimental data on the ratio (t_i/t_i^0) have been described by a universal curve (Fig.3) in dependence of parameter $B = K_\varepsilon l/\sqrt{\kappa_1 t_i^0}$ which was derived from theoretical solution of the problem. The dependence of ignition time on the parameter B is approximately linear in the range of very small values of B.

4.3. INFLUENCE OF HEAT LOSSES INTO ENVIROMENT ON THE IGNITION TIME

Theoretical investigations allow to reveal several effects which are not obvious in the course of examination of the simple problem. Thus, upon the solution of three layer conjugate problem of thermal conductivity

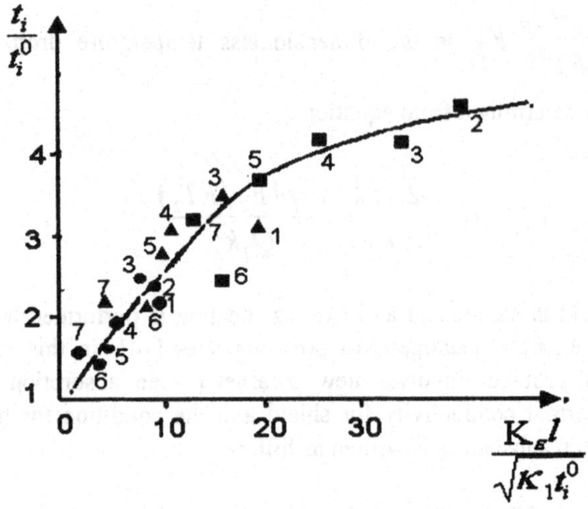

Figure 3. The data on ignition by constant heat flux of pressed nitrocellulose shielded with polyimide layer of different thickness l : ● - 0.004 cm, ▲ - 0.009 cm, ■ 0.021 cm. q = 167 W/cm² (1), 142 (2), 105 (3), 59 (4), 41 (5), 36 (6), 25 (7).

$$c_3\rho_3 \frac{\partial T_3}{\partial t} = \frac{\partial}{\partial x}\lambda_3 \frac{\partial T_3}{\partial x}, \quad x<0, \tag{26}$$

$$T_3\big|_{x=0} = T_1\big|_{x=0}, \quad \lambda_1 \frac{\partial T_1}{\partial x}\bigg|_{x=0} = \lambda_3 \frac{\partial T_3}{\partial x}\bigg|_{x=0} \tag{27}$$

it was found that taking into consideration the heat loss from shield surface to environment by the conduction mechanism [9,14] does not change the qualitative dependence of ignition time on parameter B. Two interesting formulae (16) reflect this fact:

$$\frac{t_a}{t_a^0} = 1 + \sqrt{2\pi}\frac{\gamma K_\varepsilon}{\sqrt{\theta_0}}\sqrt{\frac{t_a^0}{t_a}} \cdot \mu_0 \quad \text{and}$$

$$\frac{t_a}{t_a^0} = 1 + \sqrt{2\pi}\frac{\gamma K_\varepsilon}{\sqrt{\theta_0}}\sqrt{\frac{t_a^0}{t_a}} \cdot \mu_\infty, \tag{28}$$

Parameters μ_0 and μ_∞ are defined as

$$\mu_0 = \frac{1}{K_\varepsilon^2} \cdot \frac{K_\varepsilon^2 - B_\varepsilon^2}{1 + B_\varepsilon} \quad \text{and} \quad \mu_\infty = \frac{1}{K_\varepsilon^2} \cdot \frac{K_\varepsilon^2 + B_\varepsilon}{1 + B_\varepsilon} \tag{29}$$

They correspond to absolutely transparent and opaque shield material, respectively and reflect the complex influence of third medium. Here $B_\varepsilon = \sqrt{(c_3\rho_3\lambda_3)/(c_2\rho_2\lambda_2)}$, and index "3" relates to environment medium properties. It has been shown that transparent thermally thin shield (with $K_\varepsilon \approx B_\varepsilon$) does not affect the ignition time. At the same time, provided that the conditions $B_\varepsilon > K_\varepsilon$ and $l << \sqrt{\kappa_1 t_a}$ are met, the thermally thin transparent shield decreases the ignition time in relation to the case of non-shielded material.

4.4. IGNITION UNDER INTERRUPTIVE HEATING AND THROUGH REACTING SHIELD.

Some interesting results were obtained numerically when solving ignition problems with finite time of external heat source action [12, 15]. Numerical solution of the problem of energetic material ignition through shield with high thermal conductivity has been obtained using boundary condition

$$-\lambda_2 \left.\frac{\partial T_2}{\partial x}\right|_{x=0} = q - c_1\rho_1 l \left.\frac{\partial T_1}{\partial t}\right|_{x=0}, \quad T_2 = T_1$$

and additional condition

$$q = \begin{cases} q_0, & t < t_e, \\ 0, & t \geq t_e. \end{cases}$$

The solution gives critical exposure time t_{cr}. If inequality

$$t_e \geq t_{cr}$$

is fulfilled, the exothermic reaction starts after time t_e due to heat accumulated in the shield. If inequality

$$t_e < t_{cr}$$

is fulfilled, ignition does not occur. For certainty, the time t_{cr} was chosen in calculations as exposure time when induction period after interruption is twice longer than the ignition time at constant heating.

The results of numerical calculations can be presented in convenient form using known parameter B:

$$\frac{t_{cr}}{t_i^0} \cong 0{,}25\left(1+\sqrt{1+3.2B}\right)^2$$

The shield between the heat source and reagent can affect essentially the ignition characteristics if the shield material is able to decompose at high temperatures [16].

If shield is thermally thin one may neglect a difference in temperature between free surface and interface and formulate the simple problem.

The condition on the free surface of shield has a form [17]

$$-\lambda_2 \frac{\partial T_2}{\partial x}\bigg|_{x=0} = q - c_1\rho_1 l(t)\frac{\partial T_1}{\partial t}\bigg|_{x=0} - Q_1\rho_1\frac{dl}{dt}$$

Let thickness of shield changes according to the pyrolysis law (E_p is the energy of activation)

$$\frac{dl}{dt} = -k_p\, exp\left(-\frac{E_p}{RT_1}\right), \quad T_1 = T_2$$

The simple analytical estimates allow to divide the ignition time on three parts:

$$t_{q\Sigma} = t_d + t_{qi} + t_h,$$

where t_d is the time of shield decomposition at the temperature T_p, t_{qi} is the induction period of ignition and t_h is the time of inert heating up to the pyrolysis temperature T_p:

$$T_p = -\frac{E_p}{R}\left[ln\left(\frac{q}{\rho_1 Q_p k_p}\right)\right]^{-1}.$$

The relation between stages may be different for various regimes that correspond to various shield properties.

A similar problem formulation is valid for the case of reaction excitation through liquid layer under action of high heat flux when evaporation of the layer proceeds in kinetic regime [19]. An example is the ignition of wood impregnated by moisture. In this case, the boundary condition can be written in the form of Eq.(17) with supplement conditions (18-20):

$$-\lambda_2 \frac{\partial T_2}{\partial x}\bigg|_{x=0} = q - c_l \rho_l h \frac{dT_1}{dt}\bigg|_{x=0} + Q_{ev}\frac{dm}{dt}, \quad (17)$$

$$\frac{dm}{dt} = -\rho_l V\, exp\left(-\frac{Q_{ev}\mu}{RT_1}\right), \quad m \neq 0; \quad (18)$$

$$\frac{dm}{dt} = 0, \; m = 0; \quad (19)$$

$$m(0) = m_0, \; m_0 \leq m_\infty = h\rho_l \quad (20)$$

Here h is the characteristic size of pores, m is the mass of liquid in pores, V is the speed of sound, μ is the molar mass of liquid, m_∞ is the maximal possible mass of liquid in pores, c_l and ρ_l are the specific heat and density of liquid, correspondingly.

The results of numerical solution of problem with boundary conditions (17) – (20) can be represented by straight line in coordinates $\frac{t_i}{t_i^0}$ and $\left(\frac{lK_c}{\sqrt{\kappa_2 t_i^0}}\right)$. The value $l = m_0 / \rho_l$ is the effective thickness of shield in this case.

If thermally thin shield decomposes exothermically and corresponding heterogeneous reaction proceeds in kinetics or diffusion control regime [17], we can write the kinetic equation and boundary condition in the form:

$$\frac{dl}{dt} = -\frac{k_1\, exp\left(-\frac{E_1}{RT_1}\right)}{1 + \frac{k_1}{\beta_1} exp\left(-\frac{E_1}{RT_1}\right)}, \quad T_1 = T_2,$$

$$-\lambda_2 \frac{\partial T_2}{\partial x}\bigg|_{x=0} = q - c_1\rho_1 l(t)\frac{\partial T_1}{\partial t}\bigg|_{x=0} + Q_1\rho_1\frac{dl}{dt},$$

where β_1 is the reaction rate in diffusion control regime, k_1 is the reaction rate in kinetic regime.

The results of numerical calculations show that heterogeneous reaction serves as an additional source of heat for energetic material. The thickness changes slowly until certain moment of time but when the heterogeneous reaction enters into diffusion control regime, the shield burns very fast. The burning time of shield decreases with parameter β_1. It has been revealed that shield with heterogeneous reaction on the external surface favors the ignition of energetic material in comparison with inert shield. Furthermore, it was found that the time of ignition through chemically reacting shield is less than the time of ignition without shield for large interval of change of parameter β_1.

4.5. THE ROLE OF THE INTERFACE THERMAL RESISTANCE

The assumption on ideal thermal contact between substances hardly can be met in particular cases. The role of this factor can be examined in the first approximation analytically [11 ,12, 9].

Let us consider the inert layer with small constant thickness and negligible heat capacity which is placed between shield and energetic material. In this case, we have the condition

$$T_2(l,t) - T_1(l,t) = \Delta \frac{\lambda_1}{\lambda} \frac{\partial T_1}{\partial x}\bigg|_{x=l} \tag{21}$$

instead of equality of temperatures at the interface between substances, which is hold in the case of ignition by hot plate. Here Δ is the thickness of inert layer, λ is its thermal conductivity coefficient, the ratio Δ/λ is its thermal resistance. The new non-dimensional parameter appears:

$$h = \Delta\lambda_2 \Big/ \left(\lambda\sqrt{\kappa_2 t_*}\right)$$

Analytical solution can be presented in the form

$$\frac{t_q}{t_q^0} = 1 + \frac{\theta_0 \lambda_2}{\sqrt{\pi \kappa_2 t_q^0}} \left[\frac{1}{\lambda_1} + \frac{\Delta}{\lambda}\right],$$

The thermal resistance of all layers for multi-layer shield can be summarized if their total thickness is small

$$\frac{t_q}{t_q^0} = 1 + \frac{\theta_0 \lambda_2}{\sqrt{\pi \kappa_2 t_q^0}} \sum_{(k)} \left(\frac{l_k}{\lambda_k}\right).$$

When radiant flux is used as external energy source, the transparency of additional inert layer may be of importance. Upon radiative heating, the absolutely transparent layer reduces the heat loss into shield while opaque layer may serve as additional barrier for external flux.

In the first case, the condition for temperature drop has a form of Eq.(21), and in the second case (opaque layer) it takes a form of Eq.(22):

$$T_2(l,t)-T_1(l,t)=\Delta\frac{\lambda_2}{\lambda}\frac{\partial T_2}{\partial x}\bigg|_{x=l} . \quad (22)$$

In particular case of thermally thin shield, $\gamma << 1$, approximate analytical solution leads to formula

$$\frac{t_a}{t_a^0}=1+\gamma K_\varepsilon\sqrt{\frac{2\pi}{\theta_0}}\left[1\pm\frac{1}{\sqrt{\pi t_a^0}}\frac{hK_\varepsilon}{1+K_\varepsilon}\right]\sqrt{\frac{t_a^0}{t_a}},$$

where sign minus corresponds to absolutely transparent layer.

The role of this process may be studied for more complicated formulation of the problem taking into consideration the dynamics of detachment of shield from energetic material. We do not discuss this problem here because it needs special examination.

5. Conclusions

The results of the study can be summarized as follows.

1. The mathematical models for ignition of shielded energetic materials are formulated. The problem is studied for shield materials of different thickness and various properties: inert, reactive, transparent, opaque, double-layer.

2. Analytical approaches to solve the problem are developed and numerical solutions are obtained. Comparison with original experimental data showed their good agreement with theoretical predictions.

3. The approach developed has to be improved in the future in order to take into account the effects of gas release during heating of energetic materials: breakdown of ideal thermal contact with shied layer, change of density in preheat layer of energetic material, etc.

4. Similar approaches can be used for analysis of ignition of shielded pure fuels. In this case the model should be modified in order to analyze destruction processes in the shield material and diffusion of oxidizer to the fuel.

6. References

1. Vilyunov, V.N. and Zarko, V.E. (1989) Ignition of Solids, "Elsevier Science Publishers", Amsterdam, Oxford, New York, Tokyo, 442 pp.
2. Merzhanov, A.G. and Averson, A.E. (1971) The present state of the thermal ignition theory: An Invited Review, *Combustion and Flame*, **16**, No.1, 89-124.
3. Zeldovich, Ya.B. (1963) To the theory of ignition, *Doklady Akademii Nauk*, **130**, No.2, 283-285.
4. Averson, A.E., Barzykin V.V., and Merzhanov A.G. (1968) Approximate method of solution of the problems in thermal theory of ignition. *Doklady Akademii Nauk*, **178**, No.1, 131-134.
5. Dik, I.G. and Zurer A.B. (1982) Application of the integral methods in the problems of thermal ignition theory, *Combustion, Explosion, and Shock Waves*, **18**, No.4, 16-22.
6. Burkina, R.S. and Vilyunov, V.N.(1980) *Asymptotic relations in the problems of ignition theory*. Tomsk State University, Tomsk.
7. Lyubchenko, I.S. (1987) Thermal theory of ignition of reacting substances, *Uspekhi chimii*, **15**, No.2, 216-240.
8. Grishin, A.M. (1966) On ignition of reacting substances, *PMTF*, No.5, 25-30.
9. Knyazeva, A.G. (1989) *Ignition of shielded opaque condensed substances*. PhD Thesis, Tomsk.
10. Knyazeva, A.G. Approximate estimations of gasless composition ignition characteristics under conjugate heat exchange conditions. (Izhevsk, the Institute of Applied Mechanics, UB.RAS, p.202-207) Intern.Conference on Combustion ICOC 93, Moscow-St.Petersburg.
11. Dik, I.G. and Knyazeva, A.G. (1989) *Approximate calculation of ignition characteristics under conjugate heat exchange*. VINITI, **6441-B89**, Tomsk, 27pp.
12. Knyazeva, A.G. and Dik, I.G. (1990) Ignition of condensed substance by hot plate with inert shield at the interface, *Combustion, Explosion, and Shock Waves*, **26**, No.2, 8-18.
13. Dik, I.G. and Knyazeva, A.G. (1980) Ignition of condensed substances through a screen. *Combustion of condensed systems*, Chernogolovka.
14. Knyazeva, A.G. (1996) Approximate estimations of ignition characteristics of a propellant by radiant flux through shields with different properties, *Combustion, Explosion, and Shock Waves*, **32**, No.1, 26-41.
15. Dik, I.G., Zurer A.B., and Knyazeva, A.G. (1989) On ignition of condensed substance by pulse of thermal flux through an opaque screen with high thermal conductivity, *Combustion, Explosion, and Shock Waves*, **25**, No.6, 3-9.
16. Librovich, V.B. (1963) On ignition of solid propellants, *PMTF*, **6**, 74-79.
17. Knyazeva, A.G. and Dik, I.G. (1989) *On modelling of ignition of condensed substance through thermally degrading barrier*, VINITI, **6442-B89**, Tomsk, 23 p.
18. Knyazeva, A.G. and Dik, I.G. (1989) Ignition of condensed substance through a screen with different physical and chemical properties, *Macroscopic Kinetics and Chemical Gas Dynamics*, Tomsk, 127-132.
19. Knyazeva, A.G. and Slivka, L.N. (1989) On ignition of condensed substance by hot body through a screen , VINITI, **6443-B89**, Tomsk, 24 p.

APPLICATION OF HIGH ENERGY MATERIALS FOR COMMERCIAL USE – THE INDIAN SCENE

H. SINGH
High Energy Materials Research Laboratory,
PUNE 411 021, INDIA.

1. Introduction

After independence in 1947, India has created a vast science and technology (S&T) infrastructure in the country. Investment in S&T research has increased from 0.5% of GNP in 1977 to about 1 % of GNP in 1997. S&T manpower has increased from 1.17 million in 1970 to 3.9 million in 1997 and country is witnessing a changing process catalyzed by the wave of globalization and all the sectors are engaged in value addition.[1]

Just as a creative force can be used for destructive purposes, a destructive force can also be tamed for constructive role. However, High Energy Material's (HEM's) are generally associated with destructive roles and their important contributions to industry and economy is very often lost sight off. High explosives are finding increasing applications in extracting ores and coal from mines, for building roads, city sub-ways, airports, canals and harbours. Metals can be formed and welded, diamonds made from graphite, ordinary rocks compressed, stones blasted inside the bladders of patients. One of the most innovative use of explosives is in the field of explosive welding or cladding. This technique is employed to bond dissimilar metals particularly those which are difficult to weld by conventional methods, especially in chemical and ship building industries. Explosives can demolish buildings, where conventional demolition happen to be hazardous and time consuming. By drilling a large number of holes in the lower walls of the building to be demolished and charging them with carefully selected explosive, the whole structure can be made to collapse gently on its own foundation and the entire process is over in a few seconds. Recent developments in detonator technology includes the use of semi-conductor micro-chips. When the chip receives the electric signal, it triggers an explosive charge thousand times faster than a conventional electric detonator. Explosive power can also be used to form metal sheets into complex shapes. Typical examples are nose cone of a rocket and the hemispheric plates for thick wall steel reactor tank. Thus experts all over the globe agree that HEM Industry is one of the acknowledged technological leader and has always thrived on technological innovation and creativeness.

V.E. Zarko et al. (eds.), Prevention of Hazardous Fires and Explosions, 265–272.

development involves powder ingredients which react chemically and can be pressed to make new solids with useful properties. One such example is gallium arsenide, which can be made by mixing and synthesizing two elements, gallium and arsenic through explosive shock. Titanium and aluminium, which normally resist alloying, react as powder under explosive shock to yield new alloys that find application in the aerospace industry. Explosives driven generators are also on the anvil, which can produce short electrical pulses lasting for short time but having extra ordinary high power.

Today there are 30 civil explosive factories in India manufacturing different types of civil explosives including NG (nitroglycerine) based permitted, slurry and emulsion explosives (both cartridged as well as site mixed) and allied items like detonators, detonating cords, safety fuse, flexible linear shaped charges (FLSC) etc. India is not only meeting its full requirement of civil explosives but is also exporting both explosives and explosive technologies to other developing countries. Today country's requirement of civil explosives is of the order of 0.4 million tones. Table-1 presents the details of different types of civil explosives produced in India during 1991-92 and 1992-93.[2] The fire works manufactured during 1993 were 24,000 tonnes (value – 145 crores) and blasting explosives worth 73 million rupees were exported to various foreign countries.

Table – 1 : Production of explosives during the year 1992-93

Description	1991-92	1992-93
Gunpowder	38.0 MT	26.0 MT
Nitrate Mixture	125059.0 MT	12900.0 MT
Nitro-Compound	46496.0 MT	45817.0 MT
Total:	171594.0 MT	158744.0 MT
Safety Fuse	99.0 Million Meters	103.0 Mn. Meters
Detonating Fuse	63.0 Million Meters	59.0 Mn. Meters
Detonators	292.0 Million Numbers	292.0Mn.Numbers

Western countries have tested utilization possibilities of energetic materials, particularly military explosives for commercial use. Technically feasible concepts to convert energetic components to chemical raw materials failed in practice, in view of the fact that chemical compounds, resulting from the conversion processes are not able to compete with common materials; if purity, availability in commercial amounts and return on investments are taken into account [3]

This manuscript presents the details of application of high energy materials (HEM's) and other related technologies for commercial use in India.

2. Civil Applications

2.1 COAL MINING

60 % of commercial explosives produced in India are used by mining industry. Permitted and conventional explosives have been extensively used for gassy and open pit mines. In majority of coal mines, particularly located in Bihar (one of Indian State), there is a continuos evolution of methane gas, which forms an explosive mixture with air. By suitably reducing the flame temperature(T_f) and duration during explosion, the ignition of explosive mixture (methane + air) is reduced drastically.

One of the major problem faced by Indian coal mining industry is the extraction of coal through mines which are on fire during summer ($+40^0$–$+48^0$C ambient temperature). To meet this requirement thermally stable compositions have been developed and evaluated. Development of site mixed slurry (SMS) explosive with suitable pump trucks has considerably increased their application in open pit mining of large diameter and deep bore holes. More than 100 holes are routinely fired in a single blast for best performance by sequential firing. To obtain higher water resistance, particularly for bore holes having watery discharge, emulsion explosives have been developed. These are more water resistant, in view of oxidizer solution covered by continuous fuel, lower channel desensitization and resistance to low temperature desensitization. Better detonating properties of emulsion explosives are another unique feature of emulsion explosives. This property has been attributed to presence of droplets of Super Saturated Oxidizer phase leading to better intimacy of oxidizer and fuel and short reaction time.[4,5]

Both smokeless power and composite propellant based explosives have been used for many commercial applications including mining and cheap source of fuel. 15 –40 % of granular smokeless powder (1 mm size) has been used in watergels. Shock and bubble energy of propellant containing slurries (10 – 40 %) was found to be much superior than the conventional slurry explosives containing hexamine, AN, Guargum, crosslinker and water. Inclusion of shredded composite propellant (10 – 20 %) in slurry explosives results in an energetic explosive material, which has higher shock and bubble energy.[6]

2.2 SPACE APPLICATION

For space exploration, Meteorological studies, sounding rockets, Satellite Launch Vehicles (SLV), Augmented Satellite Launch Vehicles (ASLV), Polar satellite launch vehicle (PSLV) and Geostationary Satellite launch vehicle (GSLV) with different payload composite solid propellants based on PBAN (Polybutadiene acrylic acid, acrylonitrile) , CTPB (Carboxy Terminated Polybutadiene) and HTPB (Hydroxy Terminated Ploybutadiene) and liquid propellants have been used.

The largest solid motor booster developed by ISRO (Indian Space Research Organization) has been used for the first stage of PSLV, which has been designed for launching 1000 kg class Indian Remote Sensing Satellite. Recently booster motor with 138 Tones of propellant has been statically evaluated which gave peak thrust of 415 Tones and burnt for 110 sec. duration. This improved rocket motor will be used to place 1200 kg Indian remote sensing satellite (IRS) into 817 km Polar sun-synchronous orbit during 1998.

2.3 DETONATORS

Different types of initiatory explosives, essentially developed for military applications, have been used for commercial detonators and cap compositions. Service Lead Azide (SLA) has been extensively used as initiatory compound for detonators. However, due to its limitation of the formation of hydrozoic acid on aging and ultimately formation of sensitive Copper azide with copper tube based detonators, which has been responsible for many unfortunate accidents all over the globe, SLA has been replaced by a new and safe initiatory compound Basic Lead Azide (BLA). BLA has better hydrolytic and thermal stability, flash pick up, high bulk density and flow property.[7,8] A large number of igniferous detonators for various applications are being manufactured in India using BLA as main ingredient.

2.4 OIL EXPLORATION

One of the important factors affecting the economy of many countries nowadays is oil. For oil exploration, it is necessary to fix the location of actual oil bearing strata, after drilling of exploratory holes. This is achieved by firing a set of explosive devices, called shaped charges. RDX based shaped charges with copper /steel as core material have been used in India. Steel tube containing shaped charges are lowered to the required depth in the well and then fired electrically from the surface. The jet formed from the shaped charges bore holes into soil upto a depth of 2 – 3 meters. If the strata in which charges are fired contains oil, it rushes to the surface. β-HMX based shaped charges with more effective liner materials have been developed to obtain superior performance.

2.5 METAL CLADDING

In cladding, an explosively driven metal plate hits another kept at a short distance. The extreme temperature and pressure produced with high energy impact bonds the plates together with practically a metallurgical bond. The flyer plate and base plate are separated by a distance greater than half the flyer plate thickness in order to allow this plate to achieve its maximum impact velocity. Dissimilar metal plates have been bonded using this technique.(fig.1) Explosive welding of tubes (100) to plate has been successfully used for the manufacture as well as repair of heat exchangers. This technique increases service life of heat exchangers and gives much stronger bond than that obtained in seasoned welding.

2.6 METAL WORKING

HEM have been extensively used in Metal working industry. Operations like explosive farming and explosive sizing have proved to be of high value to aircraft and missile industries. Use of explosive rivet in aircraft construction is well known. Another interesting application of explosives in metal working is the explosive hardening. This results in change of engineering properties of metal. (TS, elongation, yield strength etc). This is achieved by detonating a thin layer of plastic sheet explosive in contact with the metal surface to be hardened. Yield strength was improved by 100 % and T.S. was improved by 40 % using this technique.

2.7 AGRICULTURE

Explosives have been used to clear land of trees, stumps and boulders. This is achieved at a much faster rate and cheaper cost than getting it done manually. Swamp drainage, stream diversion , sub soiling and tree planting are other operations, in which explosives have been used in India. Blasting is the quickest and most economic method of breaking up boulders, reducing them to a size that can be handled easily. For tree planting as a part of social forestry and to take care of environmental pollution, particularly in rocky soil, explosives not only help in making hole easily but also loosen the ground over an area considerably in excess than generally obtained by conventional methods.

2.8 CIVIL ENGINEERING

In large scale excavation work, explosive technique similar to those used in quarrying are employed. Demolition work is a branch of engineering in which explosives have been found useful in dismantling and clearing of structures and buildings. Explosives have been extensively used to demolish structures made of stone, brick, concrete, steel and timber. For cutting action flexible linear shaped charges have also been used, whereas for demolition both conventional and silent explosives have been used. «ACCONEX» is a non-explosive demolition agent developed by HEMRL. It is a special type of expanding cement which produces cracks while setting. This cement when mixed with 25 – 30 % water forms a slurry. The slurry is poured in pre-drilled holes of about 65 – 70 % of size of boulder/rock/target. The slurry sets in about 15 minutes. With passage of time it develops high expansive stresses due to presence of special silicates in the composition. The phenomenon of demolition occurs with crack initiation. Propagation of cracks from hole increases in number. The process of cracking gets complete between 24 and 72 hours, depending upon nature/size of target and temperature. The effect of ACCONEX is that it demolishes rocks/concrete structures without any noise or adverse effect on neighboring structures. This is advantageous when demolishing in densely populated and built-up areas. It does not cause any pollution as no gases are liberated.

2.9 HIGH ALTITUDE FUEL

Conventional fuels such as firewood, coal etc are not suitable at high altitudes because of difficulty in ignition and low heat due to lower oxygen content and low temperature. A gel based fuel developed earlier, was not found very attractive, because of its high degree of inflammability and toxic combustion products.

A solid fuel containing mixture of wood flour and energetic material like double base propellants with binder is being used for heating purposes to provide adequate heat energy to warm food or to prepare tea/coffee. These fuels are highly cost effective as the ingredients particularly, the propellants can be used from the waste/life expired material, available in plenty from the production agencies and reduce disposal problem to a very large extent.

Compositions based on small pieces of waste double base (DB) propellants gelled with acetone and then mixed with fine saw dust as major ingredient have been formulated, pelletised and finally coated with wax and evaluated for essential parameters such as total burn time, flame temperature and ash content. These formulations were also studied in detail for their ignitability and sustained burning behavior under low atmospheric condition (an altitude of 3000 meter and subzero temperature) and for actual performance of boiling the water in an aluminium vessel on a specially designed foldable stove. Fuel performance has been satisfactory. The toxicity level in terms of CO and hydrocarbon was found within the acceptable limits. No inconvenience was felt by the personnel using this fuel. It is very safe for storage and transportation.

A promising High Altitude Fuel (HAF) composition containing around 25 parts of DB propellant and 10 BSS size saw dust gave density of 0.80 gm/CC, flame temperature of 1050^{0}C and ash content of 3 %.

2.10 PILOT SEAT EJECTION SYSTEM

The ever increasing speeds of combat aircraft- mainly the fighters creates difficulties to allow the pilot to bail out in an emergency. FLSC (flexible linear shaped charges), when detonated cause separation/severance without producing any shock or vibration are used for this purpose. They can be used for canopy severance system, stage separation of space crafts as well as to obtain clean cutting action. FLSC with explosive loading from 0.8 gm/meter to 120 gm/meter (RDX based) normally in lead cover have been developed and used for various applications. FLSC developed for canopy severance system for an advanced aircraft can cut a 7 mm thick acrylic sheet.[9] (Fig.2).

A few FLSC's have been developed for both military and civilian applications. A miniature detonating cord with an explosive loading of 0.8 g/m designed to cut the canopy material and an explosive transfer line that can transfer the explosive impulse from one point to another without affecting the surroundings are amongst the important explosive components for In-flight Egress System of Canopy Severance system. On receiving the pressure impulse from the seat ejection cartridge, the in-flight egress system initiates and cut the canopy within 2 milliseconds.

2.11 EXPLOSIVE DETECTION KIT (EDK)

The increasing use of explosive for criminal and terrorist activities targeted at chosen civilian installations and populated areas is leading to large scale destruction and heavy casualties. This has necessitated a method for screening of suspected objects. DRDO has developed an inexpensive, handy, user-friendly kit to meet field requirement of fast detection and identification of explosives at micro levels. Its operation does not need a knowledge of chemistry or any advanced training,. Explosive detection kit is meant for detection and identification of explosives planted in vulnerable civil areas by police/Home ministry. Kit can detect and identify explosives based on any combination of nitroesters, nitramines, nitrotoluene (TNT), dynamite even up to 0.0001 g or black powder. Reagents used are ready to use type. Testing can be carried on hands, besides on porcelain plate. The kit is being exported to other countries. (Fig.3 EDK).

2.12 AIR RE-GENERATING COMPOSITION

It is a chemical mixture, which regenerates air inside a confined space by liberating oxygen and absorbing carbon dioxide simultaneously and thus maintains the breathable air within restorable limit. The active ingredient of this composition is Potassium super oxide (KO_2). This composition already developed for Indian Navy in the form of thin sheet to regenerate air inside submerged submarine and in the form of granules for self contained breathing apparatus by divers during underwater operations etc. can be used in sealed battle tank against NBC warfare and for high altitude applications and in primary health centres, toxic gas chemical plant operations, manned space craft, medical oxygen in inaccessible places, underwater habitant/mineral explorations, underground public shelter/control rooms, oxygen for gas cutting in inaccessible environments etc. This composition has been developed and manufactured on bulk scale by HEMRL and technology has been adopted to meet commercial requirements.

2.13 RIOT CONTROL AGENTS (NON- LETHAL CHEMICALS)

Non-lethal weapons (NLWs), disabling munitions (DMs), non-lethal disability techniques (NDT), less than lethal weapons (LLWS) or low collateral damage munitions (LCDMs) are assuming high importance for riot and mob control and to help law reinforcing agencies. These are smeant to prevent normal operation of personnel and equipments and cause temporary human impairment. The technologies range from laser weapons to relatively simple compounds. Low energy laser weapon, Isotropic radiators, high power microwave, Infra sound and polymer agents have been considered for this purpose. Polymer leased super adhesives can be applied directly on equipments, vehicles or facilities to alter or deny their use. e.g. polymeric agents can foul air breathing internal combustion or jet engines or cooling system for power plants, communication gear and facilities. Polymer agents could be employed to glue a person to almost anything that one may think including another person.[10]

Any chemical used by police force must be non-toxic in normal concentration, yet it must produce a rapid and incapacitating physiological reaction that is readily reversible, when the affected individual is removed. Chemical agents, which have been extensively used to create physiological effect include lachrymators e.g. chloroacetophenone (CN), o-chloro benzidine malonitrile (CS), dibenzo1,4 – oxazepaone (CR). CS is more effective than older agent CN. Its effects are short lived and quicker, CN based multichaseer Tear gas grenades have been developed by HEMRL, which have been productionised. One unit of tear gas grenade cover 15 meter diameter area and action lasts upto one minute. Irritant smokes when enter nose cause sneezing followed by severe headache, nausea and physical weakness . Such smokes are being considered for riot control applications.[11]

Thus, India has used a large number of High Energy Materials and related technologies for civilian applications and has worked out a number of plans and programmes to use defence related technologies for civilian applications including composite materials, electronic based devices and non-conventional metal/metal alloys for heart patients (stent and pace makers), tooth implantation, saline water purification and a host of similar applications.

3. References

1. Garg, KC, Dutt Bharvi, «Out look on Indian S&T »National Institute of Science & technology and development studies, Delhi – Nov'97 (Rep- 226:97).
2. Annual Report, Commercial Explosives Department (Ministry of Industry), 1992.
3. Schubert H, Plans, programs and challenges in the destruction of conventional weapons: Dismantlement and destruction of chemical, nuclear and conventional weapons, 103-184 (1997).
4. Cook M.A., The science of Industrial Explosives. Graphic Service and supply Inc. USA(1974).
5. Singh H., Development of new explosive products to meet changing needs of India. National seminar on Civil explosive technology (1992).
6. Machacek O; Eck G.R. Waste propellants and smokeless powders as ingredients in commercial explosives. 23rd International conference of ICT (1992).
7. Singh H., High energy materials research and development in India J Propulsion and Power Vol.11(4), (1995).
8. Surve, RN; Studies on the preparation and properties of Basic Lead Azide. Ph.D. Thesis, Pune University (India) ,(April '1992).
9. Indian Defense Technology – A DRDO publication 1996
10. Evancoe, P, Non-lethal technologies enhance warrior's punch. National Defense (Dec 1993)
11. Proceedings of 16th Int. pyrotechnics Seminar, Sundbyberg, Sweden (1991).

MODELLING OF FIRE EFFECTS ON EQUIPMENT ENGULFED IN A FIRE

E.PLANAS-CUCHI, J. CASAL
Centre d'Estudis del Risc Tecnològic (CERTEC)
Department of Chemical Engineering
Universitat Politècnica de Catalunya
Diagonal 647. 08028-Barcelona. Catalonia, Spain

1 Introduction

Amongst the different major accidents which can happen in process plants and in the transportation of hazardous materials, fire is the one which has its effects and consequences restricted to the shortest distances (except for smoke dispersion). Thermal radiation from all types of fire is limited usually to distances much shorter than the effects of explosion or toxic releases. Nevertheless, fire can lead to important major accidents: its thermal effects on containers, storage tanks or process equipment can originate further loss of containment of toxic or flammable substances, thus enlarging the scale of the accidental scenario.

When fire affects the aforementioned equipment, the situation is significantly different from that found when there is a fire in a building and, consequently, the management of the emergency usually follows different directives. If there is a fire in a process plant, the eventual direct effects on people (deaths or injured persons) will take place probably at the beginning of the accident; victims will be associated to an explosion, a flash fire or something similar which does not give any possibility of escape over a certain area. After the first step of the accident, people who have survived will have had probably the possibility of evacuating the plant. From that moment, any eventual further victims will correspond to the emergency brigades, which will be the only people in the plant.

Instead, in the case of fire in a building it is possible that during considerable time the efforts of firemen be devoted to rescue the people trapped in the building. During the control and extinction of fire in a building, fire-fighters will face what could be called "classical" risks (smoke, collapse, etc.). However, in a process plant or in a storage area the situation can be much more complicate. Probably fire will have thermal effects on some equipment, heating it and thus creating a supplementary danger. If there is flame impingement on the surface of this equipment, the situation can be really dangerous, as heat fluxes will be much larger and the risk of further explosions or fire is high. The heat flux, which can be of the order of 100 kW/m^2 in the case of pool fire, can reach values up to 350 kW/m^2 in the case of jet fire impingement.

Furthermore, these situations can have some impredictable aspects. For example, the BLEVE of a tank can take place after a very short time of flame impingement (the first BLEVE in the accident of Mexico City occurred after 59 s of fire ignition), or after half an hour, or after several hours, or it may not occur even after hours of flame effects. A boil-over may occur after many hours of fire-fighting.

V.E. Zarko et al. (eds.), Prevention of Hazardous Fires and Explosions, 273–286.

Fire is the most frequent accident in process plants and in transportation of hazardous materials. A historical analysis carried out on 5325 accidents and 6168 entries[1] gave the results in Table 1 (the sum of the percentages is greater than 100, as more than one of the hazards considered may exist in any given accident).

Table 1. Types of accidents

Type	*Number of entries*	*%*
Loss of containment	3022	51
Fire	2603	44
Explosion	2133	36
Gas cloud	719	12

Recently, another historical analysis of fire accidents in process plants and in the transportation of hazardous materials[2] (7029 entries corresponding to 6099 accidents that occurred up to the end of 1993, taken from the databases MHIDAS) has given the following distribution concerning the place or the activity in which the accident took place (Table 2).

Table 2. Origin of the accident

Origin	*Number of entries*	*% of total*
Process plant	779	28.4
Transport	747	27.2
Storage plant	574	20.9
Loading/unloading	193	7.0
Warehouse	191	7.0
Domestic/commercial	164	6.0
Waste storage	28	1.0
Unknown	69	2.5
Total	2745	100.0

The most important contribution corresponds to accidents in process plants (28.4%), followed by accidents in transportation (27.2%) and in storage plants (21%). Approximately 50% of fire accidents occur in process or storage plants. These three large contributions are followed by loading/unloading (7%); although these operations are usually considered delicate and dangerous, the frequency of accidents in them is significantly high. Concerning the type of fire accident, the aforementioned historical analysis gave the data summarised in Table 3. It must be taken into account that most materials handled by the process industries are fluids, and that amongst these most are liquids. As many of these liquids are flammable, especially in certain processes, it is relatively frequent for the loss of containment of one of them to give rise to a flammable pool.

Table 3. Distribution of fire accidents according to the general type of fire.

Type	*Number of entries*	*% of type*
Undetermined fire	2413	87.4
Pool fire	112	4.1
Tank fire	111	4.0
Flash fire	98	3.5
Jet fire	15	0.5
Fireball	11	0.4
Fire storm	1	0.1
Total	2761	100.0

Taking into account both the frequency of fire accidents and the severity of their consequences, the knowledge of the effects of fire on equipment is an important aspect to increase the safety of certain industrial activities.

2. Fire "secondary" effects: an example

As stated before, the effects of fire heating on equipment can lead to further loss of containment of hazardous materials, thus enlarging the scale of the accident: from a distance of 50-100 m reached by thermal radiation, the accident can cover much greater distances if explosions or toxic releases occur. As an example, a significative case will be briefly described in the next paragraphs.

In 30 January 1996 a ship, with a cargo of diverse chemicals, entered the harbour of Barcelona (11.20 am.). There was a small fire on board, in a container containing sodium hydrosulfite. The ship had found bad weather and the container moved; probably it became cracked and some water entered in it, reacting with the chemical. When the firemen arrived at the pier, it was already known that one of the containers engulfed by the fire contained hexane drums. More containers were located near the fire: two of them containing plaguicides, two containing monomer and one containing toluene diisocyanate. It must be pointed out that the separation criteria (keeping flammable products apart from those that can produce toxic gases) were not followed, infringing the corresponding rule (IMDG). However, all these containers — except the one containing hexane — could be removed in spite of the fire.

The firemen were not able to extinguish the fire, although three ships were also pumping water on the containers. The hexane drums container was then involved in the fire: at 13.18 h the first explosion took place, caused by the bursting of a hexane drum. A series of explosions took place from time to time, increasing the fire and giving rise to fireballs At 16.27 h one of the explosions opened the doors of the hexane container, resulting in the enlargement of the flames.

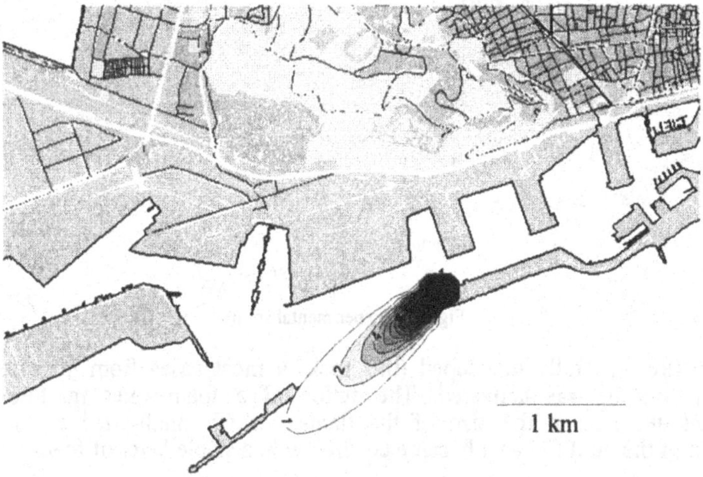

Figure 1. Dispersion of the SO_2 cloud originated by the fire.

As a result, the fire gave rise to a large smoke plume from 11.30 until 19.30 (when the fire was finally extinguished). This plume extended to the harbour inlet, as the wind was essentially NE. The smoke did not affect either the harbour buildings or the town, being dispersed along the evacuated pier and over the sea. This smoke was a mixture of gases. The main toxic component was SO_2; the estimation of the toxic dispersion was therefore based on this gas. The simulation performed[3] for the atmospheric situation allowed the estimation of the variation of concentration as a function of distance (Figure 1).

This simulation has shown that even if the wind had blown towards the town (all other meteorological conditions remaining constant) the emission of gases would not have posed any danger for the population. However, this accident — originated by a relatively small fire — produced a significant alarm in the population and gave important problems to the harbour authority.

3. Experimental tests

A series of large-scale fire tests were conducted at the facility of SP in Borås. The tests were carried out within a specially constructed rig comprising a process tank . Two sides were closed by bulkheads and a series of deluge manifolds were installed. 18 thermocouples were installed on the tank surface (14 on the inner surface and 4 on the external surface), and one radiometer was set up at 6 m (Figure 2). Eight tests were carried out in which a horizontal cylindrical tank (diameter = 1.2 m, length = 3 m) was completely engulfed in a pool fire. Four tests were performed with hexane (with a pool surface of 4 m^2) and four with kerosene (12 m^2). A detailed description of this experimental set-up can be found elsewhere[4].

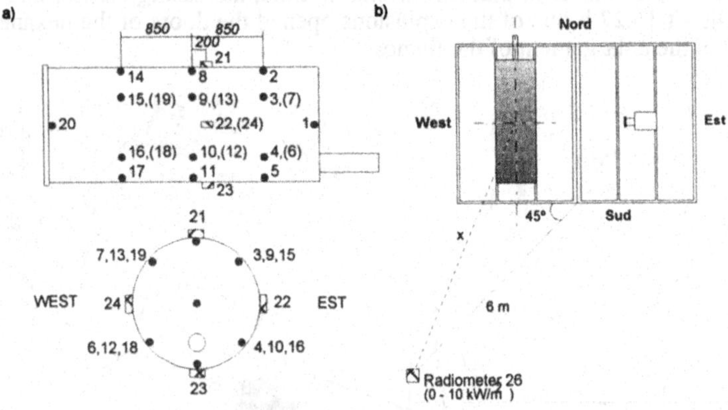

Figure 2. Experimental set-up.

Once the fire was fully developed (this took in most cases from 40 s to 60 s) an extinction system was activated. The action of a foam-water mixture caused a significant decrease in the size of the flames, which finally led to the complete extinction of the pool fire as it became covered with a stable layer of foam.

4. Experimental results

The evolution of the fire during its development and during the extinguishing process can be seen in Figure 3, where the temperature of one of the thermocouples has been plotted versus time. This diagram establishes in a relatively clear way the existence of five stages[4] in the evolution of temperature at the diverse points of the tank wall.

The first one, which corresponds to the initial development of the fire, is characterised by a rapid increase in temperature; in the plot of temperature versus time a practically straight line is obtained with a steep slope (the heating rate depends on the location). In the second stage, which corresponds to a fully developed pool fire, temperature continues to increase but at a slower rate and with a trend which is no longer linear; in this stage, a considerable wind appeared in all the tests, increasing flame turbulence and modifying its shape (Figure 4).

The third stage initiates with the start-up of the cooling/extinguishing system. The temperature starts to decrease — its maximum value having been reached at the transition between these two steps — gradually, at an approximately constant rate.

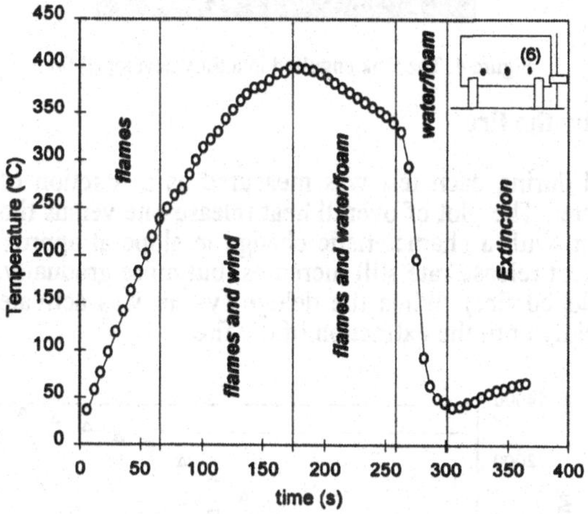

Figure 3. Stages of the temperature evolution at the lower tank laterals (thermocouple 6; hexane pool-fire, 4 m^2).

The fourth stage corresponds to a situation in which, due to the blanqueting action of the foam over the pool, the magnitude of the flames has decreased considerably and they no longer reach the point in the tank where the temperature is being measured. Now the heat balance is clearly influenced by the dominant mechanism of the foam/water cooling, with a relatively weak heat input by radiation; thus, temperature decreases abruptly, following an approximately linear trend. Finally, once the fire is extinguished, gentle heating is observed due to radiation from rig walls.

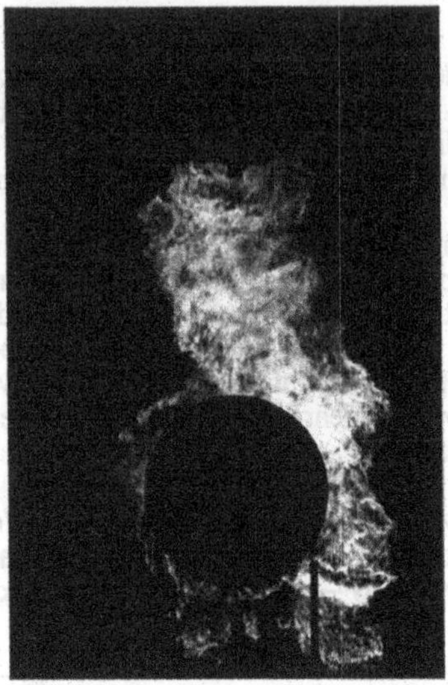

Figure 4. The tank engulfed in a fully developed fire.

5. Heat released by the fire

The heat released during each test was measured as a function of time by using an adiabatic calorimeter. The plot of overall heat release rate versus time (Figure 5) shows a steep rise at first, with a characteristic change in slope at approximately 30 s; from this moment the heat release rate still increases, but more gradually, up to a maximum value (fully developed fire). When the deluge system was activated, the heat release rate decreased rapidly until the extinction of the fire.

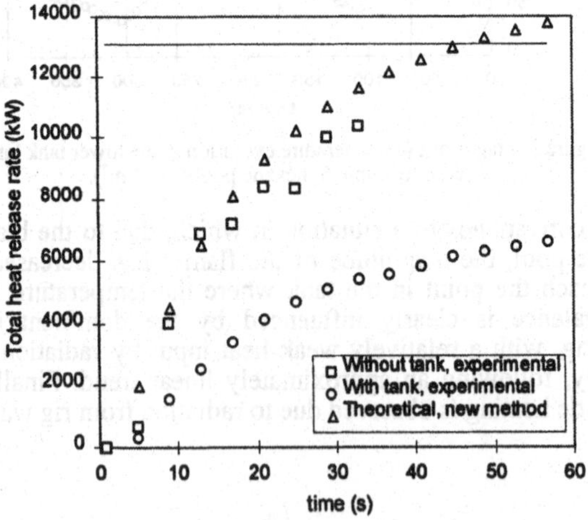

Figure 5. Total heat release rate as a function of time for tests number 13 and 9 (4 m^2, hexane, with and without tank respectively). The horizontal line corresponds to the theoretical heat release rate at steady state, calculated according to Babrauskas's equation

The experimental values were significantly lower than the theoretical values which should be expected in a pool fire without obstacles. The existence of the tank is in fact an obstacle for the flames. When there is some equipment engulfed in the fire, this equipment has a hindering or obstructing effect on the fire. Thus, both *a "hindering factor"* (ξ) and an *"efficiency of hindered combustion"* (η_{obst}) were defined[4]:

$$\xi = \frac{\textit{Heat released in the combustion of a pool with an obstacle}}{\textit{Heat released in the combustion of the same pool without any obstacle}}$$

$$\eta_{obst} = \eta \cdot \xi = \frac{\textit{Heat released in the combustion of a hindered pool}}{\textit{Heat released in the complete combustion of all the fuel evaporated in a free pool}}$$

The hindering effect of equipment engulfed in fire can be observed from the data plotted in Figure 5, which shows experimental data corresponding to the heat release rate for the pool fire with and without a tank engulfed in it. Furthermore, the maximum theoretical heat release rate (η = 100%) has also been plotted, calculated according to the method proposed by Planas-Cuchi et al.[4]. To check this method, the following data can be considered: the heat release rate at steady state (pool fire of hexane, 4 m^2) calculated from the combustion rate predicted by Babrauskas equation[5] is 13045 kW; by applying the new method — which can be used even during the unsteady state — a value of 13807 kW is obtained at t = 60 s (after reaching the stationary state). The small difference (5.5% with respect to the highest value) confirmed the validity of the method. Figure 5 shows that for a free surface pool fire, the steady-state would be reached after approximately 45 s, with a combustion efficiency of approximately 88%.

For pools with a diameter of 1 m or more, the radiation mechanism is the only one which is really important in the heat transfer from the fire to the pool. Therefore the rate at which heat is transferred to the pool is given by:

$$\frac{Q}{\left(\pi d^2/4\right)} = \sigma \cdot F \cdot \left(T_f^4 - T_a^4\right) \cdot \left(1 - e^{-Kd}\right) \qquad (1)$$

The theoretical combustion rate can be obtained from this expression and from the heat required to evaporate the fuel:

$$\dot{m}''_f = \frac{\sigma \cdot F \cdot \left(T_f^4 - T_a^4\right)}{\Delta H_v + c_p \cdot \left(T_b - T_a\right)} \cdot \left(1 - e^{-Kd}\right) \qquad (2)$$

6. Flame shape and temperature

To calculate the radiation from a fire it is necessary to know the shape of the flames. For hydrocarbon pool-fires most authors have assumed the fire to be a cylindrical body or, if the pool is a rectangle or a square, a parallelepipedical body; these bodies can be tilted in case of wind. Tunç et al.[6] improved this modelization by assuming a more realistic shape defined by axisymmetric contours obtained from photographs. Planas-Cuchi and Casal[7] proposed another shape (Figure 6) much more adequate for rectangular pool-fires, obtained from video recordings of experimental tests. The analytic flame contour corresponds to the following expression:

$$\left\langle \begin{array}{ll} g(y) = 0.449 + 0.734 \cdot y - 0.383 \cdot y^2 + 0.043 \cdot y^3 & \forall\, z > 0 \\ h(y) = -0.449 - 0.734 \cdot y + 0.383 \cdot y^2 - 0.043 \cdot y^3 & \forall\, z < 0 \end{array} \right\rangle \tag{3}$$

Concerning to flame temperature, it is usually assumed to be constant with time and height for a given fire[8,9]. However, the experimental tests effectuated in the facility previously described (Figure 1) have demonstrated that this parameter changes with time, as fire develops, until reaching a maximum value for the fully developed fire.

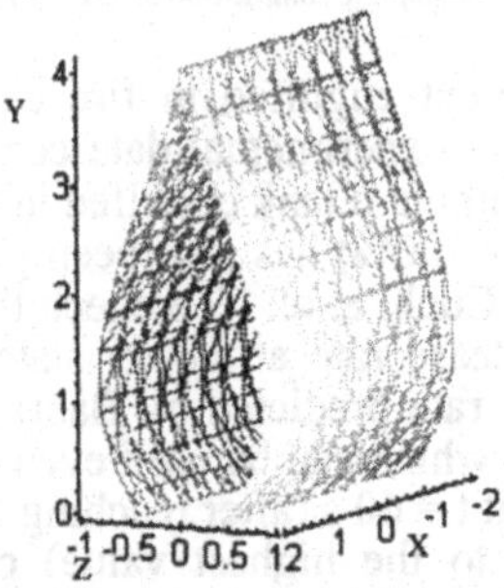

Figure 6. Flame contour observed for a 4 m² pool-fire.

Furthermore, flame temperature decreases with height. Thus, for an hexane pool-fire with a surface of 4 m², the variation of flame temperature as a function of height and time can be described by the following expression:

$$T_f(t,h) = \frac{t}{0.000851 \cdot t + 0.021 \cdot h + 0.0034} + 290 \tag{4}$$

7. Heat flux to a surface engulfed by the flames

Reference 7 shows a detailed way to calculate the radiation incident on the cylinder surface, the results are given by equation (5). Unfortunately there is not experimental data available to compare and validate the results of the proposed model. Nevertheless as the results obtained here will be applied to the model for the prediction of the temperature evolution of the equipment, this could be an indirect way to do this validation.

$$Q = \int_{V_i} \int_{A_j} \frac{K \cdot \sigma \cdot T_f^4 \cdot \cos(\phi_{ij}) \cdot \exp(-K \cdot S_{ij})}{(S_{ij})^2} dV_i dA_j \tag{5}$$

The integral can be solved numerically using a multidimensional gaussian quadrature. For a point placed along the length of the cylinder, $cos(\phi)$ and S take the following values:

$$\cos(\phi) = \frac{y - h_3 - r}{S} \tag{6}$$

$$S^2 = (y - h_3 - r)^2 + (x - a)^2 + z^2 \tag{7}$$

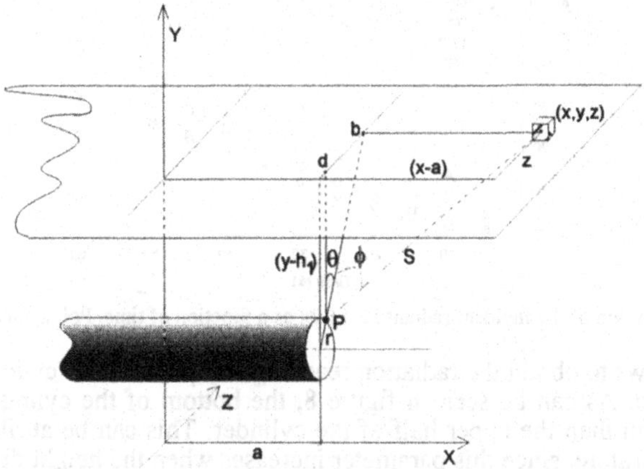

Figure 7. Scheme showing the radiation incident on a point located around the circumference of the cylinder[7].

For a point placed around the circumference of the cylinder, the following expressions apply (Figure 7):

$$\cos(\phi) = \begin{cases} \dfrac{(z - r \cdot sin(\theta))}{S \cdot sin(\theta)} & \text{for } S \cdot \cos(\phi) > \overline{Pb} \\ \dfrac{(h_3 + r \cdot \cos(\theta) - y)}{S \cdot \cos(\theta)} & \text{for } S \cdot \cos(\phi) \le \overline{Pb} \end{cases} \tag{8}$$

$$S^2 = (y - h_3 - r \cdot \cos(\theta))^2 + (z - r \cdot sin(\theta))^2 + (x - a)^2 \tag{9}$$

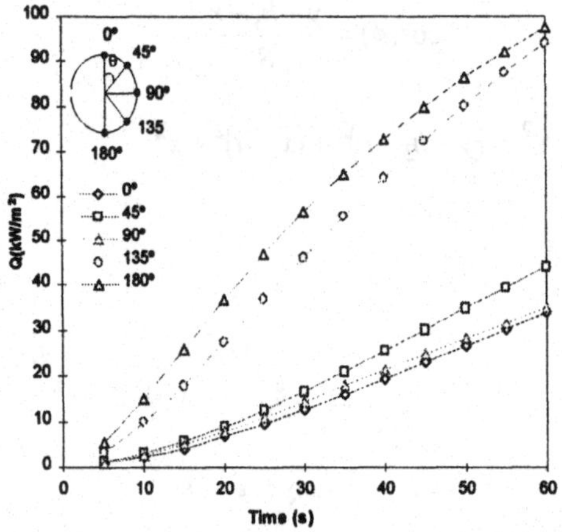

Figure 8. Evolution of the incident radiant heat flux as a function of time. Points located at a = 1m.

This model allows to obtain the radiation reaching any point of the cylinder surface as a function of time. As can be seen in figure 8, the bottom of the cylinder has a higher incident radiation than the upper half of the cylinder. This can be attributed mainly to the flame temperature, since this parameter increases when the height diminishes and in equation (5) appears at the four power.

8. Modelling temperature of the cylinder surface

To find the evolution of temperature as a function of time on the diverse points of the cylinder surface, the method of finite elements has been used. The nodes distribution has been taken in such a way that all the measuring points in the experimental test have a corresponding node:

Table 4. Correspondence between nodes and thermocouples.

Node	*Thermocouple*
0	20
3	14
5	8
8	15,19
10	9,13
18	16,18
20	4,6
23	17
25	11

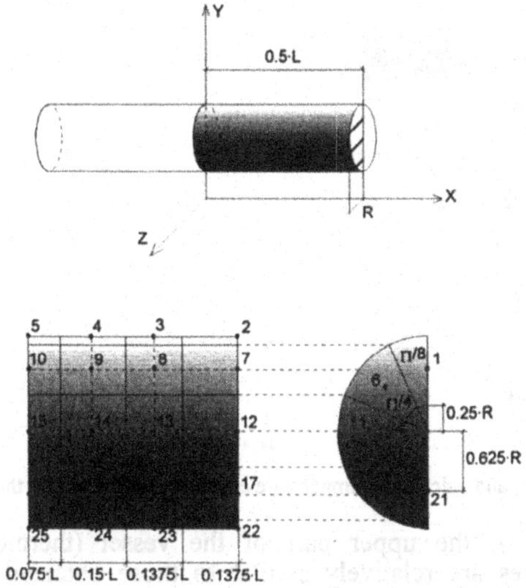

Figure 9. Scheme showing the part of the cylinder used for simulation. Nodes distribution in this part.

To represent the general equation of the implicit method that has been used in this case, it is necessary to imagine a node i surrounded by N nodes. It is also necessary to make a time discretisation then this method can be expressed by the following equation:

$$q_i + \sum_{j=1}^{N} K_{j\leftrightarrow i}\left(T_j^{(p+1)} - T_i^{(p+1)}\right) = \rho_i \cdot c_i \cdot V_i \cdot \frac{\left(T_i^{(p+1)} - T_i^{(p)}\right)}{\Delta t} \tag{10}$$

The cylindrical tank used in the experimental work was made of carbon steel, with a diameter of 1.2 m, a length of 3 m and a wall width of 4 mm. Its inner surface was not insulated. In order to apply equation (10) it is necessary to define some geometrical parameters and also the conductances $K_{j\to i}$.

9. Solution of the equations system

From the initial moment t=0, in which the temperature at each node is assumed to be the atmospheric temperature, the equation corresponding to the implicit method is written for each node for the moment $t=t+\Delta t$. Thus, a system of 26 equations and 26 unknown parameters is obtained; each equation gives the temperature at one node at the instant n as a function of the temperature of the surrounding nodes in the same instant and of the temperature of that node at the previous moment n-1. This system must be solved for each time increment taken. It was solved by using a C routine which applies a triangular factorisation method (LU decomposition).

10. Results

In general, the results obtained tend to underestimate the evolution of temperature with time on any point of the tank, especially on the lower part.

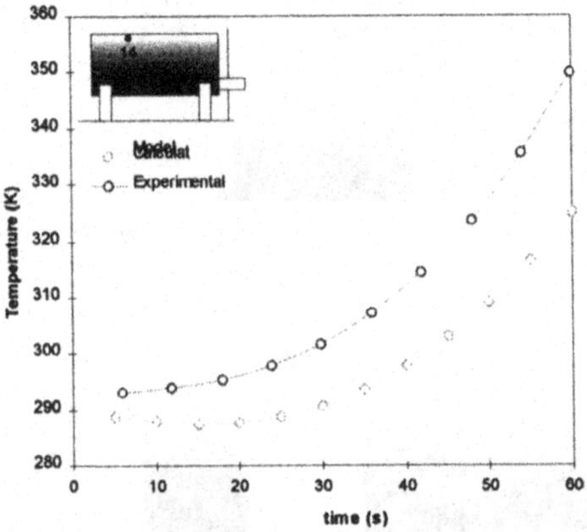

Figure 10. Experimental and calculated temperature evolution with time. for thermocouple 14.

Figure 10 shows that on the upper part of the vessel (thermocouple 14), the experimental temperatures are relatively similar to those obtained from the model (differences of 25 K, the maximum temperature being 350 K). On the lower part of the vessel, temperatures are significantly higher than those obtained from the model; the differences are approximately 230 K (the maximum temperature being 650 K).

If the values reached by the model 60 s after ignition are compared to the experimental values, for every thermocouple, it can be observed that both sets of values (figure 11) show practically the same trend with temperatures increasing from the upper to the lower part of the vessel and with temperatures from the front part of the vessel being slightly higher. Nevertheless, it must be noticed that some points have important differences (thermocouples 5, 11, 17), all of which correspond to points located on the lower part of the vessel.

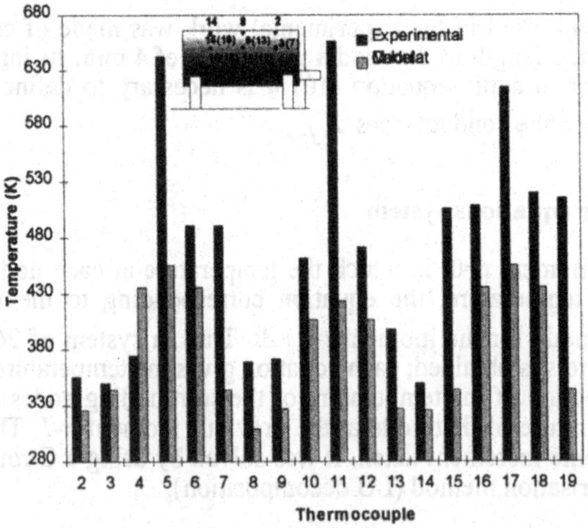

Figure 11. Temperature reached by the thermocouples on the vessel surface (60 s after ignition). The numbers in brackets correspond to thermocouples located on the west side of the vessel.

11. Conclusions

Fire accidents are the most frequent major accidents in industrial installations; amongst them, pool-fires are most common. Although thermal radiation is important only over rather reduced distances, its effects on other equipment can lead to severe consequences. Their mathematical modelling is very interesting to predict these effects.

The results of a series of experimental tests have allowed a better characterisation of pool-fires, especially in their first step (development of the fire). This information has been introduced in a new model for the prediction of the temperatures reached by equipment engulfed in the fire. Although the results from this model are somewhat lower than the experimental values, the trend in temperature distribution on the cylinder surface is very similar. Further research is now being developed at UPC to improve this model.

12. Nomenclature

A	surface area, m^2
a	x coordinate of the point under consideration, m
c_i	specific heat of the node, $J \cdot kg^{-1} \cdot K^{-1}$
c_p	specific heat of the generated gases, $kJ \cdot kg^{-1} \cdot K^{-1}$
d	pool diameter, m
F	view factor (dimensionless)
h	height, m
ΔH_v	heat of vaporisation, $kJ \cdot kg^{-1}$
K	mean absorption coefficient of the combustion gases, m^{-1}
$\dot{m}''_f$	combustion rate, $kg \cdot m^{-2} \cdot s^{-1}$
$\dot{Q}$	heat transfer rate, kW
q_i	heat reaching the node surface by radiation, $W \cdot m^{-2}$
r	distance between the centre of the cylinder and the point of interest, m
S	absorbing path length, m
T	temperature, K
T_a	room temperature, K
T_b	boiling temperature of the fuel, K
T_f	flame temperature, K
V	volume, m^3
ϕ	angle between the connecting vector and the vector normal to the surface element, rad
θ	angle between the vertical vector and the vector between the centre of the cylinder and the point of interest on the circumference of the cylinder, rad
ξ	hindering factor (dimensionless)
η	combustion efficiency (dimensionless)
η_{obst}	efficiency of hindered combustion (dimensionless)
σ	Stefan-Boltzmann constant ($=5.67 \cdot 10^{-12}$ $kW \cdot m^{-2} \cdot K^{-4}$)
ρ_i	node density, $kg \cdot m^{-3}$

13. References

1. J.A. Vílchez; S. Sevilla; H. Montiel; J. Casal; *«Historical analysis of accidents in chemical plants and in the transportation of hazardous materials»* J. Loss Prev. Process Ind. Vol. 8, No. 2, pp. 87-96, 1995.
2. E. Planas-Cuchi; H. Montiel; J. Casal; *«A survey of the origin, type and consequences of fire accidents in process plants and in the transportation of hazardous materials»*Trans IchemE, Vol. 75, Part B, pp. 3-8, 1997.
3. E. Planas-Cuchi, J.A. Vílchez, X. Pérez-Alavedra, J. Casal. "Effects of fire on a container storage system. Case study". J. Loss Prev. Process Ind. (in press).
4. E. Planas-Cuchi; J. Casal; A. Lancia; L. Bordignon; *«Protection of equipment engulfed in a pool fire»* J. Loss Prev. Process Ind., Vol. 9, No. 3, pp. 231-240, 1996.
5. V. Babrauskas. *Fire Tech.* 1983, 19, 251

6. M. Tunc; A. Karakas; *«Three-dimensional formulation of the radiant heat flux variation on a cylinder engulfed in flames»* Journal of Heat Transfer – Transactions of the ASME, Vol. 107, pp. 949-952, 1985.
7. E. Planas-Cuchi; J. Casal; *«Modelling temperature evolution in equipment engulfed in a pool fire»* Fire Safety Journal (in press), 1998.
8. J.J. Gregory; N.R. Keltner; R.Jr. Mata; *«Thermal measurements in large pool fires»* J. Heat Transfer – Transactions of the ASME, Vol. 111, pp. 446-454, 1989.
9. M. Tunc; J.E.S. Venart; *«Incident radiation from an engulfing pool-fire to a horizontal cylinder»*Fire Safety Journal, 8, pp. 81-87, 1984.

MATHEMATICAL MODELING OF CATASTROPHIC EXPLOSIONS OF DISPERSED ALUMINUM DUST

A.V. FEDOROV, V.M. FOMIN, T.A. KHMEL'
Institute of Theoretical and Applied Mechanics SD RAS, 630090, Novosibirsk, Russia

1. Introduction

Technologies using finely dispersed powders of combustible substances (metal particles) as working bodies are widely used in industry. The study of explosibility and detonation ability of these gas mixtures is necessary to prevent industrial explosions with catastrophic consequences. Thus, very important is the development of mathematical models verified with experimental data, and the study of the conditions of emergence and propagation of detonation waves under the influence of various gasdynamic and physical factors.

Review of the results on heterogeneous detonation of an aluminium dust in oxygen is given in the present paper. The stationary and non-stationary problems of detonation wave (DW) propagation are investigated on the basis of the mathematical model developed in Refs. [1,2]. This model is in agreement with the experimental data [3] by the correlation between the DW velocity and initial particle concentration. The combustion process is described with the equation of Arhenius type taking into account incomplete burning of the particles. An analysis of the flow types is given in the form of Chapman-Jouguet (CJ) regime, strong DW, and weak DW in one-velocity two-temperature approximation and with regard for the differences between gas and particle velocities. Stability with respect to small and finite disturbances of the all stationary DW types supported by the piston motion is shown by means of numerical modeling of non-stationary equations. The problem of interaction of the DW and adjacent rarefaction wave (RW) is investigated at each detonation regime and self-sustained regimes are determined. Some new results are concerned with the initiation problem of heterogeneous detonation within the framework of two-velocity two-temperature mathematical model.

2. Governing System of Equations

The propagation of plane detonation waves within the framework of the two-velocity two-temperature model of mechanics of reacting heterogeneous media is described by the following system of equations:

V.E. Zarko et al. (eds.), Prevention of Hazardous Fires and Explosions, 287–299.

$$\frac{\partial\rho_1}{\partial t}+\frac{\partial(\rho_1 u_1)}{\partial x}=J, \qquad \frac{\partial\rho_2}{\partial t}+\frac{\partial(\rho_2 u_2)}{\partial x}=-J$$

$$\frac{\partial\rho_1 u_1}{\partial t}+\frac{\partial(\rho_1 u_1^2+p)}{\partial x}=-f+Ju_2, \quad \frac{\partial\rho_2 u_2}{\partial t}+\frac{\partial(\rho_2 u_2^2)}{\partial x}=f-Ju_2 \tag{1}$$

$$\frac{\partial\rho_1 E_1}{\partial t}+\frac{\partial[\rho_1 u_1(E_1+p/\rho_1)]}{\partial x}=-q-fu_2+JE_2,$$

$$\frac{\partial\rho_2 E_2}{\partial t}+\frac{\partial(\rho_2 u_2 E_2)}{\partial x}=q+fu_2-JE_2$$

$$p=\rho_1 RT_1, \quad E_1=\frac{u_1^2}{2}+c_{v,1}T_2, \quad E_2=\frac{u_2^2}{2}+c_{v,2}T_2+Q.$$

$$J=\frac{\rho}{\tau_\xi}(\xi-\xi_\kappa)\exp(-E_a/RT_2)\max(0,sign(T_2-T_{ign})), \tag{2}$$

$$f=\frac{\rho_2}{\tau_u}(u_1-u_2),\ q=\frac{\rho_2 c_{v,2}}{\tau_T}(T_2-T_1). \tag{3}$$

Here p is the pressure, T is the temperatures, Q is the heat released in the chemical reaction, ρ_i, u_i, E_i, $c_{v,i}$ are the macroscopic density, velocity, total energy per mass unit, and heat capacity of the i-th phase, respectively. The subscripts 1, 2 refer to the gas and the particles, the subscript κ indicates the final state. The combustion law (2) corresponds to that in Refs.[1, 2], where $\xi=\rho_2/\rho$ is the relative mass concentration of particles and $\rho=\rho_1+\rho_2$, E_a is the activation energy, ξ_κ is the fraction of unburned particles, T_{ign} is the ignition temperature, and τ_ξ is the characteristic combustion time. The characteristic times of velocity (τ_u) and thermal (τ_T) relaxation processes in the general case are variable and depend on the flow parameters.

3. Stationary Detonation Problem. One-Velocity Approach

Let us consider the process of a DW stationary propagation in a mixture of oxygen and small aluminium particles. The governing equations (1-3) at one-velocity two-temperature approach in front-fitted system can be reduced to two ordinary differential equations

$$\frac{dM}{dx}=\frac{(\gamma_1-1)(1+\gamma_1 M^2)^3}{2A_0^2\, M(M^2-1)}\left[\dot{\xi}(c_{v2}T_2+Q)+\xi c_{v2}\dot{T}_2\right],$$

$$\frac{d\xi}{dx}=-\beta(\xi-\xi_k)\exp(-E_2/RT_2), \quad A_0=c_0\frac{1+\gamma_1 M_0^2}{M_0}. \tag{4}$$

Here $\dot{T}_2 = dT_2 / dx$, $\dot{\xi} = d\xi / dx$, $\beta = \tau_T / \tau_\xi$, $dx = dy / u\tau_T$, $M = M_f$, (M_f, M_e are the Mach numbers, γ_1 and γ are the frozen and equilibrium adiabatic exponents [6]. The indexes f and e stand for frozen and equilibrium values, respectively.

We formulate a problem for (4): find M, ξ (discontinuous and continuous functions) and constants c_0, β such that a solution of the boundary value problem

$$\dot{M}, \dot{\xi} \to 0, \quad M \to M_{fk}, \quad \xi \to \xi_k < \xi_0 < 1 \text{ at } x \to +\infty,$$
$$M = M_0, \quad \xi = \xi_0 \text{ at } x = -0; \tag{5}$$

exists in the form of Chapman-Jouguet solution (CJ) ($M_{fk} < 1, M_{ek} = 1$), or in the form of solution of strong ($M_{ek} < 1$) or weak detonation ($M_{ek} > 1$) type.

Singular Points. The internal singular point arises when the numerator and denominator of the first equation from (4) are equal to zero. It is the sonic point in terms of frozen sound velocity. For the values of Q used in accordance to experimental data [3] this point is always a saddle. The study of the type of final equilibrium state in general case $u_k \neq c_e$ shows that at $u_k > u_f$ or $u_k < u_e$ the final point is a stable node and at the interval $u_k \in (u_e, u_f)$, $\lambda_1 \cdot \lambda_2 < 0$ it is a saddle, where u_e, u_f are defined from initial state [6].

Existence of Stationary Solutions. The influence of the parameter $\beta = \tau_T / \tau_\xi$ on the flow behaviour is investigated. A chart of detonation regimes (Fig. 1) in the plane (M_{e0}, β) is constructed on the basis of the numerical calculations. Domain I (strong detonation regimes) is bounded below at $\beta \in (0, \beta_m)$ by the line $M_{e0} = M_{CJ}$ (normal CJ regimes) and the curve $\beta = \tilde{\beta}(M_{e0})$, which have the asymptote $M_{e0} = M_{e*}$ at $\tilde{\beta} \to \infty$. Weak detonation regimes with final saddle point are determined on the manifold $\beta = \tilde{\beta}(M_{e0})$ at $M_{CJ} < M_{e0} < M_{e**}$; weak and strong detonation regimes with internal singular point are determined on this manifold at $M_{e0} > M_{e**}$. There are no stationary solutions in domain II.

4. Stationary Detonation Problem. Two-Velocity Approach

The governing equations including the equation of particle burning, the equations of particle velocity and temperature relaxation processes, and the equations of state in form (1-3) are written by the following way:

$$\rho_1 u_1 + \rho_2 u_2 = c_1, \quad \rho_1 u_1^2 + \rho_2 u_2^2 + p = c_2,$$
$$\rho_1 u_1 E_1 + \rho_2 u_2 E_2 + p u_1 = c_3, \tag{6}$$

$$\frac{d(\rho_2 u_2)}{dx} = -J, \quad u_2 \frac{du_2}{dx} = \frac{1}{\tau_u}(u_1 - u_2), \quad u_2 \frac{dT_2}{dx} = \frac{1}{\tau_T}(T - T_2), \tag{7}$$

$$p = \rho_1 R T_1, \quad E_1 = \frac{u_1^2}{2} + c_{v1} T_1, \quad E_2 = \frac{u_2^2}{2} + c_{v2} T_2 + Q. \tag{8}$$

Here, the constants c_1, c_2, c_3 are found from the initial state parameters.

The following boundary conditions are specified. The initial state is set for $x < 0$; at $x = +0$ the gas parameters are determined from the frozen shock wave conditions with regard for $u_2 = u_0, T_2 = T_0, \xi = \xi_0$; at $x \to +\infty$ the equilibrium conditions for burning-out are satisfied $u_2 = u_1, T_2 = T_1, \xi = \xi_K$. In the CJ regime, the final velocity of the mixture equals the equilibrium sound velocity $u_K^2 = c_e^2 = \gamma p / \rho = \gamma(1-\xi)RT$. The completely frozen speed of sound equals to the propagation velocity of small disturbances in the gas $c_f^2 = \gamma_1 p / \rho_1 = \gamma_1 R T_1 > c_e^2$.

The properties of solutions to (6-8) depend on two relaxation parameters $\alpha = \tau_T / \tau_u$ and $\beta = \tau_T / \tau_\xi$ and will be analysed throughout the entire plane (α, β).

Final Singular Point. In weak detonation regimes the final velocity is higher than c_e. The DW velocity being changed from the CJ velocity to a certain critical value, $M_{CJ} < M_0 < M_*$ (here M_0 is the Mach number of DW velocity), the final state remains subsonic with respect to the frozen speed of sound ("dispersion" interval). Thus, in "dispersion" interval the final state is always unstable (saddle, saddle-focus). In weak detonation regimes beyond the "dispersion" interval the final points are stable (node or node-focus).

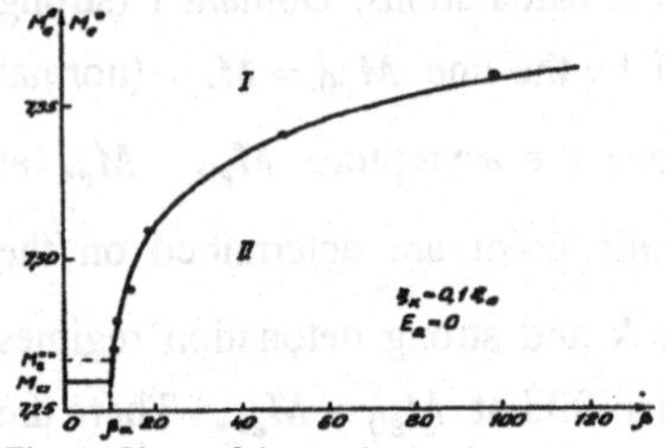

Fig. 1. Chart of detonation regimes in one-velocity model.

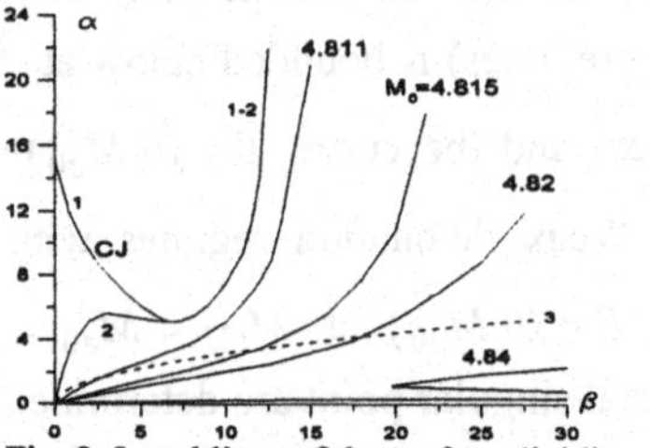

Fig. 2. Level lines of the surface dividing stationary regimes in two-velocity model.

Steady Regimes. Not all final states can be achieved because a "choking" line can appear in the solution at $M_0 < \tilde{M}(\alpha, \beta)$. The surface $M_0 = \tilde{M}(\alpha, \beta)$ separates the detonation regimes in the space (M_0, α, β). The solutions above the surface correspond to the strong detonation regime. Regimes (strong and weak) with an internal singular point are realised on the surface $M_0 = \tilde{M}(\alpha, \beta)$ that separates stationary regimes from the unsteady ones. The surface is presented in Fig. 2 by the level lines for various M_0. The CJ solutions exist in the domain located above curve 1. The curve has an asymptote $\beta \to \beta_m$, $\alpha \to \infty$. Beginning with a certain $M_0 > M_{CJ}$, the level lines are split into two branches and then move away from the point (0,0). At $M_0 \geq M_{**}$ steady-state solutions (strong detonation regimes) exist within the entire plane of relaxation parameters.

Within the region of steady regimes the final singular points turned out to be always stable at CJ and strong detonation regimes. In weak detonation regimes they are stable at $M_0 > M_* > M_{CJ}$ and unstable at $M_{CJ} < M_0 < M_*$ («dispersion» interval)

Flow structure Behind the Front. The initial parameter values were taken as in [1,2] in accordance to [3]. The characteristic time of combustion varied from 0.04 ms to 0.1 ms. Strictly speaking, the characteristic times of thermal and velocity relaxation are variable quantities,

$$\tau_T = \frac{d^2 \rho_{22} c_{v2}}{6\lambda_1 Nu}, \tau_u = \frac{4d \ \rho_{22}}{3c_D \rho_{11}|u_1 - u_2|} \tag{9}$$

(ρ_{22} is the microscopic density of particles, λ_1 is the heat conductivity of the gas, d is the particle diameter, $Nu = 2 + 0.6\,\mathrm{Re}^{1/2}\,\mathrm{Pr}^{1/3}$, c_D is the particle drag coefficient). The computations were carried out both for constant values of α, β and with regard for their dependence on the flow parameters.

A specific feature of the flow is the presence of a narrow layer of elevated concentration of particles (ρ-layer) (Fig.3, α=6, β=6, τ_ξ=0.1 ms) and a characteristic inflection in the gas temperature profile (Fig. 4, lines 1-3). At first, the temperature of gas increases behind the DW front, which is related to the heat release due to friction forces. Then some portion of heat is consumed on particles heating, and the gas temperature slightly decreases.

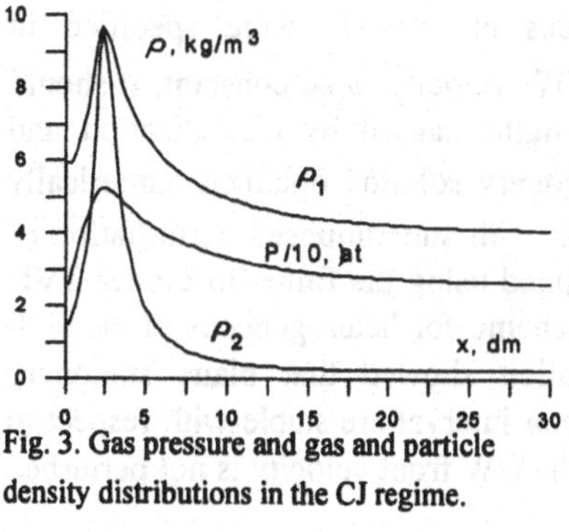

Fig. 3. Gas pressure and gas and particle density distributions in the CJ regime.

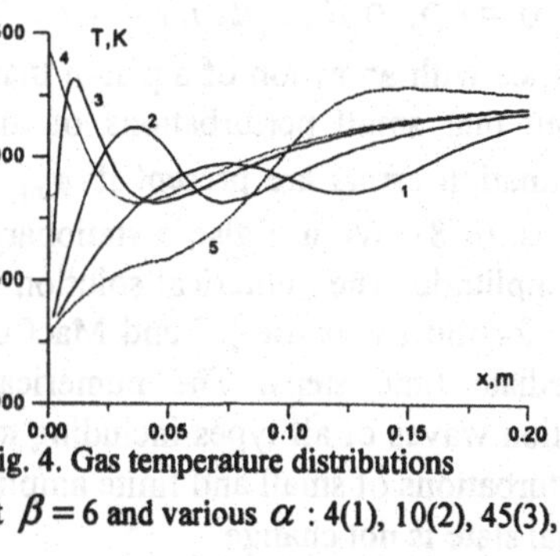

Fig. 4. Gas temperature distributions at $\beta = 6$ and various α : 4(1), 10(2), 45(3), 1(5).

When the particles achieve the ignition temperature, combustion begins and the temperatures of gas and particles increase. The presence of a local maximum of gas temperature at the stage of ignition delay and a local minimum at the moment of ignition is observed for $\alpha > 1$. Line 4 corresponds to the one-velocity model. At $\alpha \leq 1$ the *T(x)* dependence in the relaxation zone is monotonic without local maximum (line 5 in Fig. 4 for α =1, τ_ξ=0.01). The ρ-layer appears under conditions when $\alpha / \beta \geq 1$ for

α, β from 0 to 1 (the characteristic time of velocity relaxation is smaller than the characteristic time of combustion); or $\alpha^2/\beta \geq 1$ for $1<\alpha \leq 7$ (the characteristic time of velocity relaxation is smaller than the mean geometric value of the characteristic times of thermal relaxation and combustion), or at $\alpha > 7$. The functional dependence of the ignition delay zone width l_{ign} on the relaxation parameters has the form $l_{ign} = l_0\beta/\sqrt{\alpha} = u_0\sqrt{\tau_T\tau_u}F_0$, where F_0 depends only on the initial state parameters and on T_{ign}. For $T_{ign} = 900$ Ê (used in the basic computations in agreement with experimental and calculated results [4]), $F_0 = 0.495$.

The flow parameters were calculated using formulae (9) for τ_u, τ_T. The CJ regime doesn't exist here. As the detonation velocity is increased, a weak regime with an internal singular point is formed, its final state being fully supersonic. It turned out that constant values α=4.6, β=20 give the same width of ignition and combustion zones as in the case of use Eq.(9). Despite the differences in the relaxation zone, caused by velocity non-equilibrium and attendant variability of τ_u, τ_T, the typical flow features are the same: a ρ- layer and $T(x)$ - inflection are observed.

5. Investigation on Stability of Stationary Detonation Regimes

Stability of the stationary solutions has been studied by direct numerical modelling of non-stationary equations. The following Cauchy problem is posed for Eqs. (1-3)

$$t = 0, \quad \phi = \begin{cases} \phi_0, & 0 < x < l, \\ \phi_{ini}(x), & -L < x \leq 0, \end{cases} \tag{10}$$

where $\phi = (\rho_i, \rho_i u_i, \rho_i E_i), i = 1,2$. The values at $x = -L$ were specified in accordance with an action of a piston that retain DW velocity to be constant. It should be noted that small perturbations of all wavelengths caused by discretization and approximation errors are present if ϕ_{ini} is a stationary solution obtained numerically from Eqs. (6-8). We use also a stationary solution with superimposed perturbation of finite amplitude. The numerical solution was obtained using the finite-difference TVD scheme for the gas phase [5] and MacCormack scheme for heterogeneous phase with intermediate time steps. The numerical integration showed that plane stationary detonation waves of all types including «dispersion» interval are stable with respect to 1D perturbations of small and finite amplitude, if the DW front velocity is not perturbed and final state is not change.

6. Interaction with Adjacent Rarefaction Wave

Initial - Boundary Value Problem. Let us now assume that the piston is instantly removed, and a centred rarefaction wave (RW) is formed in the gas phase, while the parameters of particles retain their values determined by the final state. The problem involves the study of the following processes: a) evolution of a non-equilibrium rarefaction wave formed in a two-phase medium, and b) its interaction with the flow behind the front in various detonation regimes. Corresponding Cauchy problem for the Eqs. (1-3) is formulated as follows

$$t=0,\quad \phi_i=\begin{cases}\phi_{i,0}, & x_D<x<+\infty,\\ \phi_{i,st}(x), & x_K<x\le x_D,\\ \phi_{i,R}(x), & 0\le x\le x_K.\end{cases} \tag{11}$$

Here $\phi_{i,0}$ is the initial state ahead of the front, $\phi_{i,st}(x)$ is the steady-state solution to the problem (6-8), $\phi_{1,R}(x)$ is the solution describing the centred rarefaction wave in the gas phase assuming the pressure at point $x=0$ to be equal to the value ahead of the DW front. At the left boundary $x=0$ «soft» boundary conditions were posed.

The Region of Existence of CJ Regimes. In the CJ regime the real DW front structure, consisting of the shock wave, combustion zone and adjacent rarefaction wave, appears to be stable and self-sustained. Figure 5 shows the distribution of gas and particle temperatures with the time step of 0.12 ms for $\alpha=6$, $\beta=6$, $\tau_\xi=0.002$ ms. Non-equilibrium in the rarefaction wave zone occurs at the initial stage. It is seen that the DW amplitude, velocity, and the distance between the frozen shock wave front and the point of merging with the rarefaction wave are retained. Let us consider the strong regimes with α,β upper the curve 1 in Fig. 2 and detonation velocity higher than the CJ velocity. For M_0=5.5 and initial pressure 68.0 at (u_{DW}= 1770 m/s) the flow stabilisation and approaching the CJ regime (Fig. 6, pressure distributions) occurs for about 2 ms. Thus, for α,β in the region of existence of CJ regimes the strong DW becomes weaker and transforms into the CJ wave.

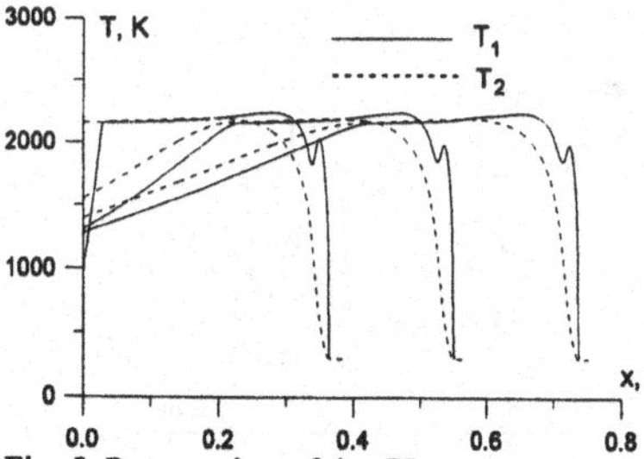

Fig. 5. Propagation of the CJ wave with an adjacent RW.

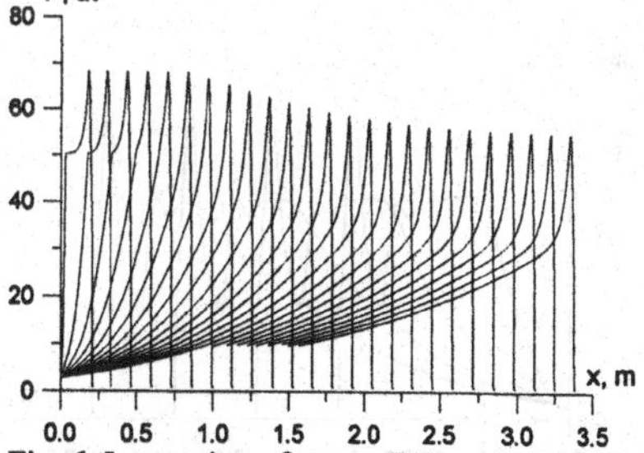

Fig. 6. Interaction of strong DW with the RW. approaching the CJ regime.

The Region of Existence of Weak Regimes with a Supersonic Final State. For a more illustrative representation of effects in the region of DW and RW merging, these regimes were studied for model mixture [6-8] with a more appreciable dispersion of the speed of sound in the final equilibrium state than aluminium and oxygen. For this mixture, a steady-state solution of weak detonation is determined for $\alpha = 10$, $\beta = 25$, M_0=5,7411 (u_{DW}=1786 m/s) and characterised by p_{max}=70.02 at (the temperature profile is presented in Fig. 7, branch W). The density distributions characterised by a narrow layer of elevated concentration of particles are shown in Fig. 8 with a step of 0.1 ms. It is seen that the DW is separated from the RW, and the flow structure behind the front remains unchanged. Thus, the detonation flow with a supersonic final state is self-sustained.

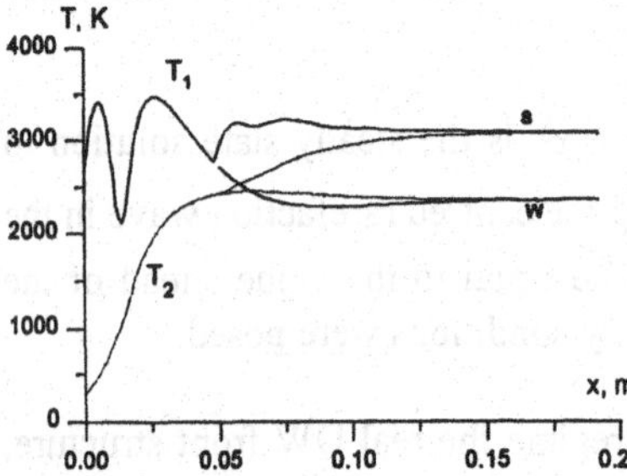

Fig. 7 Stationary solutions with internal sonic point

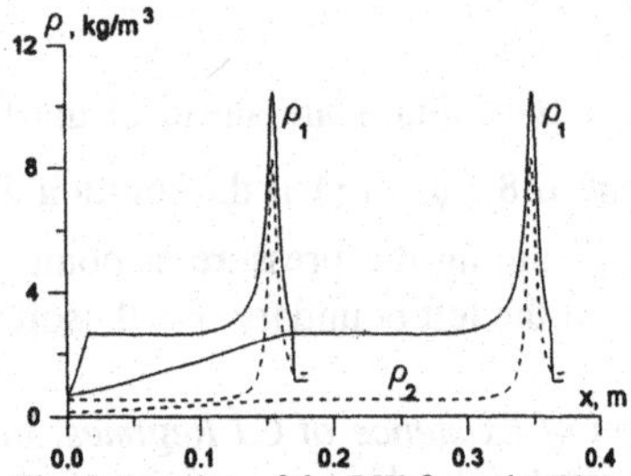

Fig. 8. Lagging of the RW from the DW front in weak detonation regime.

In the strong regime with the same values of relaxation parameters, the interaction between the DW and RW takes place, but the CJ regime does not exist. An approaching to weak detonation regime with a certain detonation velocity higher than the CJ velocity can be expected. The calculated results for initial data corresponding to the branch S in Fig. 7 are presented in Figs. 9a-9c. The sonic (with respect to the frozen speed of sound) points marked in pressure profiles. At the first stage (Fig. 9a, time step 0.053 ms) the rarefaction wave passes through the final equilibrium state zone and reaches the compression wave zone behind the sonic point. At the second stage (Fig. 9b) the compression wave becomes weaker, lags behind the sonic point and the DW front, and the formation of a flow region behind the sonic point corresponding to the weak detonation profile is observed.

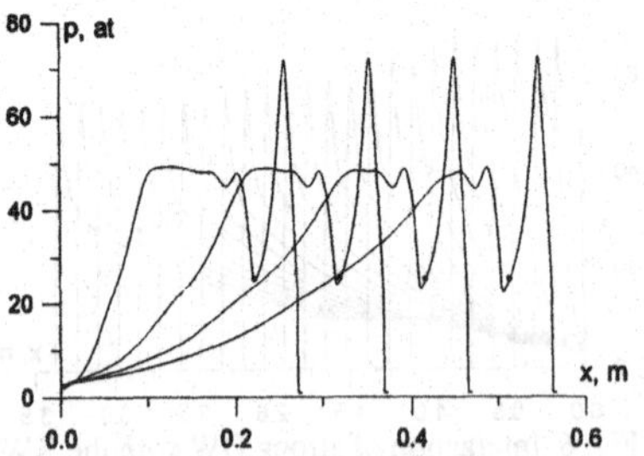

Fig. 9a

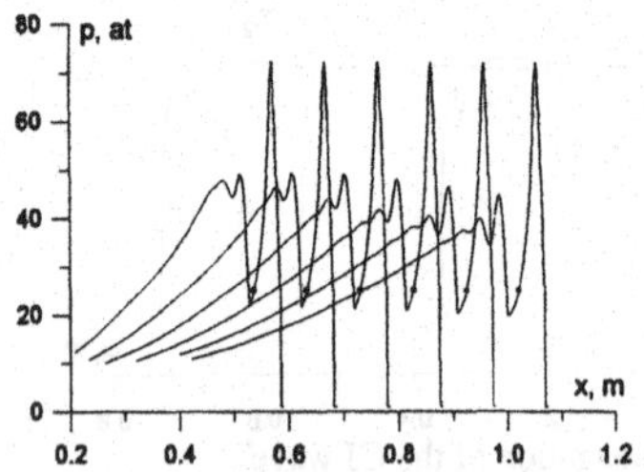

Fig. 9b

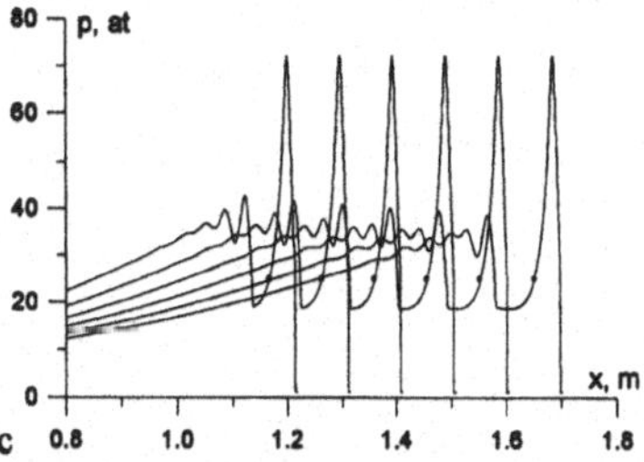

Fig.9c

Fig. 9. Interaction of strong DW with internal sonic point and RW (a, b, c).

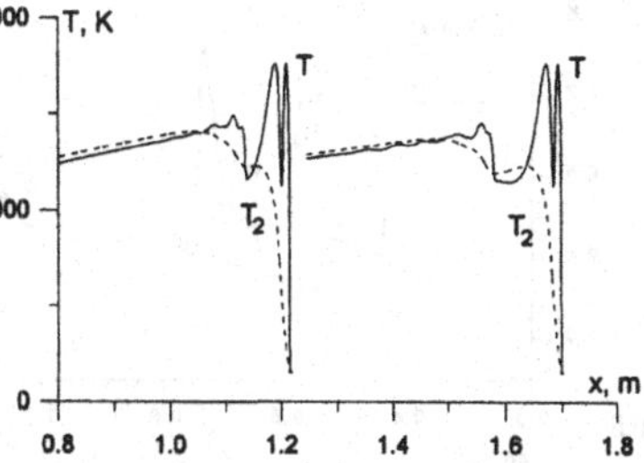

Fig. 10. Approaching weak detonation regime

These non-equilibrium processes lead to emergence of an oscillatory pressure profile in the region of rarefaction wave merging with the compression wave. The third stage is shown in Fig. 9c. The flow behind the sonic point reaches the final equilibrium state of weak detonation. The compression wave lags behind the DW front and gradually attenuates. The flow structure behind the DW front up to the sonic point is the same for all three stages.

Figure 10 shows the temperature distributions corresponding to the first and last curves in Fig. 9c. The lagging compression wave transforms here to an attenuating shock wave, which was also observed in the one-velocity approximation [7, 8]. The steady flow behind the front well corresponds to the stationary solution (Fig. 7, branch w). Thus, the waves interaction results in the self-sustained weak detonation regime with an internal sonic point.

The Region of Existence of Weak Regimes with a Transonic Final State. This final state is unstable (a saddle in the velocity-equilibrium model or a saddle-node in the two-velocity formulation). However, with a piston support the DW of this type proved to be stable both to small and finite perturbations. A one-velocity study of the interaction of this DW with the RW revealed the DW attenuation and its splitting into a shock wave and lagging rarefaction front, which caused a loss of stability [7, 8]. The results of computations for system of Eqs.(1-3) with the initial conditions (11) (the model mixture with $\alpha = 10$, $\beta = 10.0454$, M_0=5.2, u_{DW}=1618 m/s, in the final state $M_{e\kappa} = 1.168$, $M_{f\kappa} = 0.872$) showed a steady propagation of the structure consisted of the DW front with adjacent RW. Figure 11 shows the velocity profiles corresponding to time

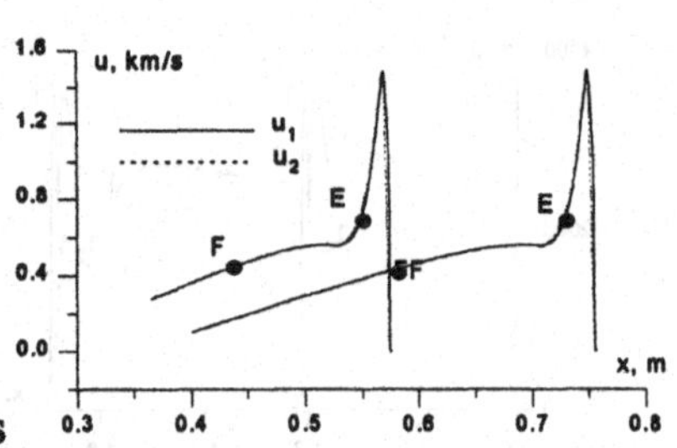

moments

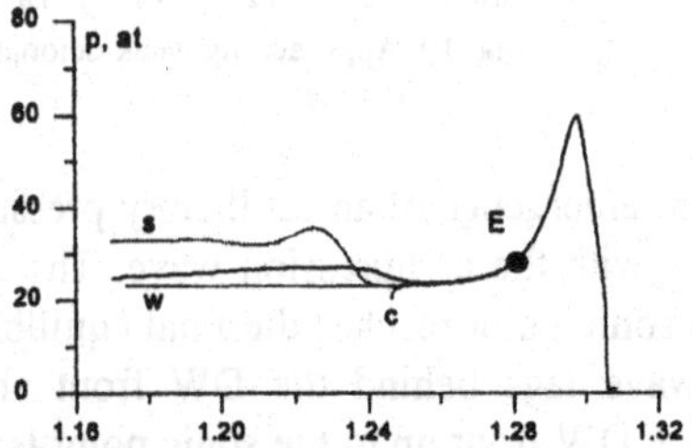

Fig. 11. Self-sustaining propagation of weak DW with transonic final state in self-sustaining regime. «dispersion» interval.

Fig. 12. Result of strong DW and RW interaction within the

Comparison with stationary solutions.

0.316 ms and 0.428 ms. The letters E and F denote, respectively, the equilibrium and frozen sonic points in the DW front-fitted system.

It is seen that the wave structure from the frozen SW front to point E is unchanged, while the interaction, that occurs in the merging region, blurs the fore front of the rarefaction wave but does not destroy the final equilibrium state. At the section EF the gas velocity with respect to the front is smaller than the frozen speed of sound but larger than the equilibrium one. Thus, the weak regime with a transonic final point in the velocity- and temperature-non-equilibrium approximation is self-sustained.

For initial data corresponding to strong regime for the same values of relaxation parameters, the calculations revealed DW attenuation and formation of a flow that also corresponds to the weak branch of solution. Figure 12 shows the pressure profiles reduced to the same front position in the model mixture: the lines s and c correspond to steady-state solutions for M_0=5.2+10^{-8} (strong detonation) and M_0=5.2-10^{-8} (solution with the flow choking line), the line w - M_0=5.2 (weak detonation, final point is saddle-node). The intermediate line shows the result of DW interaction with the RW for t = 0.76 ms. The curves merge up to the sonic (with respect to equilibrium speed of sound) point E and further in the region of stabilisation of the final state of weak detonation. The difference in the scenario of DW/RW interaction, as compared with a fully supersonic weak regime, is that there is no complete separation of the RW from the DW here, but an expanding section of transitional flow is formed, which remains unsteady and blurs the fore front of the rarefaction wave. Thus, the interaction between a strong DW and rarefaction wave in the dispersion interval of relaxation parameters (transonic final state) leads again to a self-sustaining weak detonation regime.

7. Initiation of Detonation

Let us consider semi-infinite space divided with a diaphragm. Suppose that the mixture between the boundary and diaphragm is characterised by high values of pressure and temperatures (assuming the particles have burnt). At the moment $t = 0$ the diaphragm destroys. The initial-boundary-value problem for system of Eqs.(1-3) is formulated by following way

$$t = 0, \phi = \begin{cases} \phi_l, 0 < x < l, \\ \phi_0, l < x \leq \infty \end{cases}; \; t > 0, \; d\phi / dx = 0, x = 0 \tag{12}$$

where ϕ_l corresponds to the state of mixture between the boundary and diaphragm. Numerical investigations using the method described above show that in ideal formulation (absence of viscous and heat conductivity effects) the initiation of detonation takes place only under the conditions for particle ignition immediately behind the front of the leading shock wave formed as a result of diaphragm destroying. For α =1 and β=20 the detonation arise at T_l = 2000K, p_l =100 at and doesn't arises at T_l = 2000K, p_l =80 at. The transition of the combustion front adjoining the shock wave to detonation front is shown in Figs. 13 as gas (solid lines) and particle (bold solid lines, Fig. 13b) temperature distributions. On the first stage the leading SW amplitude increases (Fig. 13a) and then the detonation front structure is formed (Fig. 13b).

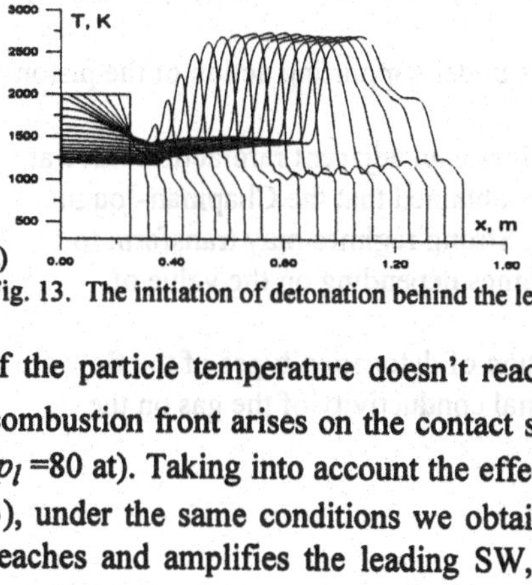

a)

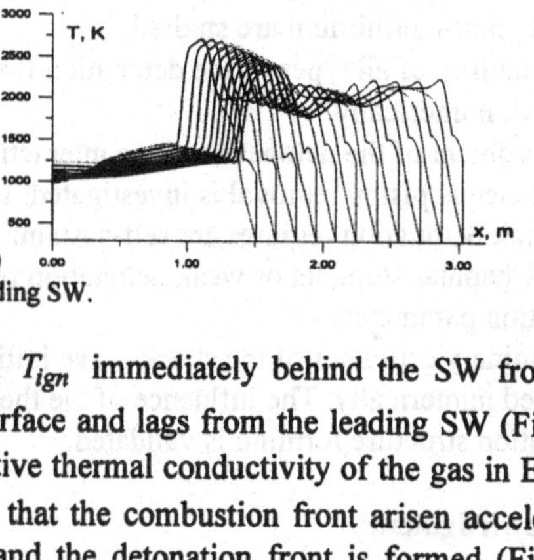

b)

Fig. 13. The initiation of detonation behind the leading SW.

If the particle temperature doesn't reach T_{ign} immediately behind the SW front the combustion front arises on the contact surface and lags from the leading SW (Fig. 14, p_l =80 at). Taking into account the effective thermal conductivity of the gas in Eqs.(1-3), under the same conditions we obtain that the combustion front arisen accelerates, reaches and amplifies the leading SW, and the detonation front is formed (Fig. 15, p_l =80 at).

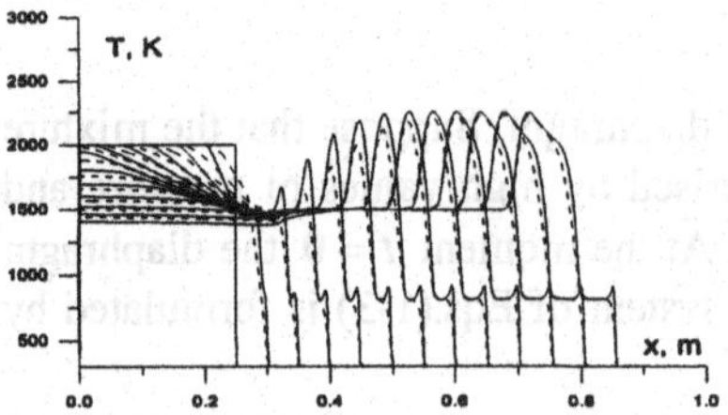

Fig. 14. The combustion front lagging.

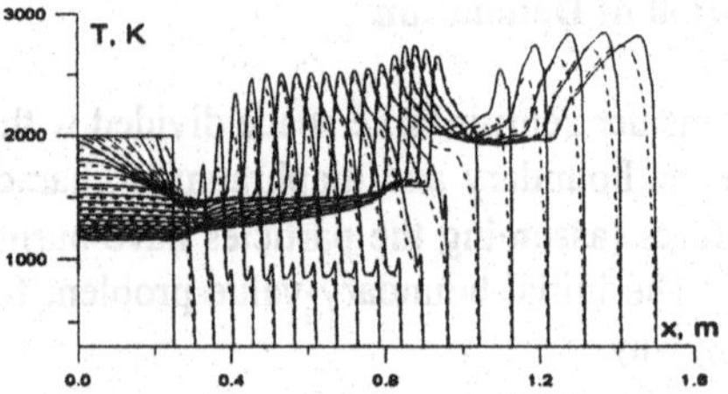

Fig. 15. The combustion front acceleration.

8. Conclusions

The following results are obtained in course of mathematical simulation of the detonation regimes in oxygen-aluminium mixture on the basis of one-velocity two-temperature approach and with regard for different velocities and temperatures of the components.

The types of final singular points depending on the relaxation parameters in various detonation regimes are determined by qualitative analysis of the final steady state.

The regions of existence of different steady regimes: Chapman-Jouguet, weak and strong detonation, are determined numerically in dependence on the detonation velocity and relaxation parameters.

The features of the detonation wave structure (ρ - layer and T - inflection) caused by velocity non-equilibrium are studied.

The stability of all types of the detonation flows under supporting action of the piston is shown numerically.

The problem of the detonation wave interaction with an adjacent rarefaction wave at instantaneous piston removal is investigated. It is obtained that the Chapman-Jouguet and weak detonation regimes are self-sustaining. Strong regimes may transform to either Chapman-Jouguet or weak detonation regimes depending on the value of relaxation parameters.

Preliminary estimates of the shock-wave initiation of detonation types of the flow are obtained numerically. The influence of the thermal conductivity of the gas on the detonation structure forming is validated.

Acknowledgment

The work was supported by the Russian Foundation for Basic Research (grant 96-01-01886).

9. References

1. Eremeeva T., Medvedev A.E ., Fedorov A.V., Fomin V.M. (1986) To the theory of ideal and non-ideal detonation of aerosuspensions, *Pre-print N 37-86*, Institute of Theoretical and Applied Mechanics SD AS USSR, Novosibirsk (in Russian).
2. Fedorov A.V. (1992) Structure of the heterogeneous detonation of aluminium particles dispersed in oxygen// *Combustion, Explosion, and Shock Waves*, **28**, p. 277.

3. Strauss W.A. (1968) Investigation of the detonation of aluminium powder-oxygen mixtures,*AIAA J.*, **6**, N 12, 1753-1761.
4. Borisov A.A., Khasainov B.A., Veyssiere B. (1991) On the detonation of aluminium dispersion in air and oxygen, *Chemical Physics*, **10**, p. 250 (in Russian).
5. Wang J.C.T., Widhoph G.F. (1987) A High-Resolution TVD Finite Volume Scheme for the Euler Equation in Conservation Form, *AIAA Paper*, N 87-0538.
6. Fedorov A.V., Khmel' T.A. (1996) Types and stability of detonation flows of aluminium suspension in oxygen, *Combustion, Explosion, and Shock Waves*, **32**, p. 181.
7. Fedorov A.V., Fomin V.M., Khmel' T.A. (1996) Real detonation waves in oxygen-aluminium mixtures, *Pre-prints of the 7-th Int. Colloquium on Dust Explosions*, Bergen, Norway, 23-26 June, 1996. pp. 4.33-4.46.
8. Fedorov A.V., Khmel' T.A. (1997) Interaction of detonation and rarefaction waves in aluminium particles dispersed in oxygen, *Combustion, Explosion, and Shock Waves*, **33**, N 2, 211-218.
9. Fedorov A.V., Fomin V.M., Khmel' T.A. (1997) An investigation of steady detonation of aluminium particles-oxygen suspension with regard for particles slipping,*Conference Proceedings 16-th ICDERS*, Aug. 3-8, 1997. University of Mining and Metallurgy, AGH, Cracow, Poland, 303-307.

CHARACTERISTICS AND APPLICABILITY OF RADIOTHERMAL LOCATION SYSTEM FOR THE PURPOSE OF FIRE DETECTION IN CHERNOBYL NPP AREA

I.I. ZARUDNEV, Y.P.BURAVLEV,
V.V.GAZHIYENKO, and Y.A.GORBUNOV
ICSC-World Laboratory, Ukraine Branch,
Turguenevskaya 32-a str., Kiev 252054, Ukraine

1. Introduction

The upper layer of a terrestrial surface of the Chernobyl 30km zone of alienation contains a lot of radioactive substances. The possible fires in this zone may promote spread of these substances on hundreds kilometers and increase of a radiating pollution level of the populated areas and large territories containing pastures, cultivated arable land and drains of waters.

In this connection the problem of fire-prevention monitoring of a zone of alienation with the purpose of detection and providing means for fight against arising fires at an early stage of their development is urgent.

The given problem is supposed to be solved by development of radio engineering system of remote detection of fires and notification of the appropriate service. The paper presents description of opportunities and specific features of design of the system developed.

2. Structure and Purpose of System

The system is intended for remote detection of fires in Chernobyl zone at an early stage of their origin; determination of coordinates of the flared up areas; the appropriate notification of fire-fighting sub-units; monitoring of dynamics of detected fires.

The system is supposed to be constructed from:

- four stations of observation of the high efficiency with radius of action up to 15 km and more (depending on a terrain profile);
- three stations of observation of medium efficiency with radius of action of the order of 10-15 km;
- two movable stations of observation with radius of action up to 15 km;
- the central data acquisition station combined with one of the high efficiency stations;

V.E. Zarko et al. (eds.), *Prevention of Hazardous Fires and Explosions*, 301–304.

- the means of delivering of the information on the detected fires to appropriate institution

The measuring equipment is based on radio direction finders, which are supposed to be placed on towers available in a zone of alienation. The equipment should work in autonomous mode. The exception makes the central station, where there is an on duty guard team and mobile stations. The mobile stations are represented by motor vehicles with telescopic towers that ensure lift of the equipment on height up to 15 m. Seasonally and depending on weather conditions the mobile stations are in a hangar with readiness of moving to the specified position, or in the certain places of a zone on the dominant heights.

3. Methods of Operation System

The following methods are used:

radio-direction finding of objects with increased temperature;

triangulation of the found out sources of radiation;

discrimination of sources produced by a fire between sources of radiation of a natural and artificial origin.

The listed methods, excepting last one, are well known. They are used for a long time in radio-direction finding, radio astronomy, in monitoring a terrestrial surface from the artificial satellite. As for the methods of recognition of fires, they are investigated in less detailed way in comparison with other specified methods.

Applying of all listed methods in Chernobyl zone is specific. The specificity consists in the following:

The terrestrial surface is observed from towers under rather small elevation, therefore noise of the sky, presence of laid fogs, crown of trees masking action have a significant effect on the measured signal;

In a zone of observation there is a number of thermo-microwave contrast objects (inhabited and industrial buildings);

At use of two and three-centimeter wavelength band for the various reasons the presence of interfering influence of various radio electronic means is possible.

4. Used Frequency Range

By various organizations of Kiev for various purposes the radiometers in two and three-centimeter, millimeter and infra-red ranges of electromagnetic waves were developed and investigated. On the basis of the analysis of these results in view of specificity of their use in Chernobyl zone the following conclusion was formulated: the system of monitoring should be based on radiometers of 2-cm wavelength band. The radiometers in this frequency range have rather good characteristics in difficult conditions of observation (at presence of a fog, rain, smoke). They are of small size. Thus, it is possible to build multi channel meters and, if necessary, integrate with them radiometers working in different ranges of electromagnetic waves.

5. Novelty Aspects that Are Offered to Use in Design of Radiometric System

As is known, at construction of systems of the given class there are contradictions between providing high efficiency of the given systems achieved due to use of narrow directed beams, their cost and high opportunities under the review of space.

In the system under discussion these contradictions are supposed to be overcome at the expense of realization of the following ideas.

a. Use of technique of frequency synthesis of signals that allows approximately five times increase of direction finding opportunity for measuring stations, when building meters on a basis of interferometer.

b. The parallel review of space in elevation plane and parallel-serial review in azimuth plane with rational combination of the information from one review to another not only in each of meters, that is known, but also in their total.

The arising fires are not point-size objects. In the beginning of their history at the expense of ascending warm flows of air carrying on products of combustion, they (fires) appear vertically distributed. Then in course of developing the fires they become spatially distributed. The volumetric distribution of sources of radiation of electromagnetic waves leads to fluctuation character of signals in points of reception. Therefore it is supposed to use not simple power reception stochastic on the nature of signals, but to take into account their complex internal structure, i.e. to use results of the analysis of detailed structure of signals. It will allow to ensure authentic selection (extraction) of weak signals produced by fires on a background of interfering noisy signals.

In radiometers it is supposed to set adaptive devices of suppression of rather narrow-band powerful signals of various radio electronic means working in wide (of the order of 4-6 GHz) frequency range. That is to ensure a noise stability of system in relation to purposely put interference, whose probability of occurrence in a two-centimeter range of wavelengths cannot be neglected.

6. Prospective Use of System for Scientific Research

According to the experts, in Chernobyl zone the fire hazard period is the part of year, composing about half of year. Taking into account this fact, and also that circumstance, that not all equipment of system is continuously used in an on duty survey mode, there is a possibility to use constructed system in addition for performing various studyes.

For the specified purposes it is supposed to equip one of measuring stations. It is planned to carry out works of the following character: to estimate an opportunity of enhancement of a system effectiveness at the expense of use of the radar-tracking equipment, to carry out studying ionosphere in interests of the forecast of

meteorological conditions, to measure radiation of ionosphere and other signals of the sky.

7. Conclusion

There is basis to believe that the construction of the radar-tracking system of fire monitoring in Chernobyl zone of alienation will promote reduction of radiation pollution both in the very zone, and in adjoining areas. Besides, the development of such system will allow to carry out scientific researches directed to measurement of radiation of an environment and space.

LIMITING CONDITIONS OF FOREST FIRE SPREADING AND ELABORATION OF THE NEW METHODS TO FIGHT THEM.

A.M. GRISHIN
Tomsk State University,
Tomsk 634050, Russia

1. Introduction

It is known that forest fires greatly damage forests all over the world. In the modern world there is a tendency of growing a number of forest fires and damage caused by them. It is caused by the rise of antropogenic load and the absence of efficient methods of fighting forest fires.

In this paper concepts of mathematical theory of combustion [1, 2], general mathematical model of forest fires and method of small perturbations [3] are used for determining limiting conditions of the most dangerous crown forest fire spread.

It is shown that it is possible to determine temperature fields, component concentrations and forest fire contour at any instant of time with using a general model and numerical integration of the basic system of equations Method of small perturbations is briefly presented to determine limiting conditions for forest fire spread. Some new waterless methods of fighting forest fires are briefly described.

2. Physical and Mathematical Setting up a Problem on Crown Forest Fire Spread

Let us examine a finite size forest canopy of height h, ventilated by a wind with a constant speed u_∞. The direction of the wind coincides with the orientation of the x-axis. The composition of the multiphase, multicomponent medium in the forest canopy, as well as the scheme of physicochemical transformations during the spread of the crown fire correspond to the model given in Refs. [2,3]

At the initial moment of time a high-temperature zone $[x_L, x_R]$ $[y_L, y_R]$ (fire source), with characteristics T_g, c_g, which are maintained constant until initiation at $t = t_{ig}$ of the fire front appears near point 0 in the canopy. Here T_g and c_g is a temperature of burning forest fuels and concentration of an oxidiser in the fire source. Outside the source, the state variables of the medium are the same as in the unperturbed atmosphere.

It is known [2,3] that the basic system of equations of the forest fire theory represents a complex system of nonlinear equations in partial derivatives.

V.E. Zarko et al. (eds.), Prevention of Hazardous Fires and Explosions, 305–318.

For simplifying this system let us assume that the terms characterising convective transfer along the y-axis of mass of α - component, $\rho_5 v \frac{\partial c_\alpha}{\partial y}$, and energy $\rho_5 c_{p5} v \frac{\partial T}{\partial y}$ (ρ_5 is the density of gas phase, v is the projection of the velocity onto the y-axis, c_α is the concentration, T is the temperature) are small compared with the terms $\rho_5 u \frac{\partial c_\alpha}{\partial x}$ and $\rho_5 c_{p5} v \frac{\partial T}{\partial x}$, which characterise convective transfer of mass of the α-component and energy in the direction of the *x*-axis. In addition, let us assume that the projection of wind velocity onto the *x*-axis *u* coincides with the value of the equilibrium wind speed in the forest canopy, u_∞* [3] , and the rate of combustion chemical reactions of gaseous products of the forest fuel pyrolysis is limited by concentrations of carbon dioxide and oxygen in the forest fire front. Therefore, when setting up a problem only concentrations of oxygen (α=1), fuel gas (carbon oxide α=2) and nitrogen (α=3) are considered. For simplicity of analysis we will use the method of averaging the basic system of equations throughout the height of the forest canopy presented in Refs.[2, 3]. Thus we have the following averaged through the height two-dimensional mathematical model:

$$\frac{\partial \rho_1 \varphi_1}{\partial t} = -R_1, \frac{\partial \rho_2 \varphi_2}{\partial t} = -R_2; \tag{1}$$

$$\frac{\partial \rho_3 \varphi_3}{\partial t} = \alpha_C R_1 - R_3 M_C / M_1, \frac{\partial \rho_4 \varphi_4}{\partial t} = 0; \tag{2}$$

$$\rho_5\left(\frac{\partial c_\alpha}{\partial t} + u_\infty \frac{\partial c_\alpha}{\partial x}\right) = \frac{\partial}{\partial x}\left(\rho_5 D_{TX} \frac{\partial c_\alpha}{\partial x}\right) + \frac{\partial}{\partial y}\left(\rho_5 D_{TY} \frac{\partial c_\alpha}{\partial y}\right) + \tag{3}$$

$$+ R_{5\alpha} - c_\alpha Q - (I_\alpha^+ - I_\alpha^-)/h, \alpha = 1,2;$$

$$\sum_{\alpha=1}^{3} c_\alpha = 1, 1/M = \sum_{\alpha=1}^{3} c_\alpha / M_\alpha, \rho_5 RT / M = P_e = const; \tag{4}$$

$$\left(\sum_{i=1}^{4} \rho_i^0 c_{pi}\right) \frac{\partial T}{\partial t} + \rho_5 c_{p5} u_\infty \frac{\partial T}{\partial x} =$$

$$= \frac{\partial}{\partial x}\left(\lambda_{TX} \frac{\partial T}{\partial x}\right) + \frac{\partial}{\partial y}\left(\lambda_{TY} \frac{\partial T}{\partial y}\right) - q_2 R_2 + q_3 R_3 + q_5 R_5 - \tag{5}$$

$$- (q^+ - q^-)/h, \rho_i^0 = \rho_i \varphi_i;$$

---

* Equilibrium wind speed in a forest canopy, which is represented by a set of the tree crowns, is averaged across the height of the forest canopy, stabilized wind speed.

$$T = T_r, c_\alpha = c_{\alpha r}, \rho_i^0 = \rho_{ir}^0, x, y \in D_1, 0 \le t \le t_3; \quad (6)$$

$$T = T_e, c_\alpha = c_{\alpha e}, \rho_i^0 = \rho_{ie}^0, x, y \in D_2, t = 0; \quad (7)$$

$$x = \pm\infty: T = T_e, c_\alpha = c_{\alpha e}, \varphi_i = \varphi_{ie} \quad (8)$$

Here λ_{Tx}, λ_{Ty}, and D_{Tx}, D_{Ty} are the coefficients of turbulent thermal conductivity and diffusion in the directions of x and y, respectively; x, y, z are the Cartesian coordinates; t is the time; I_α^+ and I_α^-, q^+ and q^- are the intensities of diffusion and thermal fluxes at $z = 0$ and $z = h_3$, respectively; $z = h_3$ is the upper boundary of the forest canopy; φ_1, φ_2, φ_3 φ_4 are volume fractions of dry organic substance, water in liquid-drop state, coke fine and ash, respectively; ρ_1, ρ_2, ρ_3, ρ_4, ρ_5 are the densities of the dry organic substance, water, coke fine, ash and gaseous phase, respectively; R_1, R_2, R_3, $R_{5\alpha}$ are the mass rates of formation (disappearance) of the dry organic substance, water connected with this substance, coke fine and components of a gaseous phase, respectively; the indexes e and "g" are added to gas state variables in the unperturbed medium and in the fire source region, respectively, while the remaining symbols are taken from [3].

As in Ref.[3], we assume that the coefficients of turbulent transfer $\lambda_{\grave{o}x}$, $\lambda_{\grave{o}y}$, $D_{\grave{o}x}$, $D_{\grave{o}y}$ are expressed in terms of the coefficients of turbulent dynamic viscosity $\mu_{\grave{o}x}$, $\mu_{\grave{o}y}$, i.e., $Pr_{\grave{o}x} = Pr_{\grave{o}y} = Sc_{\grave{o}x} = Sc_{\grave{o}y} = 1$, while $\mu_{\grave{o}x}$, $\mu_{\grave{o}y}$ = const, and in the general case $\mu_{\grave{o}x} \neq \mu_{\grave{o}y}$.

We also assume that the heat and diffusion fluxes at the canopy lower boundary are equal to zero ($q^- = I^- = 0$) (see [3]), while the fluxes at the upper boundary of the forest canopy are expressed by formulas (4.2.9), (4.2.10) from Ref.[3].

Note also that in deducing the system (1)–(8) it was assumed that the process of transfer is determined by forced convection, while the effect of free convection on the process of fire spread is taken into account by the selection of the coefficients of heat and mass transfer α_0 and α_0/c_{p5} [3].

3. Analysis of Numerical Results

The method for solving the problem is described in Ref.[3]. Figures 1-3 give some results for two-dimensional spread obtained by this method. In calculations the following values of kinetic constants were used (i = 1, 2, 3, and 5 [3]): $E_i/R = 9.4 \cdot 10^3$; $6 \cdot 10^3$; $9 \cdot 10^3$, and $11.5 \cdot 10^3$ K; $k_i = 3.63 \cdot 10^4$ s^{-1}; $6 \cdot 10^5$ K$^{0.5} \cdot$ s^{-1}; 430 s^{-1}, and $1.1 \cdot 10^{15}$ K$^{2.25} \cdot$ s^{-1}. Heat fluxes were taken as $q_i = 0.3 \cdot 10^6$; $1.2 \cdot 10^7$; and 10^7 J/kg. The following values of thermophysical quantities were selected: c_{pi} = 2; 4.18; 1.1; 1; 1 kJ/(kg · K), where i = 1–5; and ρ_i = 500; 10^3; 200; 200; 1 kg/m^3, where $\rho_5 = \rho$, i = 1–5. It was assumed that $\alpha_c = 0.06$, $\nu_{\mathring{a}} = 0.7$, coefficient of kinematic turbulent viscosity $\nu_{\mathring{a}} = \mu_T/\rho_{5e} = 1$ m^2/s, $\rho_c = 0.2$ kg/m^3, $T_{\mathring{a}}/T_e = 3$, $c_{\alpha\mathring{a}}/c_{\alpha e} = 1$, coefficient of heat-exchange between near-ground layer of at atmosphere and canopy $\alpha_0 = 300$ J/(m² · s · K). For simplicity it was assumed that $\mu_{Tx} = \mu_{Ty} = \mu_T$, $\lambda_{Tx} = \lambda_{Ty} = \lambda_T$, $D_{Tx} = D_{Ty} = D_T$.

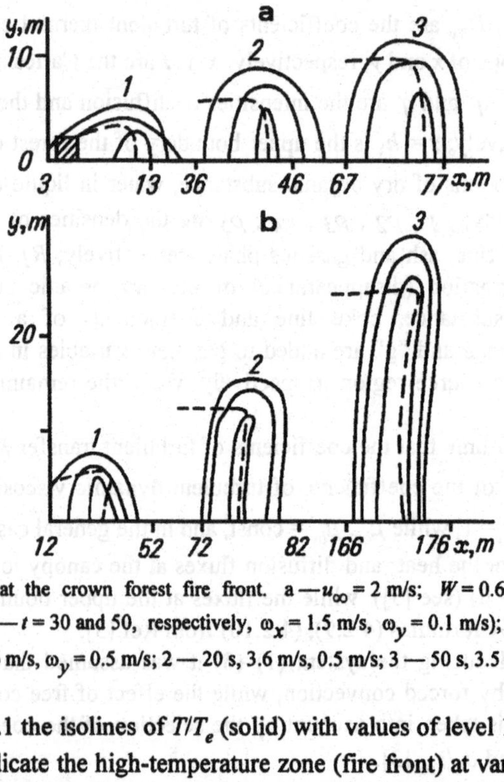

Fig. 1. Isotherms at the crown forest fire front. a — $u_\infty = 2$ m/s; $W = 0.6$ (1 — t = 7 s, $\omega_x = 1.3$ m/s, $\omega_y = 0.7$ m/s; 2, 3 — $t = 30$ and 50, respectively, $\omega_x = 1.5$ m/s, $\omega_y = 0.1$ m/s); b — $u_\infty = 5$ m/s, $W = 0.4$ (1 — $t = 8$ s, $\omega_x = 3.6$ m/s, $\omega_y = 0.5$ m/s; 2 — 20 s, 3.6 m/s, 0.5 m/s; 3 — 50 s, 3.5 m/s, 0.5 m/s).

In Fig.1 the isolines of T/T_e (solid) with values of level 1.5, 3, and 4, read from the outside, indicate the high-temperature zone (fire front) at various moments of time. This drawing illustrates the development of the fire front from the initial source (cross-hatched section). It can be seen that with time, as a result of the transfer of heat from the source, the neighbouring part of the canopy is heated and ignited. The time of ignition in the given calculation $t_{ig} = 9.8$ s. A high-temperature zone forms in the canopy that shifts in the direction of the wind (x-axis), and spreads in a transverse direction (y-axis). In doing so the velocity of the leading edges in both directions depends on time. As a result of the spread of the front, a wedge-shaped burn-out portion forms in the canopy in which $\rho_c = 0$. The burn-out part (see Fig. 1, a) is situated below and to the left of the boundary marked by the broken line, which corresponds to the temperature of pyrolysis, while above and to the right is the unburned portion of the canopy.

The process of spread of the crown fire front presented in Fig. 1 b differs from that examined above. The boundary of the burn-out part, marked by a broken line, and the front of the fire as a whole, bend in the direction of movement of the front. It is curious that at moment $t = 50$ s, the flanks of the front outstrip its central part.

It is worthwhile to explain the causes of the existence of the two substantially different types of forest fire contours: the convex (see Fig. 1a) and the concave-convex (see Fig. 1 b). To aim on end, Figs.2a and 2b give isolines of concentrations of oxidant

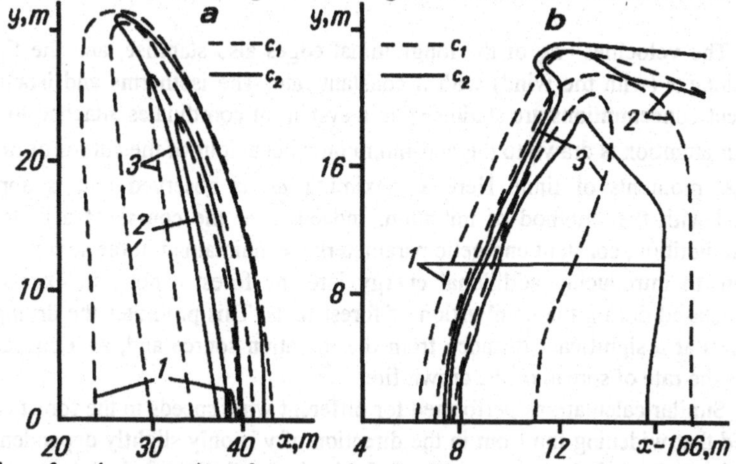

Fig. 2. Lines of equal concentrations of oxygen and combustible gas. a — $u_\infty = 2$ m/s; $W = 0.3$, $t = 30$ s (1 — $c_1 = c_2 = 0.15$; 2 — 0.1; 3 — 0.05); b — $u_\infty = 5$ m/s, $W = 0.4$, $t = 50$ s (1 — $c_1 = c_2 = 0.1$; 2 — 0.05; 3 — 0.02).

and combustible gas corresponding to the isotherms in Figs. 1a and 1b. It can be seen that ahead of the fire front, in the direction of the x-axis, a zone of maximum oxidant gradient forms, and hence the central part of the fire contour outstrips its flanks (see Figs. 1b and 2b). Fig. 2b shows isolines of oxidant and combustible gas that correspond to the isotherms in Fig. 1 b at $t = 50$ s. Comparing these data we can see that ahead of the center of the high-temperature zone in the direction of x, a region of reduced concentration of oxidant forms. It develops because the wind removes combustion products from the fire front, and the oxidant is displaced by them. Hence in this case the flanks spread faster than the central part of the front.

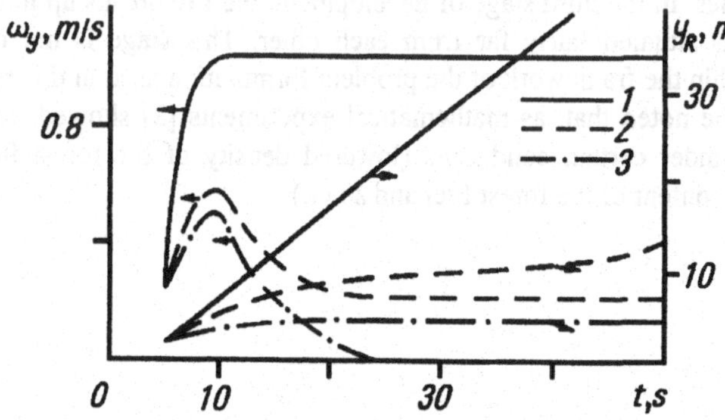

Fig.3. Projection of the contour of the crown forest fire front y_R and the rates of spread in the transverse direction ω_y when $u_\infty = 2$ m/s. Values of W: 0.3 (1), 0.5 (2), and 0.67 (3).

Fig. 3 gives the characteristics of fire front spread in the direction of y. As the moisture content increases, the conditions of spread of the fire deteriorate, and at high values of W, the fire front in the direction of y for all practical purposes does not spread.

The velocities ω_x of the longitudinal edges also stabilise, and the fire front moves along x (with the wind) with a constant rate. The isotherms and isochores of component concentrations are stationary in a system of coordinates attached to the fire front. Our attention is drawn to the non-monotonic behaviour of the curves of rate ω_y at the initial moments of time. Here a maximum ω_y is observed that is apparently connected with the method of initiation. Indeed, over the course of a certain time preceding ignition, constant energetic parameters are maintained in the source, which is equivalent to introducing additional energy into the forest canopy in excess of the energy released during the combustion of forest fuels. This promotes the drying of the forest fuels at a significant distance from the initiation source and, as a consequence, increases the rate of spread of the crown fire.

Similar calculations performed for different wind speeds in the forest canopy u_∞ showed that widening the front in the direction of y is only slightly dependent on u_∞. The results presented above, demonstrate the basic feasibility of obtaining contours of forest fires based on consideration of physicochemical processes at the fire front, diffusion of oxidant, and the forest fire type.

Mathematical experiments showed that there is diffusional-thermal instability of the forest fire front manifested as a distortion of the initial rectangular contour of the forest fire even when the forest-pyrological properties of forest massifs are uniform. In the other words, the contours of the forest fires large enough are unstable absolutely.

This conclusion agrees with the results of observations of actual forest fires [5]. According to these data, a large forest fire has three stages of development. The first stage is characterised by a closed contour having the shape of an ellipse elongated in the direction of the wind. In the second stage of the fire, its contour splits into separate sections, which points also to the absolute instability of the contour. Qualitatively, this effect is in agreement with the data presented above from theoretical and numerical studies. In the third stage of development, the fire breaks up into a series of separate sections situated fairly far from each other. This stage is not described mathematically within the framework of the problem formulation used in this section.

It should be noted that, as mathematical experiments [3] showed, forest fuel combustion stops under certain conditions (lowered density of the forest fuel layer, increased moisture content of the forest fuel and so on).

4. Diffusion-Thermal Instability (DTI) of the Forest Fire Front and Limiting Conditions of Its Spread

To solve the problem of suppressing forest fires the limiting conditions of their spread should be considered. These conditions can be determined as a result of mathematical experiments based on general mathematical model of forest fires [3]. Since the problems of the forest fire theory contain a lot of determining parameters, application of the small perturbation method and simplified (linear) mathematical models of the forest fire spread are more efficient and acceptable.

To analyze the diffusion-thermal instability of a two-dimensional crown forest fire front in the horizontal plane it is expedient to use the formulation and method proposed in section 4.3 Ref.[3] for studying the diffusion-thermal instability of a one-dimensional fire front in the vertical plane *x*, *z*. Applying the method of averaging throughout the height of the stand (see section 1.1[3]) and using simplified structure of the front of a crown forest fire (see Fig. 78 [3]), we obtain differential equations identical to equations (4.3.1), (4.3.2) from [3], where instead of the second independent spatial variable *z*, the independent variable *y* is used (the *y*-axis lies in the horizontal plane and is perpendicular to the *x*-axis, oriented in the direction of the wind). For the given problem, in the equations mentioned above, which are identical to equations (4.3.1), (4.3.2) [3], all the state variables are averaged throughout the height of the forest canopy. In addition, in studying diffusion-thermal instability, limitations on the wave length of the perturbation associated with the finite height of the forest canopy (see also [4]) are relaxed. Taking into account the above mentioned, to analyse the stability of a crown forest fire front in the horizontal plane *x*, *y* we can use the characteristic equation (4.3.29) [3].

Assuming in (4.3.29) [3] $\Omega = 0$, after bringing the equation of neutral hyper surface into dimensionless form, we obtain:

$$\Phi(z_1, z_2, L, \overline{K}, \overline{u}, \alpha_j, \overline{b}, d_j, q, Pe) = 0. \tag{9}$$

Here, as in section 4.3, [3] $L = \rho_{53}c_{p5}D/\lambda$ is the Lewis number; $\overline{K} = 2K\lambda/(\rho_{53}c_{p5}\omega)$ is the wave number; $\overline{u} = u/\omega$ is the wind speed; $a_j = 4bD/(\rho_{5j}c_{p5}\omega^2)$, and $b = 4b\lambda/(\rho_{53}c_{p5}\omega)^2$ are the coefficient of mass and heat emission, respectively; $q = q_2/q_5$ is the ratio of thermal effects; $Pe = \rho_{53}c_{p5}\omega l/\lambda$ is the Peclet number;

$$\overline{d}_j = 1 + \left[\sum_{i=1}^{4} \rho_i c_{pi} \varphi_{iH} / (\rho_{53}c_{p5})\right]_j$$ is the total volume heat capacity; and index *j* corresponds to the various zones of the front.

The form of function Φ is easy to construct from (4.3.29) [3] taking into consideration the fact that the value of $\rho_{53}c_{p5}\omega/l$ where l is the thickness of forest fires front, was used as the scale for α_{ik} and β_{ik}. Assuming that two parameters are variable, while the rest are constant, we can obtain the boundaries of the domain instability in the various planes of the determining parameters.

Equation for short-wave length perturbations is significantly simplified and is converted into the expression examined in Ref.[5] (see section 4.3 [3]).

Numerical calculation of the boundaries of the instability domain according to equation (9) showed that the form of the neutral curves is qualitatively preserved. Specifically, for the neutral curve $z_1 = z_1(\overline{u})$ for variable $\rho_{5j} \neq$ idem, W_j,≠idem there is a shift of the maximum in the direction of the smaller values of $\overline{u}$ compared to the case ρ_{5j} = idem, W_j,=idem examined in Ref.[5] (see also section 4.3 [3]).

Of interest is the relationship between the position of the boundaries of the region of stability, and the dimensionless wave length of the perturbation $\overline{\lambda}_B = \lambda_B / l$.

Figure 4 shows neutral curves $\overline{\lambda}_{B*} = \overline{\lambda}_{B*}(Pe)$ for various values of $\overline{b}$ and $\overline{u}$. The region of stability is located under the corresponding curve. The critical wavelength λ_{B*} can also be regarded as a certain limiting transverse dimension H_* of the burn-out zone of forest fuel, the exceeding of which leads to distortion of the leading edge of the fire front. As was shown in sections 4.1, 4.2 [3], the parameter

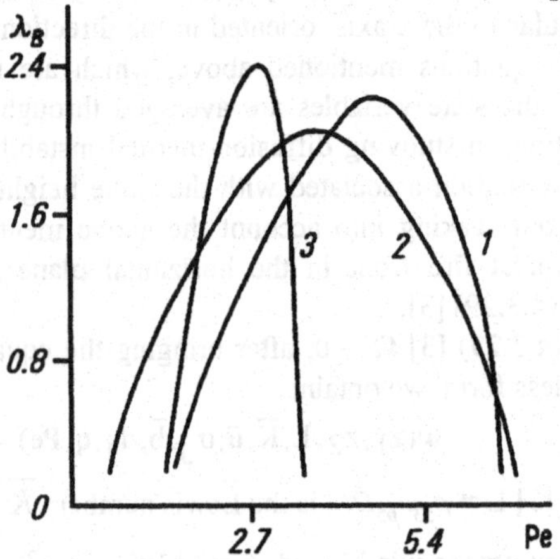

Fig. 4. Diagrams of stability in the plane of variables $\overline{\lambda}_B$—Pe. L=1, $\overline{d}_j = 1.4$, $\overline{q} = 0.3$, 1—$\overline{u} = 1.2, \overline{b} = 0.005$, 2—$\overline{u} = 1.2, \overline{b} = 0.05$, 3—$\overline{u} = 1.3, \overline{b} = 0.005$

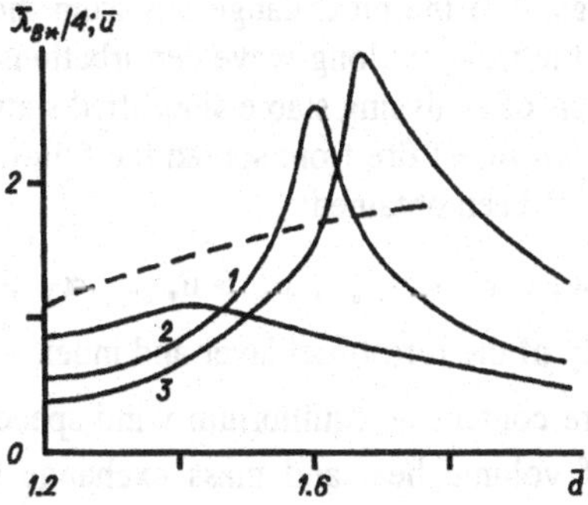

Fig.5. Critical transverse dimension of burn out area of forest fuel (solid curves) and $\bar{u}$ (broken curves) as functions of dimensionless volume heat capacity $\bar{d}$. L=1, b=0.05, $\bar{q} = 0.3$, z_1=3, Z_2=3, Pe=3. Values of $\bar{u}$: =1.4(1), 1.6(2), 1.7(3).

$\bar{u} = u/\omega$ for crown fires at fixed forest fuel supply and moisture content is practically constant at 3 m/s < u < 9 m/s, i.e., ω is a linear function of wind speed. Hence, the Pe number is unequivocally linked to wind speed u. It can be seen that the relation $\bar{\lambda}_{v*}(\mathrm{Pe})$ is non-monotonic, and stability occurs in a certain range $u_1 < u < u_2$. The values of u_1 and u_2 can be treated as the lower and upper limits of fire spread with respect to wind speed. From an analysis of Fig. 4 it follows that with increasing $\bar{\lambda}_v$ the region of stability narrows significantly, i.e., during the spread of a fire, long-wave length perturbations are the most dangerous. A rise in heat and mass exchange with the near-ground layer of atmosphere leads to an expansion of the zone of instability (see Fig. 4, curves 1, 2), while when speed $\bar{u}$ increases, the region of stability with respect to wind speed narrows (see Fig. 4, curves 1, 3).

Fig. 5 gives the effect of dimensionless volume heat capacity $\bar{d}_j$ on the position of the boundaries of the region of instability for various values of the parameter $\bar{u}$. The results are obtained for d_j = constant = $\bar{d}$ (d is the linear function of the volume supply of forest fuels). The non-monotonic nature of the neutral curve $\bar{\lambda}_{v*} = \bar{\lambda}_{v*}(\bar{d})$ testifies to the presence of limits to fire spread with respect to forest fuel supply. The physical meaning of the lower limit is completely clear, while the appearance of the upper limit is associated with significant consumption of energy for heating of the phytomass, leading to decrease in the maximum temperature and the rate of combustion.

The broken line in Fig.5 shows the relationship $\bar{u} = \bar{u}(d)$ for which the values of d correspond to the maximum $\bar{\lambda}_{в*}$. Qualitatively, this curve corresponds to the relation $\bar{u} = \bar{u}(d)$, obtained in Section 4.2 [3].

As follows from Figs. 4, 5 the most dangerous from the point of view of the development of distortion in the front are long-wave perturbations.

In Ref.[3] on the basis of analysing stable simplified stationary solutions of the problems connected with crown forest fire front spread the following critical conditions for crown forest fire cease have been obtained:

$$\rho_c < \rho_*\,,\quad w > w_*\,,\quad u_\infty < u_{*\infty}\,,\quad u_\infty > \overline{u}_{*\infty}\,,\quad \alpha_V > \alpha_{V*}\,, c < c_* \qquad (10)$$

where ρ_* is the critical density of the forest fuel layer and index * stands also for critical values of forest fuel moisture content w, equilibrium wind speed u_∞ in the forest fuel layer, the coefficient α_V of volume heat and mass exchange between a forest fuel element and the medium and concentration c, while the underscore of and overbear on u indicate the lower and upper limits of spread of the forest fire with respect to wind speed.

It should be stated that for surface forest fires, the lower limit of spread with respect to wind speed is lacking, since they also spread in the total absence of wind.

Note that only the basic limiting conditions are enumerated above. As in the classic theory of combustion [1], there are so-called "combined" limiting conditions in the theory of forest fires as well, when forest fuel combustion ceases owing to the action of certain factors unfavourable to fire spread. To determine the critical values of ρ, w, u_∞ α_V quantitatively [3,4], the perturbation method and mathematical experiments were used above, and the theoretically obtained limiting conditions on spread were checked by semi full-scale experiments.

As the mathematical and physical experiments showed, even if only one of the conditions (10) is fulfilled, combustion ceases and the process of forest fire spread stops. Knowing the limiting conditions of fire spread, we can realise them in the vicinity of the fire front by means of special actions, and in this way provide fire suppression. In other words, from the point of view of the theory of forest fires, any new method of fighting forest fires represents one or another specific realisation of the limiting conditions of forest fire spread.

5. A New Concept of Forest Fire Fighting

According to experimental and theoretical studies, the results of which are presented in [2, 3], the fronts of both crown and surface forest fires have a complex structure that includes zones of heating, drying and pyrolysis of forest fuels, combustion of gaseous pyrolysis products, and final combustion of solid products. The processes of combustion are limited by the inflow of oxygen and gaseous combustible pyrolysis products, that is diffusion controlled. The spread of the front of a crown forest fire is a multistage process, which is essentially limited by the formation of gaseous pyrolysis products, their mixing with the air oxygen, and subsequent combustion. If the structure of the fire front is destroyed, then, as experiments have shown (these will be described below), its spread ceases.

Its most sensitive part is the zone of pyrolysis and mixing of combustible pyrolysis products with oxygen. Indeed, this part of the front contains easy ignitable mixture, and hence a sufficiently small pressure pulse is enough to explode (detonate) this mixture and terminate flame combustion. The same effect may be obtained, if combustible gas resulting from the forest fuel pyrolysis were blown off in a direction opposite to the fire spread. It is precisely because of the effect of the explosion of the combustible mixture in the zone of pyrolysis of the crown forest fire front that we are able to destroy the structure of the front and stop its spread .

This type of approach constitutes the content of a concept of fighting forest fires, the essence of which consists in applying relatively small energy inputs to the most vulnerable part of the forest fire front in order to realise the conditions for its suppression.

Destruction of the structure of the front is also achieved by applying the familiar method of "beating" to suppress surface forest fires [6]. In this case thanks to mechanical action, the last of the limiting conditions (10) is realised. As a result, a small energy input using freshly cut branches leads to a reduction in the concentration of combustible gas in the pyrolysis zone and a cessation of gas phase combustion at the surface forest fire front.

Taking into consideration the basic requirements placed on forest fire fighting methods, the new concept of fighting forest fires can be formulated in the following manner: the purposeful use of a new theoretical and experimental data on the structure, mechanism and limiting conditions of forest fire spread in order that, through the application of relatively small, ecologically safe energy inputs to structural elements of the forest fire front, the conditions for its suppression to be realised.

Within the framework of this concept, the need for further refinement and development of a quantitative theory of forest fire spread becomes clear. Development of the theory will make it possible to find the limiting conditions of forest fire spread, and on that basis, create new methods, means, and technologies for fighting various types of forest fires. In particular, in [2, 3] by means of mathematical experiments which are considerably more safe and suitable than natural, a mechanism of suppressing forest fire front with high moisture content of the forest fuels (forest screens) or with low amount of forest fuels (forest breaks) was investigated.

For realisation of this conception at the Tomsk State University the explosive substances and special devices were used. Some methods of localisation and suppressing of forest fires worked out at the Tomsk State University and registered as patents of the Soviet Union and Russian Federation are given in the monograph [3].

Figure 6 shows a photograph of a forest strip following the explosion of a 2.4 kg PZhV-20 charge in the presence of a combustion front. It can be seen that the thin twigs and needles are dropping off, while combustion is ceasing. In the number of cases, the trees in simulated forest strip were wrenched out of the earth; this did not occur at explosions in actual forest massifs, however, since the plants are more firmly attached to the soil by their roots than freshly cut trees that have been dug into the soil.

Note also that for all practical purposes the wind speed does not affect the degree of forest fuel (branches) drop-off caused by the shock wave.

The observations of the pressure increment and the degree to which needles were broken show that in a burning forest strip the intensity of the shock wave is increasing. The blast intensification effect can be explained in the following way. Forest fuel combustion is a multi stage process. According to the results of Refs.[2, 3], initially the forest fuel is heated and then dried. After that, with increasing temperature, pyrolysis takes place accompanied by the release of gaseous and solid combustible products. Finally, both of these products burn. As estimates have shown [2, 3], the effect of blast intensification in the presence of a fire front can be caused by an acceleration of the process of chemical transformation of gaseous pyrolysis products as a consequence of the action of the shock waves, and then by the influence of energy release from the chemical reactions on the amplitude of the spreading shock waves.

Fig.6 Photographs of consequences of explosion in burning strip made up of small, freshly cut pines (a), charge (b), and consequences of explosion in a natural forest massif made up of young pines (c). a — the fire front ceases to spread; trees have been knocked down; b — trees have not been knocked down; needles and thin twigs have dropped off.

Fig.7. Interaction of shock waves with a surface forest fire front: a - surface forest fire and a charge PZhV-20; b - moment of the charge explosion; c - consequences of the explosion.

To investigate the interaction of shock waves with a surface forest fire front, a few experiments were carried out in September 9, 1993 using special methods at the proving ground of Kolpashevo operative department of Tomsk base of aviation forest protection. The proving ground was a forest meadow covered with a dry grass. The forest fuel consisted of a dry grass of 1993, grass tatters of past years, pine twigs and needle cast. A surface forest fire front was formed artificially by igniting a layer in several points. The edge of the surface forest fire was 3.5 m. The wind was weak, 0.5-1 m/s at the height of 2 m, and a surface fire front spreaded mainly in the wind direction. According to the classification [5], average amplitude fires with 0.6-0.75 m high flames occurred.

Standard charges of PZhV-20 with mass of 4-8 kg were used for experiments. Both fire explosion and charge initiation by electrodetonators were used. As the result of charge explosion on the ground immediately near the forest fire front, the combustion and spreading of the surface forest fire stopped. A small ditch (mineralised strip) of 0.4-0.6 m wide and 0.2-0.35 m deep, was formed near the front edge of the forest fire. Behind the surface forest fire front, some sources of forest fuel smoldering were observed that were extinguished by a forest satchel sprinkling device. Figure 7 shows the surface forest fire front and a charge PZhV-20 in front of it (Fig. 7, a), the moment of PZhV-20 charge explosion (Fig. 7, b) and the consequence of the explosion — mineralised strip (Fig. 7, c).

Thus, stall of the flame torch and removal of forest fuels, and the formation of mineralised strip take place during surface forest fires extinguishment with strand charge explosion. Mechanism of the flame front destruction consists in the explosion of gas combustion products of pyrolysis due to the influence of a primary shock wave generated by PZhV-20 explosion and blowing of gas products of forest fuel pyrolysis by a gas flow, which appears following the shock wave. The flow velocity at first instants of time is higher than the speed of sound, and therefore forest fuel elements are blown far into the fire rear.

Acknowledgment

The work was made under financial support of the Russian Foundation for Basis Research (Grant 96-01-00011).

6. References

1. Zeldovich, Ya. B., Barenblatt, G.I., Librovich, V.B., Makhviladze, G.M. (1980) *Mathematical theory of combustion and explosion*, M: Nauka.
2. Grishin, A.M. (1996) General mathematical model of forest fires and its applications,*Fizica Gorenija e Vzriva* , **32**, N 5, 45-63.
3. Grishin, A.M. (1997) General mathematical modelling of forest fires and new methods of fighting them. (Edited by F. Albini) Tomsk: Publishing House of the Tomsk State University.
4. Grishin, A.M., Zelensky, Ye.Ye., Shevelev, S.V. (1989) Stability of forest fire front spread, *Reactive media mechanics*, Novosibirsk: Nauka, 3-22.
5. Valendik, E.N. (1990) *Fighting large forest fires*, Novosibirsk, Nauka, Siberian Branch.
6. Instructions for detecting and suppressing forest fires. (1976) Moscow, Gosleskhoz SSSR.

VORTEX POWDER METHOD FOR EXTINGUISHING A FIRE ON SPOUTING GAS - OIL WELLS

D.G. AKHMETOV, B.A. LUGOVTSOV, V.A. MALETIN
Lavrentyev Institute of Hydrodynamics SB RAS,
Novosibirsk 630090, Russia

1. Introduction

Fires on open spouting gas-oil wells are among the most complicated industrial emergencies and cause enormous economical and environmental harm. Some idea of the fire on a well spouting under abnormal conditions can be gained from the following data: flow rate of powerful spouts may reach 10–20 million m^3 of gas or 10000 tons of oil per day, the height of a burning plume may be as much as 80–100 meters, heat release in such a plume is about several million kW. Near the well mouth where the works on emergency elimination are performed, extremely severe conditions arise for people to stay — high level of thermal radiation, strong vibrations and a noise level, exceeding the pain threshold of the human body. These factors considerably hinder efforts on eliminating emergency at the well.

Until the present time there have been two ways of extinguishing gas-oil well fires: by means of powerful water jets; by jets of fire-extinguishing powders delivered to the plume by compressed air; using gas–water jets created by aircraft turbojet engines or by explosion of a powerful concentrated charge of explosive suspended near the plume base. When the fire is extinguished using water or powder jets, they are directed below the base of the plume and by lifting the jets the flame is separated from the spout. These methods are suitable for quenching the fires on the spouts with gas flow rates up to 3–5 million m^3 per day and oil flow rates up to 3–5 thousand tons per day. However, these methods become of little efficiency in quenching more powerful burning spouts. The use of these methods requires a great deal of manpower and special equipment, complicated and expensive preparation works and large stocks of water. Therefore, the time required for elimination of the emergency at the well not infrequently protracts for many weeks and months, which leads to exhaustion of deposit resources and to the risk of destruction of the spout.

A fundamentally new method of extinguishing the fires on gas-oil spouts of practically arbitrary possible intensity has been developed at the Institute of Hydrodynamics of the Siberian Branch of the Russian Academy of Sciences in collaboration with the fire-fighting service [1, 2]. The plume quenching using this method is performed by the

V.E. Zarko et al. (eds.), Prevention of Hazardous Fires and Explosions, 319–328.

action on the plume of a vortex ring filled with a dusted fire-extinguishing powder. The vortex ring forms at explosion of a small annular charge with a layer of the fire-extinguishing powder placed over it. This method is characterized by high efficiency, moderate body of preparations and small consumption of fire-extinguishing materials.

The simplicity of realization of this method makes possible extinguishing of a burning gas-oil spout in short terms with minimum expenditure of human and material resources.

2. Combustion of Gas-oil Jets. Conditions of Flame Stabilization and Separation.

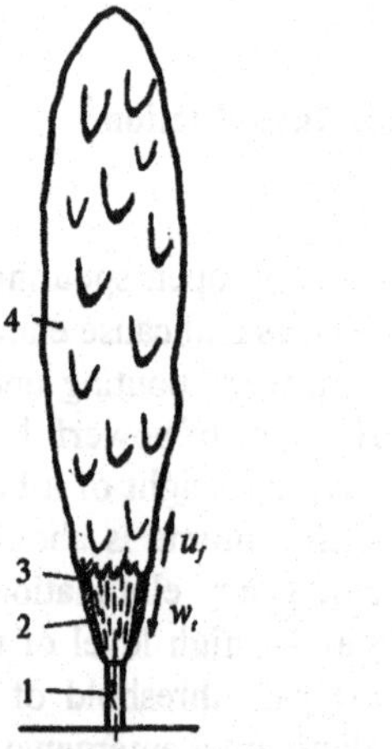

Figure 1. Diffusion plume. 1 — outlet tube, 2 — jet mixing layer, 3 — region of the flame stabilization, 4 — plume.

Characteristic features of combustion of gas-oil spouts are more conveniently considered by the example of gas jets. Precisely gas and mixed gas-oil spouts are the most intense type of spouts arising on emergencies on the wells. In actual conditions such jets are turbulent. When the jet of gas escaping from the well ignites, the so-called diffusion plume is formed which has an axisymmetric spindle-like shape (Fig. 1). Chemical reactions take place in a thin superficial layer, which to a first approximation, may be regarded as a surface where the fuel and oxidizer concentrations vanish, and diffusion flows of the fuel and oxidizer to this surface are in stoichiometric proportion. Diffusion combustion front has no propagation velocity, therefore, it cannot be stabilized the jet flowing upward. Stabilization of the flame in the jet occurs in the lowest part of the plume where a different combustion mechanism is realized. When the gas flows from the opening, a turbulent mixing layer of the gas and surrounding air is formed on the initial, not burning part of the jet surface. In this layer the gas concentration smoothly decreases in the radial direction, and the oxidizer concentration increases. In the middle part of the thickness of the mixing layer the homogeneous fuel–oxidizer mixture has composition close to the stoichiometric proportion. When such a mixture prepared for combustion ignites, the front can propagate in the mixing layer with finite velocity even upstream if the combustion rate exceeds the value of the local jet velocity. But since the jet velocity increases as the outlet opening is approached, at some height the jet velocity u_f becomes equal to the combustion rate w_t, and the flame is stabilized on the surface of the jet at this height. It appears not to be possible to calculate exactly the turbulent combustion rate w_t. However, rough estimates show that the value of w_t is of the order of the fluctuation velocities of the jet, which are proportional to the axial velocity u_m. From the experimental data [3] it follows that the maximum value of the rms fluctuations of the longitudinal velocity component is $0{,}2u_m$. Taking this value, to a first approximation, as the turbulent combustion rate, we may assume that the maximum flame propagation velocity upstream the gas jet spouting at 300–450 m/sec will be of the order of 50 m/sec.

From the above analysis it follows that extinguishing of a burning jet may be achieved by one of the two ways: either by increasing the local flow velocity u_f in the region of the flame stabilization, or by reducing the combustion rate w_t. In both cases the condition of the flame stabilization $w_t = u_f$ at a given height of the jet breaks, and the flame will be carried upward by the flow and will be ripped off the jet.

3. Evaluation of the Flow Rate of Burning Gas Spouts

In the extinguishing of the fires on the powerful gas spouts it is necessary to evaluate the burning spout flow rate, since the gas flow rate is one of the basic parameters determining the extent of the efforts and the material and technical resources that are required to eliminate the emergency. However, direct measurement of the flow rate of a burning spout is in most cases impossible, and there are no effective remote techniques for determining the flow rate of a jet. Below is shown that the flow rate of powerful underexpanded gas jets can be determined from the plume height H [4].

It is known that the height of a turbulent plume formed in combustion of normally expanded gas jets with subsonic outflow velocity does not depend on the velocity or on the flow rate of the jet and is determined only by the diameter d of the opening from which the jet flows out by the thermophysical properties of the gas, and the its temperature T at the exit of the opening. This dependence for gases with a high heating value can be written in the form [5, 6]

$$H = kd/\sqrt{T}\,, \tag{1}$$

where k is the coefficient that depends only on the individual properties of the gas. However, the most powerful and frequently encountered gas well spouts are underexpanded, i.e., the gas pressure p_* at the exit of the opening is greater than the pressure p of the surrounding medium, and the gas leaves the opening at the sonic (critical) velocity u_*. Because of this, in the initial segment of the jet there arises a system of rarefaction and compression waves, thanks to which gradual equalization of the pressure in the jet with the pressure of the surrounding medium takes place, and at some distance from the opening the flow in the jet does not differ significantly from the flow in the normally expanded jets. Observations show that when the underexpanded jets burn, the height of the diffusive plume exceeds significantly the length of the initial underexpanded segment of the jet, and in many cases the plume is lifted above the initial segment of the jet. and the gas burns only in the primary isobaric segment. Because of this the combustion of the underexpanded spout can be considered approximately as a combustion of a normally expanded gas jet, the initial diameter of which is equal to the diameter of the isobaric section of the spout (the section where the pressure in the jet first equalizes with the pressure in the surrounding medium). Therefore the dependence of the plume height, measured from this section, on the spout diameter and the gas temperature in the same section can be found from the relation (1) for isobaric jets. This approach makes possible determination of the dependence of the plume height on the spout flowrate.

To this end we shall use along with relation (1) the equations of mass and momentum flux conservation in two sections of the jet — at the exit of the opening and in the isobaric section

$$\rho_* u_* S_* = \rho u S\,, \tag{2}$$

$$\rho u^2 S - \rho_* u_*^2 S_* = (p_* - p) S_*\,, \tag{3}$$

where ρ_*, u_*, p_* and S_* are, respectively, the density, velocity, gas pressure, and the jet cross-section area at the exit of the opening; ρ, u, p, S are the same quantities in the isobaric section of the jet. Here we assume the flow to be one-dimensional and neglect air entraining from the surrounding medium at the initial segment of the jet. Invoking the equation of state of the gas and considering that the velocity of the gas outflow from the opening is equal to the speed of sound $u_* = \sqrt{\gamma R T_*}$, where γ is the adiabatic exponent and R is the gas constant, from Eqs. (1)–(3) we can obtain the dependence of the gas volumetric flow rate Q (for the gas density, referred to standard conditions $p_n = 10^5$ Pa, $T_n = 288$ K) on the plume height H

$$Q = \alpha\left(1 + \sqrt{1 - \beta/H^2}\right) H^2\,, \tag{4}$$

where $\alpha = \dfrac{\pi\sqrt{2(\gamma+1)\gamma R T_0}}{8\gamma}\dfrac{T_n p}{k^2 p_n}$, and $\beta = \dfrac{2\gamma}{\gamma+1}\dfrac{k^2 d_*^2}{T_0}$, d_* is the exit opening diameter, and $T_0 = \dfrac{\gamma+1}{2} T_*$ is the stagnation temperature. It is evident that to evaluate Q one needs also to know the values of T_0 and k. Direct measurement of T_0 is not always possible. However, thanks to the comparatively weak dependence of Q on T_0 and the small temperature variations between the various spouts, the gas flowrate can be evaluated from some average value of T_0 with an accuracy adequate for practical purposes. The dependence of k on the thermophysical properties of the gas is known [5, 6]

$$k \approx \frac{1}{c}\sqrt{\frac{(1+L)W}{c_p}\frac{\mu}{\mu_a}}\,,$$

where W is the calorific value of the gas, L is the stoichiometric coefficient; c_p is the heat capacity of the gaseous

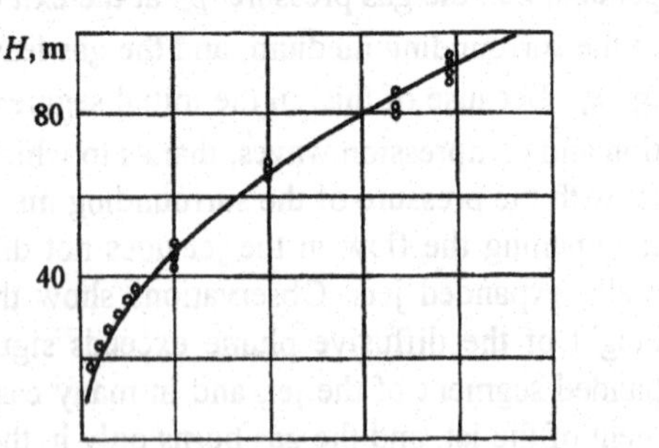

Figure 2. Gas plume height as a function of the jet flow rate.

mixture in the plume; μ and μ_a are the molecular masses of the gas and air; c is the dimensionless coefficient characterizing the turbulent jet expansion angle that depends on the ratio of the densities of the gas and air. However, because of the insufficiently accurate definition of the coefficient c the value of k can be found more reliably from Eq. (4) using the measured values of Q and H for the fixed values of α and β. It follows from experimental data that for the methane (CH_4) gaseous jet $k = 2{,}48 \cdot 10^3\ K^{1/2}$. The condition of spout underexpandedness $p_*/p > 1$, which is the condition of applicability of the formula (4) for the flowrate, is reduced to the inequality $H > \sqrt{(\gamma+1)/2}kd_*/\sqrt{T_0}$. The flow rate of the highly underexpanded spouts when $p_*/p >> 1$, the flow rate can be evaluated using the approximate formula

$$Q \approx \frac{\pi\sqrt{2(\gamma+1)\gamma RT_0}}{4\gamma}\frac{pT_n}{k^2 p_n}H^2,$$

which is obtained from Eq. (4) when $\beta/H^2 << 1$. Taking $T_0 = 293$ K and $p = p_n$ we have for methane spouts $Q = 2{,}33 \cdot 10^3 H^2$, where H is measured in meters and Q in m^3/day. Comparison of this approximate relation with the results of measurements is presented in Fig. 2. Thus, it has been shown that the plume height H is the main characteristic of the burning gas spouts.

4. Vortex Rings and Their Properties

The development of the vortex powder method of extinguishing the fires on gas-oil wells is the practical outcome of experimental and theoretical research on formation, structure and motion of vortex rings, carried out at the Institute of Hydrodynamics [7–10]. The possibility of extinguishing a plume by means of a vortex ring is related with the structure of the flow associated with its motion. In an air medium the formation of a vortex ring is associated with the impulsive motion of a finite volume of air or gas. The vortex ring formed progresses in the direction of the axis of the ring, traveling a considerable distance as compared to its own size. In its motion the vortex ring is accompanied with a certain volume of gas — the "atmosphere" of the ring, whose shape is similar to that of an ellipsoid of revolution flattened in the direction of motion. In the coordinate system attached to the vortex, the motion of surrounding medium coincides with the flow pattern around the solid

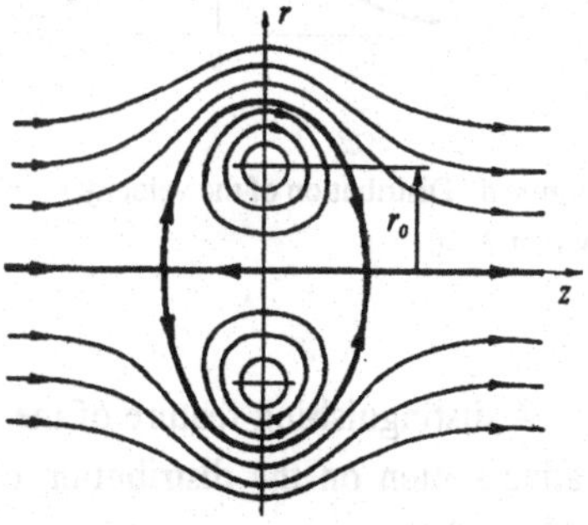

Figure 3. Streamlines of a vortex ring. Arrows indicate directions of the velocity vector. r_0 is the radius of the vortex ring.

corresponding to the "atmosphere" of the ring, and inside the "atmosphere" of the ring the motion occurs along the closed streamlines (Fig. 3). The presence of the flow region with the closed streamlines is beneficial for confinement of small particles in it, and

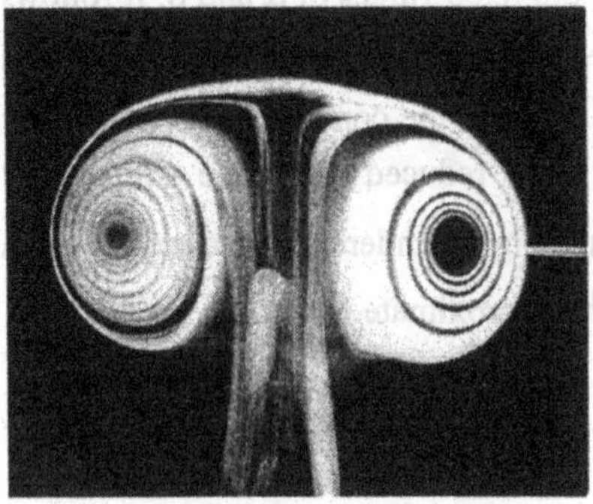

Figure 4. Laminar vortex ring.

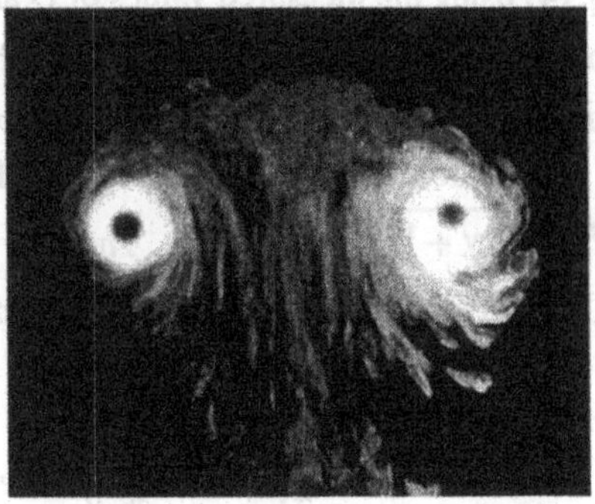

Figure 5. Turbulent vortex ring.

therefore the vortex ring can carry a dusted substance in its atmosphere, which can be introduced at the ring's formation. The structure of the air vortex rings, visualized by smoke, is shown in Figs. 4, 5.

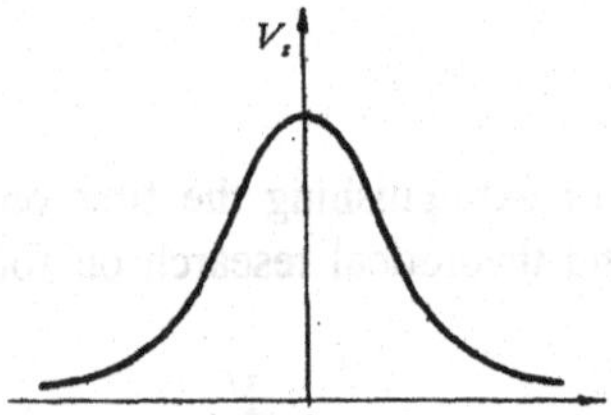

Figure 6. Distribution of the velocity V_z along the z axis at $r = 0$.

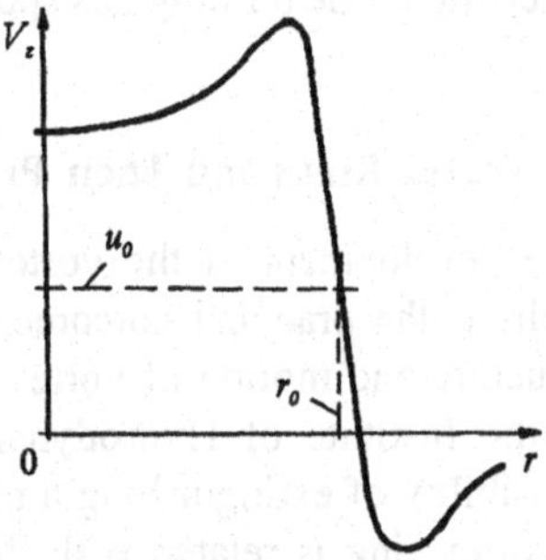

Figure 7. Distribution of the velocity V_z along the r axis at $z = 0$. u_0 is the progressive speed of the vortex, r_0 is the radius of the vortex ring.

A distinguishing feature of the vortex ring is the velocity field induced by it. Qualitative notion on the distribution of the axial component of the velocity V_z along the axis z of the vortex progressive motion and along the radial direction r, when the vortex ring moves in an immobile medium is shown in Figs. 6 and 7. These plots indicate that for $r < r_0$ the direction of the axial component of the velocity V_z everywhere coincides with the direction of the progressive velocity of the vortex ring.

5. The Mechanism of the Diffusion Plume Extinguishing by the Vortex Ring

The plume extinguishing is accomplished when the vortex ring filled with dusted fire-extinguishing powder moves along the spout axis to the top of the plume. The vortex ring is formed by detonating a small charge of explosive, placed around the well with a layer of fire-extinguishing powder put down on the charge. On explosion of such a charge, the impulsive gas-powder jet is formed, which then transforms into a mushroom -shaped circular ring containing the dusted powder and moving upwards along the axis of the spout (Fig. 8). As discussed above, the flame is stabilized on the jet with its lower edge at a certain distance from the exit section of the tube, where the turbulent combustion rate w_t is equal to the local flow velocity u_f in the mixing layer of the jet. If for whatever reason the flow velocity in the region of flame stabilization increases and becomes greater than the value of the turbulent combustion rate, the lower edge of the flame will be carried by the flow upwards to the top of the plume. A similar picture will be observed if the combustion rate decreases abruptly. Thus, to rip the flame off the jet one needs either to increase the flow velocity near the lower edge of the plume or to decrease the combustion rate.

From the conditions of stabilization and ripping off of the flame it follows that an air vortex ring moving along the plume axis can be effectively used for extinguishing the plume. The extinguishing of the plume occurs as a result of the action of the velocity filed of the vortex ring on the flame. The experiments demonstrated that air vortex rings with an initial velocity of 50–100 m/sec can be effectively used to extinguish high-speed gas plumes. From the practical viewpoint, vortex rings with relatively low velocities are more attractive for the purpose of extinguishing actual well fires since they can be obtained more easily, without the use of special devices and constructions. However, in this case the intensity of the vortex velocity field may be insufficient for the flame-out, and it will be necessary to decrease the turbulent combustion rate w_t simultaneously. To solve this problem it was proposed to use the ability of the vortex ring to transport a dusted admixture, for example, a fire-extinguishing powder. Introducing such powder into the flame substantially reduces the rates of the combustion reactions and flame propagation. This combined action on the flame considerably enhances the efficiency of the extinction process. Extinguishing of the burning spout by the vortex ring carrying the dusted fire-extinguishing powder is accomplished by the combined action of the both listed factors. First, as the vortex ring moves upwards along the flame axis, to the velocity of the spouting jet is added the velocity field shown in Figs. 6, 7, induced by the vortex ring which drifts the lower edge of the flame upwards. Second, the dusted fire-extinguishing powder carried by the vortex, enters the

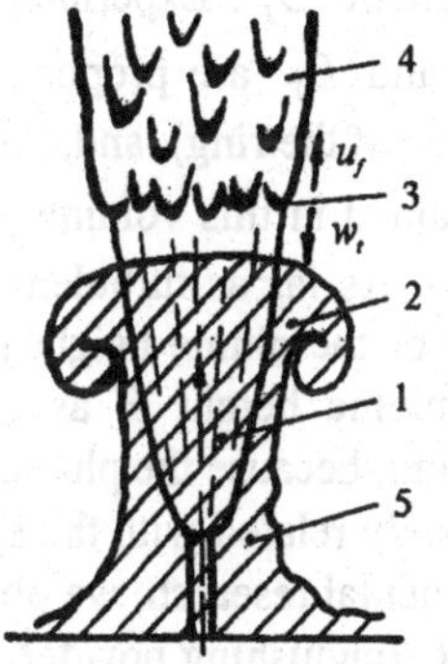

Figure 8. Diagram illustrating extinguishing of the plume by a vortex ring. 1 — jet, 2 — vortex ring, 3 — bottom edge of the flame, 4 — plume, 5 — cloud of the fire-extinguishing powder in the vortex trail.

flame and sharply reduces the combustion rate, which leads to further drifting of the flame to the top of the jet. The efficiency of the fire extinguishing is improved also by ejection of a portion of the dusted powder from the vortex ring into its trail. This trail envelops the bottom extinguished part of the spout by a powder cloud, preventing its re-ignition.

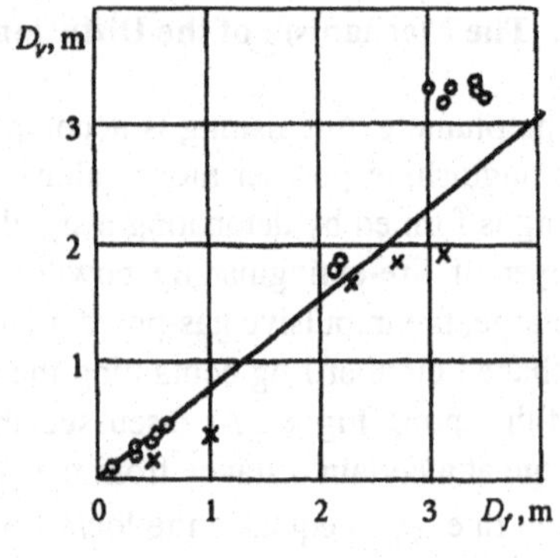

Figure 9. Diameter of the vortex ring required for extinguishing the plume as a function of the maximum plume diameter. o — the plume is extinguished, × — the plume is not extinguished.

6. Experiments and Background for Fire-Extinguishing Calculations

With the aim to establish the principles of modeling and calculating the agents required for extinguishing the fires on gas-oil spouts using the vortex ring, we performed a series of experimental investigations on extinguishing the burning methane, propane, kerosene and oil spouts. Spout flow rates in these tests varied in a wide range — from 0,01 to 70 kg/sec. The plume height varied from 2 to 50 m, and the maximum plume diameter — from 0,3 to 8 m. Experiments on extinguishing powerful oil spouts with flow rates up to 6100 tons per day were carried out on the prooving ground near the Samotlor oil field. The experiments demonstrated that the vortex powder method of extinguishing the plumes is efficient in extinguishing the fires on gas and liquid spouts of practically arbitrary intensity. The experiments revealed that the vortex ring diameter D_V, necessary for extinguishing the plume, must be approximately equal to the maximum diameter of the plume D_F. Experimental dependence of D_V on D_F is presented in Fig. 9. Since D_V and D_F are proportional to each other, the volume of the vortex ring ("atmosphere" of the ring), and, consequently, the mass of the dusted fire-extinguishing powder contained in this volume, must be proportional to D_F^3, if the powder concentration is given. As far as turbulent diffusion plumes are geometrically similar [5, 11], and the ratio of the plume height to its maximum diameter is virtually constant, we shall take the plume height H as a basis for further calculations of fire-extinguishing agent amount, because the plume height is determined easier and with greater accuracy, and is uniquely related with the spout flow rate. Based on this and using the results of the experimental research, we obtained the following formulae for calculating the mass of the fire-extinguishing powder M and the mass of explosive m, needed for extinguishing the plumes of the height H:

$$M = k_1 H^3, \quad m = k_2 H^3,$$

where k_1 and k_2 are constant coefficients.

The experiments resulted also in revealing the optimal schemes and practical procedures for placing the fire-extinguishing agents near the well mouth under the fire conditions. It was found that the deduced calculation principles and techniques for realization of the method provide assurance that the fires on compact gas-oil spouts of an arbitrary intensity will be extinguished.

7. Extinguishing Actual Well Fires Using the Vortex Powder Method

The method was utilized to extinguish a number of large-scale fires on emergency spouting wells on the territory of the former Soviet Union. The vortex powder method was first used to extinguish the fire on a powerful gas-condensate spout in Uzbekistan [2, 12]. Open spouting and fire occurred during the prospective drilling. The spout flow rate ranged up to 20 million cubic meters of gas per day, the plume height reached 90–95 m, and the maximum plume diameter was 15 m. Over a month period 8 attempts to extinguish the fire using traditional methods were undertaken. For fire extinguishing, powerful pump stations, devices with aircraft turbojet engines and other complicated fire-fighting equipment were delivered to the place of emergency. The fire-fighting was complicated by the absence of water in the region where the well was located. The water resources were accumulated in specially constructed water reservoirs. To fill these vessels and to convey the water to the well, metal pipelines 20 km long were mounted. More than 400 people took part in emergency elimination. However, despite all efforts, the attempts to eliminate the fire using conventional extinguishing methods failed. Thereafter for the first time it was decided to apply the vortex powder extinguishing method. Preparation for extinguishing by means of this method was carried out in several hours, and from the first attempt the fire was extinguished. Fig. 10 presents the picture of one of the stages of extinguishing this fire.

8. Conclusion

The highly efficient vortex powder method for extinguishing the fires on powerful gas and oil spouts, occurring at emergencies on the wells has been developed. The burning spout extinguishing using this method is accomplished by means of impulsive action on the plume by an air vortex ring filled with dusted fire-extinguishing powder. The vortex ring is formed at explosion of a small distributed charge of explosive. Advantages of this method becomes evident at extinguishing fires on extremely powerful gas-oil spouts, when traditional methods of extinguishing such spouts are low-efficient. The method allows for a drastic decrease of required fire-extinguishing time. Works on extinguishing a fire of virtually arbitrary intensity can be performed by efforts of several quali-

Figure 10. Fire extinguishing at the well in Uzbekistan. The spout flow rate was $\sim 20 \cdot 10^6$ m^3/day.

fied specialists during one day, with the use of minimum amount of fire-extinguishing agents. The method does not require delivery of bulky and costly equipment to the emergency site and its use. The method is well adapted for the use in arid desert regions and in winter conditions, because for extinguishing the fire, as such, the water is not necessary. The vortex powder method was successfully used to extinguish a number of large-scale fires in Uzbekistan and Western Siberia.

9. References

1. Akhmetov, D.G., Lugovtsov, B.A., and Tarasov, V.F. (1981) Extinguishing gas and oil well fires by means of vortex rings. Combustion, *Explosion and Shock Waves*, **16**, No. 5, 746–749.
2. Akhmetov, D.G., and Tarasov, V.F. (1983) On the extinguishing of a fire on a powerful gas spout, *Continuum Dynamics* [in Russian], Institute of Hydrodynamics SB AS USSR, No. 62, 3–10, Novosibirsk.
3. Abramovich, G.N. (1960) *Theory of Turbulent Jets* [in Russian], Fizmatgiz, Moscow, 715 p.
4. Akhmetov, D.G. (1994) Burning gusher flow rate evaluation based on diffusion plume height. *Combustion, Explosion and Shock Waves,* **30**, No. 6, 746–749.
5. Vulis, L.A., and Yarin, L.P. (1978) *Plume aerodynamics*, [in Russian], Energiya, Leningrad, 216 p.
6. Baev, V.K., Ktalkherman, M.G., and Yasakov, V.A. (1977) *Study of gaseous fuel combustion* [in Russian], Institute of Theoretical and Applied Mechanics SB AS USSR, 21–1.
7. Akhmetov, D.G., and Kisarov, O.P. (1966) Hydrodynamic structure of a vortex ring, *Zh. Prikl. Mekh. Tekh. Fiz.*, No. 4, 120–123.
8. Lugovtsov, A.A., Lugovtsov, B.A., and Tarasov, V.F. (1969) On the motion of a turbulent vortex ring, *Continuum Dynamics* [in Russian], Institute of Hydrodynamics SB AS USSR, No. 3, 50–60, Novosibirsk.
9. Lugovtsov, B.A. (1979) Turbulent vortex rings, *Continuum Dynamics* [In Russian], Institute of Hydrodynamics SB AS USSR, No. 38, Novosibirsk, 78–88.
10. Tarasov, V.F. (1973) Estimation of some parameters of a turbulent vortex ring, *Continuum Dynamics* [in Russian], Institute of Hydrodynamics SB AS USSR, No. 14, Novosibirsk, 120–127.
11. Lewis, B., and Von Elbe, G. (1968) *Combustion, flames, and explosions of gases* [Russian translation], Mir, Moscow, 592 p.
12. Kasymov, B. (1983) The fight in the Karshinskaya steppe, *Fire-fighting* [in Russian], No. 7, 18–19.

NANOSIZE ELECTRO-EXPLOSION POWDERS: ASSESSEMENT OF SAFETY IN THE PRODUCTION AND APPLICATION

G.V.IVANOV, V.G.IVANOV, V.G.SURKOV, O.V.GAWRILYUK
Institute of Petroleum Chemistry SB RAS,
634021 Tomsk, Russia

1. Introduction

Nanosize particles of metals, oxides and other compounds represent promising materials for novel technologies. They can be used as components of new types of ceramics including flexible ones, glasses, materials for electronics, dyes, abrasives, bio-filters, compounds, luminophore, adsorbent, catalysts, thermo-stable coatings, etc. [1-2]. Nanosize powders of active metals used as component of energetic materials are of the particular interest. International symposia, conferences and workshops hold on the regular base evidences the constant interest of scientists to this problem.

The review [1] cites comprehensive data about existing methods for the production of nanosized particles: evaporation and condensation of metals in fine vacuum, plasma-chemical and electrochemical processes, thermal destruction of salts in controlled conditions, etc.

Since 1970 in Tomsk the studies of the processes for the production of nanosized powders of metals and their compounds in the most extreme conditions - by the electro-explosion of conductors have been started. Usually thin foil or wire are used as these conductors. The application of wire as the initial raw material for the production of the large amounts of nanosized metal powders allows to render automatic the process of the continuous feed and explosion of conductors. During past decades the physical processes taking place at the metal wire electro-explosion has been studied in detail, the methods for simulation of the process of electro-explosion have been elaborated, automatic installations for the production of ultrafine powders (UFP) have been designed. There exists a stock of operating installations for the production of UFP providing the production of tens of kilograms per month. The methods for the wire elextro-explosion developed allows the production of nanosize particles of any metal ranging from sodium to tungsten.

The phenomenon of the conductor electro-explosion (CEE) was studied in connection with the problem of the development of special initiating agents for the auxiliary system of rockets and spacecrafts, i.e. as a reliable energy source. The first publications in the field of the CEE [3-6] were dedicated to this side of the problem. Only 3 papers [7-9] considered the CEE method as a possible way for the production of

V.E. Zarko et al. (eds.), Prevention of Hazardous Fires and Explosions, 329–340.

metal aerosols. Some scientific organisations of Tomsk first paid attention to the possibility of the production of nanosized powders with unusual properties. Nowadays we have in Tomsk scientific school in the field of the electro-explosion powders production and characterisation of their properties.

2. Manufacturing and Characterisation

The wire electro-explosion is carried out in a special chamber filled with inert gas (by argon, usually) under the pressure up to 1 MPa. The explosion is carried out by a short (less than 1 ms) high-voltage (15-30 kV) electrical pulse. The mode of the explosion and the pulse form are selected depending on the conductor properties - its specific resistance, diameter, thermo-physical characteristics. The explosion parameters are controlled by the magnitude of capacitance and inductance of the high-voltage source. The energy input into the wire during the explosion can amount two values of metal sublimation heat [10]. At the input of the current pulse into the wire a compressing electromagnetic field preventing the destruction of the wire into drops when heating the melting and boiling temperatures above up to 2-4 10^5 K, is generated. At these temperatures the energy supply cut off and an overheated material expands with the speed of 1,2-1,4 km/sec in the medium of an ionised gas (H_2, He, Ar, etc.). The simultaneous action of a powerful electric pulse, shock wave, luminous and X-radiation (recorded at the explosion) and ultrahigh rates of cooling of the particles formed (to 10^6-10^8 K/sec) provide the formation of meta-stable structures in the powders produced by this method. The metal powders have a regular spherical shape and polydisperse size distribution. However, the control of the explosion conditions for each individual conductor provides the production of the powder with the required medium size. The excess ("stored") UFP energy is assessed as some melting heats. The most remarkable property of the electro-explosion powders is their ability to keep the meta-stable structure and the excess energy obtained in the moment of their formation. Although the UFP particles have a small size (0.05-0.1 μm) and developed specific surface (5-25 m^2/g), they are thermally stable in air up to some "critical" temperature (200-450° C). Upon achievement of that temperature, they become extremely reactive. Up to now the question of the metal UFP stability, even of Ti powders, is a matter of discussions. However, it has been experimentally found that drastical increase of UFP chemical reactivity under heating is preceded by the evolution of adsorbed gases and by the exothermic heat effect. Obviously, gas film on the surface of UFP particles is formed at their cooling and is strongly bounded with the metal surface. When electo-exploding in argon, admixture gases - nitrogen, oxygen, water - are mainly adsorbed. Total gas content on the UFP particles can reach 5 % weight and more. Mass-spectrometric analysis of the gas desorption processes, when heating UFP in vacuum, reveals a complex structure of gas film and a consecutive elimination of individual layers in heating (Fig. 1). Nevertheless, it has been well established that argon, the principal gas in medium where the wire has been exploded, is the last gas evolving from the metal surface in heating.

The issue of the nature of stored energy has not been solved yet. An amorphous structure of metal particles was initially suggested but X-phase analyses have shown an insignificant deviation of the crystal lattice parameters from standard

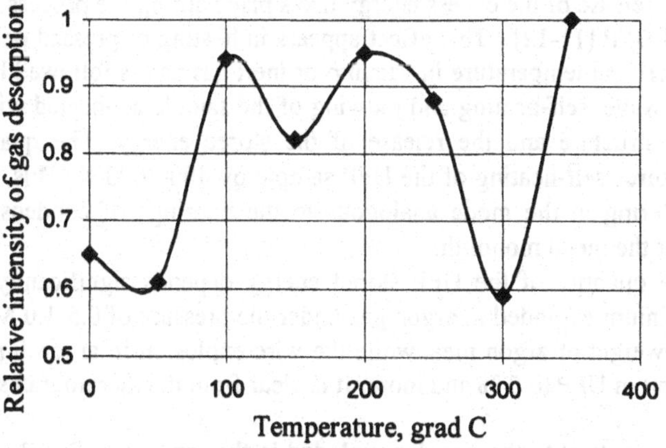

Fig.1. Gas desorption from ultra-fine aluminum powder when heating in vacuum (mass-spectrometric analysis)

values. The direct determination of the excess ("stored") energy quantity by calorimetric methods gives very contradictory results. The UFP combustion in oxygen does not allow to make precise determination of the UFP energetics because significant portion of the powder disperses and does not combust completely precipitating on the walls of calorimetric devices. Experiments carried on the DCM-2 micro-calorimeter have allowed to establish the possible limits of the stored energy. Fig. 2 shows the curve of the heat energy release in heating of the nanosize electro-explosion aluminium (Alex) in an inert

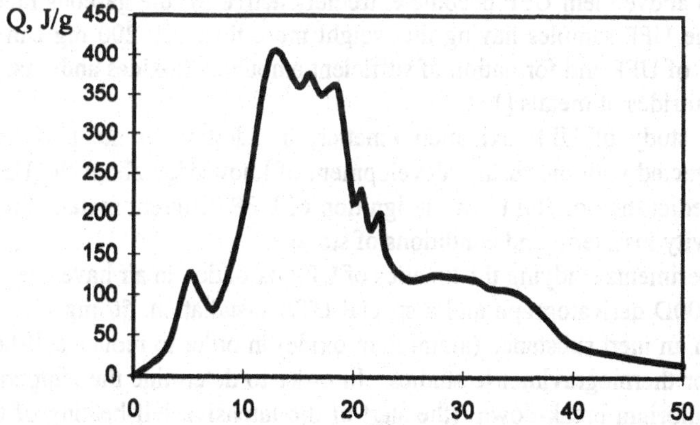

Fig.2. Heat evolution when heating 20 mg sample in a sealed vial of the differential scanning calorimeter.

medium. The maximum of the heat release occurs at 420-450 ° C for 40-50 s reaching the absolute value of 400 J/g which corresponds to the melting heat of aluminium. The integration of the heat release curve suggests that the overall UFP energy constitutes no less than 3-5 metal melting heats.

The release of the excess energy takes place during the process of self-sintering of the metal UFP [11-12]. This effect appears in heating of pressed pellet of the metal UFP to the critical temperature in vacuum or inert gas and is followed by generation of the thermal wave, self-heating and glowing of the sample connected with failure of the meta-stable structure and the release of the stored energy. The quantity of energy released ensures self-heating of the UFP sample by 400-1000 K. This is sufficient for the self-sintering in the mode analogous to the burning of condensed systems and formation of the metal monolith.

The quantity of the UFP stored energy depends significantly on the argon state. Aluminium exploded in argon gas under the pressure of 0.5-1.0 MPa contains 0,5 - 0,8 % by weight of argon meanwhile the wire explosion in argon liquid increase the argon content in UFP to 5 % and more. It is clear from the thermograms of UFP heating in air

(Fig. 3). In contrast to the powder exploded in the argon gas (Fig. 3a), UFP produced in

argon liquid shows a distinct weight loss when reaching the critical temperature of the failure of meta-stable energy-saturated structure, the exothermic effect is manifested more significantly (Fig. 3b). It is seen that this effect is not connected with the metal oxidation because in this case the weight loss occurs but not its gain.

3. Kinetics of Oxidation

The results of studies of the reactivity of metal UFP, mainly aluminium, in different oxidising media are given below. As it is stated above, metal UFP are chemically inert in air up to achievement of "threshold" temperatures individual for each metal. At the temperatures above them UFP become extremely active for the gaseous media and the heating of the UFP samples having the weight more than 100-200 mg can trigger the self-ignition of UFP and formation of sufficient amounts of oxides and, that is the most interesting, nitrides of metals [13].

The study of UFP oxidation kinetics, in addition to the pure fundamental interest connected with the further development of knowledge about the UFP structure allows to predict the possibility of the ignition of UFP different metals, the possibility of the reactivity loss, term and conditions of storage.

Experiments studying the kinetics of UFP oxidation in air have been carried out on the Q-1500D derivatograph and a special DTA installation. 50 mg UFP samples in mixture with an inert substance (aluminium oxide) in order to reduce self-heating have been used for thermogravimetric studies. In order to determine the temperature of the thermal equilibrium break-down (the start of the intensive self-heating of the sample) the samples of 100-200 mg have been used in DTA installation. A dynamic mode with the heating rates from 2.5 to 20 K/min has been used. The DTA installation owing to the heat-compensating unit and good heat insulation minimising the heat rejection from the reaction cell has allowed to determine the moment of the thermal equilibrium break-down (self-acceleration of the reaction) that usually results in the self-ignition of the UFP powder.

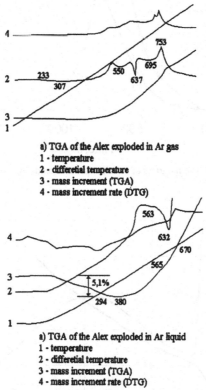

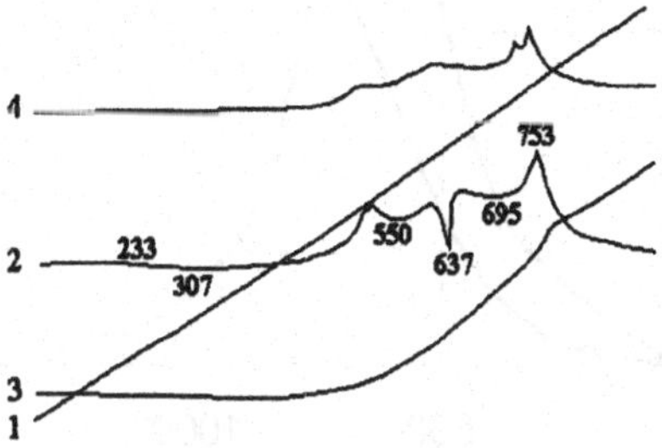

a) TGA of the Alex exploded in Ar gas
1 - temperature
2 - differetial temperature
3 - mass increment (TGA)
4 - mass increment rate (DTG)

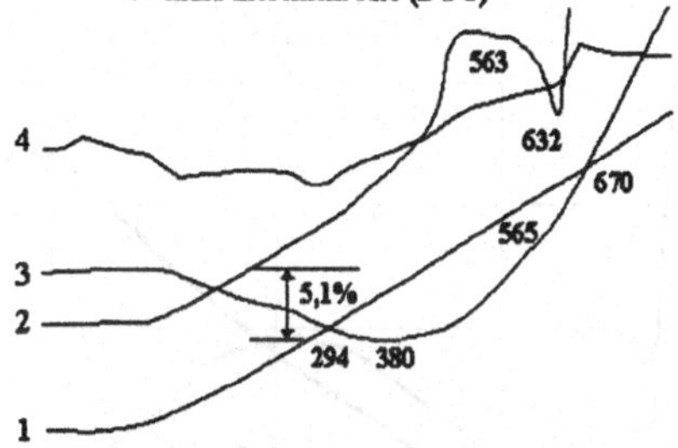

a) TGA of the Alex exploded in Ar liquid
1 - temperature
2 - differetial temperature
3 - mass increment (TGA)
4 - mass increment rate (DTG)

Fig.3. Thermograms of the ultra-fine aluminium oxidation in air at the linear heating with heating rate of 5 K/min.

The study of kinetic mechanism of UFP oxidation of six metals - Al, Zn, Mo, Fe, Cu, Sn - has allowed to make some generalisation as concerns the mechanism of the UFP oxidation and self- ignition.

Figures 4 and 5 show typical kinetic relationships of the UFP oxidation processes. Depending on the character of oxidation in air the UFP of the metals studied can be classified by 2 types:

a) oxidation has one macroscopic stage (Fig. 4); the process is characteristic for the most chemically active and high-caloric metals like Al, Zn, Mo.

b) oxidation has two stages (Fig. 5); the two-stage kinetics is observed for low active polyvalent metals forming the oxides with changeable composition - Cu, Sn.

It is remarkable that the temperatures of the oxidation with a marked rate and of the self-ignition of low-active metals are significantly lower than that of active metals.

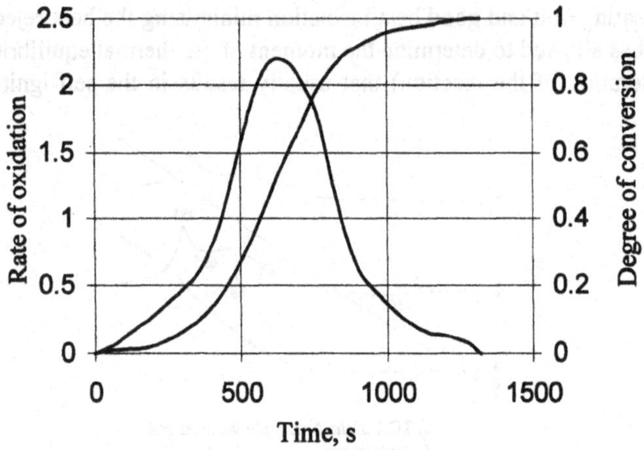

Fig.4. Thermograms of the ultra-fine zinc oxidation in air

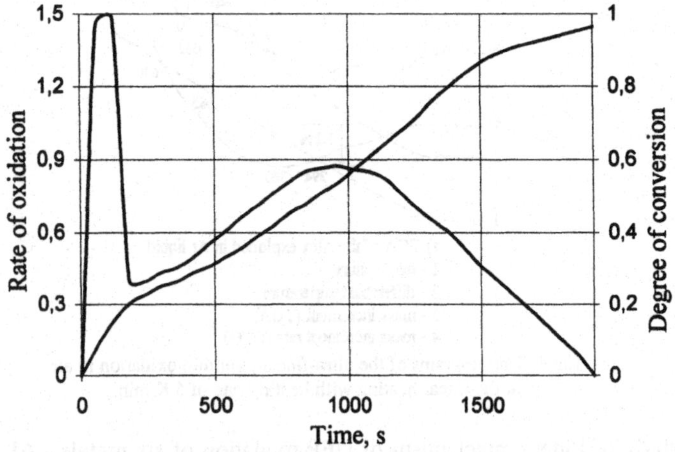

Fig.5. Thermograms of the ultra-fine iron oxidation in air

As stated above, the process of the beginning of the active UFP interaction with air depends on the temperature of gas film desorption from powder particles and on the kinetic law of oxidation.

We have found that the continuous alternation of the oxidation laws - from power function with the index n= 0, 1, 2 up to logarithmic takes place even in course of one macro-stage [14-15]. Accordingly, metals can be oxidised only by the power or logarithmic oxidation laws. General kinetic equation of the process in the non-dimension form is as follows:

$$d\eta/dt = k_0 \exp(-E/RT)\ \eta^{-n} e^{-k_1 t}$$

where η is the depth of conversion (the level of metal oxidation) determined on the basis of thermogravimetric data by the change of UFP mass in heating, t is the time, $d\eta/dt$ is the reaction rate, k_0 is the pre-exponent, E is the activation energy, R is the universal gas constant, T is the temperature, n is the degree index at the oxidation law.

At n = 0,1,2 we have the linear, parabolic or cubic oxidation law, correspondingly; k_1 is the factor for the logarithmic oxidation law. Two oxidation processes are possible: with k_1=0 according to the power law of oxidation with n = 0, 0.5, 1, 1.5, and 2 or with k_1>0 according to the logarithmic law. In course of the metal oxidation in dynamic conditions (linear heating) oxidation laws can alternate in certain temperature ranges which is connected with structural changes of the oxide film on the metal particle surface. The information about UFP oxidation kinetics (up to nowadays only qualitative assessment has been carried out), e.g. [16] has a considerable applied importance, in addition to the development of the fundamental knowledge about UFP. In particular, this allows to find specific criteria for the assessment of their reactivity and safe handling in course of synthesis.

The most common characteristic of the oxidation process of any metal UFP is, obviously, the presence of the parameters range with the linear oxidation law (n=0, oxidation rate $d\eta/dt$ does not depend on the oxidation level η, the process is analogous to the mono-molecular reaction) on the initial oxidation stage. Minimum temperatures of the thermal equilibrium break-down are connected with initiation of a strong self-heating just within this area of thermal curves. This is understandable taking into account the fast de-sorption of the protective gas film of UFP when reaching the threshold temperature.

The complete investigation undertaken and the analysis of the thermal curves of UFP oxidation have allowed to elucidate the comprehensive mechanism of powder oxidation beginning with the linear oxidation, at the absence of continuous protective oxide films in course of the gas film desorption and a strong slow-down of the process (the logarithmic oxidation) after the fast oxidation of the fresh surface (Table 1).

TABLE 1. Kinetic parameters of the metal UFP oxidation in air.

Metal	Temperature, K	Oxidation law	K_1	K_0	E, kJ/mole
Aluminium	700-830	Linear	-	$2{,}2\ 10^{6}$	155
	>830	Logarithmic	11	$4{,}0\ 10^{6}$	155
Molybdenum	580-670	Linear	-	$2{,}2\ 10^{9}$	170
	670-920	Logarithmic	16	$2{,}3\ 10^{37}$	550
Iron	530-770	Linear	-	$6\ 10^{7}$	125
	>770	Logarithmic	12,5	$8{,}2\ 10^{11}$	84
Zinc	460-620	Linear	-	$5{,}2\ 10^{5}$	100
	>620	Logarithmic	5,4	$1{,}5\ 10^{6}$	100
Tin	420-490	Linear	-	$3{,}5\ 10^{5}$	88
	>490	Parabolic	-	0,475	42
Copper	400-460	Linear	-	$9{,}5\ 10^{14}$	160
	460-510	Logarithmic	20	$1{,}0\ 10^{17}$	160
	510-560	Linear	-	$3{,}8\ 10^{-2}$	19
	>560	Logarithmic	13,5	$1{,}2\ 10^{6}$	48

However, the calculation of the heat flux balance allows to assess the minimum possible temperatures of the metal UFP self-ignition using the data obtained on the UFP oxidation kinetics within the area of the linear oxidation law (Table 2).

TABLE 2. Parameters of the initial stages of the metal UDP oxidation in air and temperatures of the thermal equilibrium break-down

Metal	η	dη/dt 10^3, s^{-1}	k_o s^{-1}	E kJ/mol	c/Q 10^5 K^{-1}	T_o calc. K	T_o exper K
Al	0,22	11,60	2,2 10^6	155	3,0	690	720
Mo	0,9	1,18	2,2 10^9	170	3,7	600	630
Fe	0,79	1,66	6,0 10^7	125	6,3	555	570
Zn	0,70	2,33	5,2 10^5	100	7,8	480	520
Sn	0,96	0,42	3,5 10^5	88	10,5	425	450
Cu	0,61	3,05	9,5 10^{14}	160	16,9	420	430

In this case under the conditions of the thermal analytical device with the linear heating rate the heat flux from the heater is as follows: $F_1 = c\,\rho\,\omega$ (J/cm³ sec)

The inherent heat evolution of the system owing to the chemical reaction at n=0 is expressed by the equation: $F_2 = Q\,\rho\,k_o \exp(-E/RT)$.

In the moment of the thermal equilibrium break-down, i.e. of the drastic increase of the reaction rate, the inherent heat evolution becomes large, $F_2 > F_1$ independently on the external heat input. The condition of the equal heat fluxes is the condition of the break-down of the thermal equilibrium and the self-supporting reaction. The formula for the determination of the minimum temperatures of the thermal equilibrium break-down is as follows:

$$T_0 = -\frac{E}{R\ln(c\omega/Qk_0)},$$

Here T_0 is the minimum possible temperature of the thermal equilibrium break-down in the UFP-air system,

ρ is the metal density,

c is the UFP heat capacity,

Q is the oxidation heat effect,

ω is the heating rate.

As follows from the formula deduced the relation c/Q have the most pronounced effect on the T_0. The comparison of calculated and experimentally observed values of the thermal equilibrium break-down temperature have been carried out using the kinetic parameters obtained in the course of thermogravimetric experiments (Table 2).

The data of Table 2 show that the calculated thermal equilibrium break-down temperatures (oxidation with self-acceleration) agree very closely with the data experimentally obtained. The copper powder, low-active in normal state, is believed to be the most flammable (the rated minimum temperature of UFP self-ignition is 160° C approximately, Table 2). UFP oxidation degree in the temperature range studied is relatively low for aluminium (22 %) and reasonably high for the other metals (70-96 %) despite sufficiently high rate of aluminium UFP oxidation. This problem calls for further investigations.

4. Chemical Reactivity

Nano-size metal powders actively react with boron, carbon, nitrogen, sulphur. In some cases the interaction goes on in the burning mode. Various compounds can be obtained by such a method.

The reaction between aluminium UFP and water beginning at 50-80° C seems to be unusual. As this takes place, hydrogen is evolved and oxide-hydroxide aluminium phases are formed (Fig. 6). If a gel-like water thinned by small polymer additives (1-3 %) is used, its mixture with the nanosize aluminium is easily ignited by an electric coil and burns in inert atmosphere as a normal solid propellant.

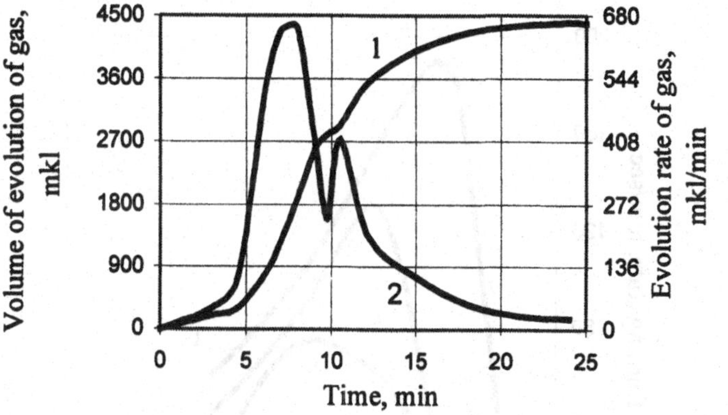

Fig.6. Total hydrogen evolution (1) and the evolution rate at the reaction of ultra-fine aluminum with water (2).

Such compositions are of interest as a source of hot hydrogen as well as for the production of ultra-fine or monolith corundum [17].

Some unusual phenomena are observed in course of burning of nanosize aluminium mixtures with nitrates and perchlorates (Figs. 7-9). Burning rate of nanosize Alex (electro-explosion aluminium) in these mixtures is sufficiently higher than that of 5 μm size commercial Al powders and reaches 100-180 mm/sec. The maximum of the burning rate is usually shifted towards the excess of aluminium content (Fig. 7-9). Alex is able to activate the burning of the conventional commercial aluminium. In the presence of Alex its burning rate increases by 2-7 times. In addition, there was observed a stable burning of triple mixtures with the polymer binder at the temperatures of liquid nitrogen.

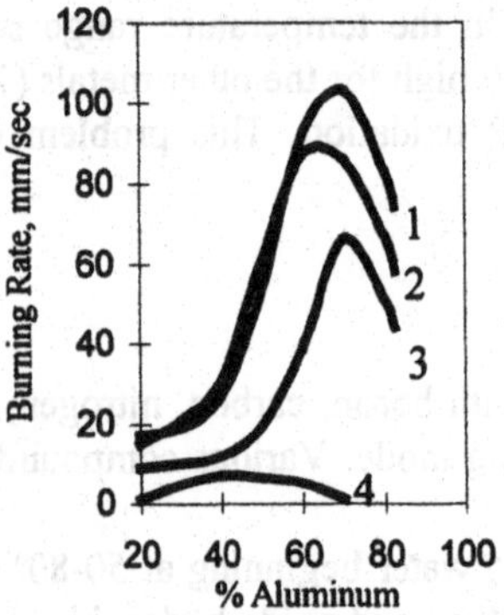

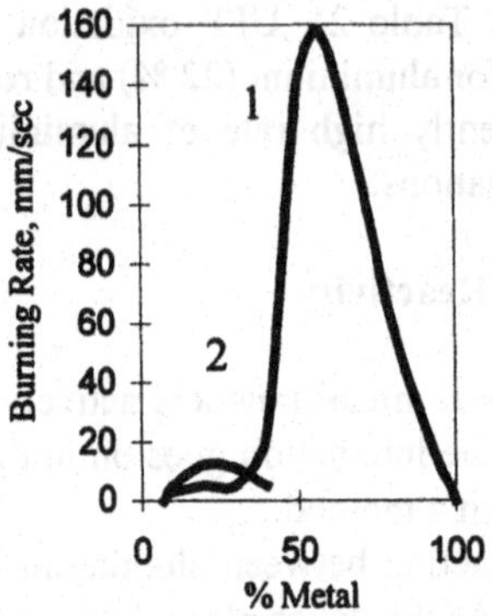

Fig.7. Burning rate versus percent of Aluminum in mixture with nitrates (P = 4 MPa): 1 - Alex + $NaNO_3$, 2 - Alex + KNO_3, 3 - 30% Alex + 70% commercial Al +$NaNO_3$, 4 – commercial Al + $NaNO_3$.

Fig.8. Burning rate versus percent of Aluminum in mixture with Ammonium Perchlorate: 1 – Alex, commercial Al.

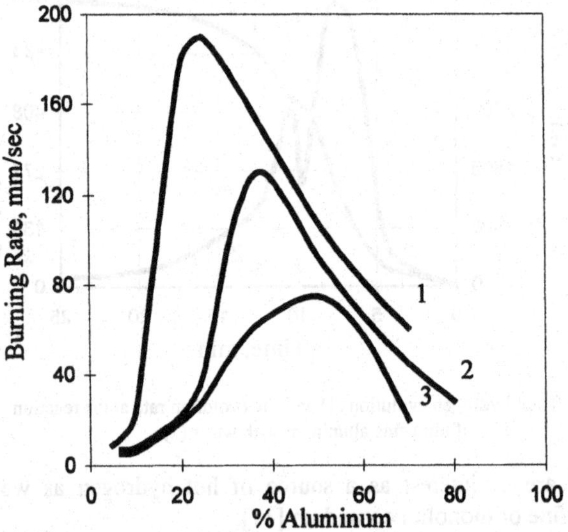

Fig.9. Burning rate versus percent of Aluminium in mixture with Potassium Perchlorate (P = 4 MPa): 1 - Alex, 2 - 30%Alex + 70% commercial Al, 3 – commercial Al.

5. Conclusions

The experimental data obtained suggest that the destruction of metal UFP metastable structure with the release of the stored energy occurs in the condensed phase near the burning surface and that the ignition of aluminium particles in contradiction with the conventional metallised propellants behaviour, takes place mostly at the burning surface. This explains a pronounced Alex effect on the burning rate and diminishing the metal particle agglomeration in burning of condensed systems.

The other advantage of Alex powders as opposed to the commercial powders is the ability to increase the detonation velocity of explosives (Fig. 10), e.g., of HMX (hexamethylene - 3- nitramine). Meanwhile the detonation velocity of the HMX-commercial aluminium mixtures monotonously decreases, that of the Alex mixtures has an extreme character and at the Alex content more than 30 % drastically increases to 7000 m/sec (Fig. 10).

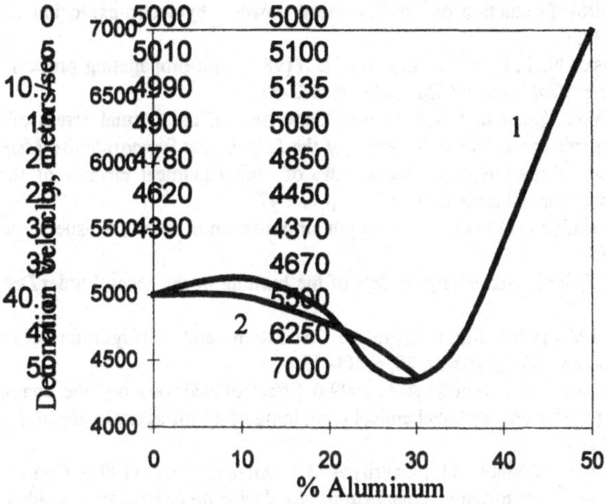

Fig.10. Detonation velocity as function of Aluminum in an explosive:
1 - Alex, 2 - commercial Al.

The above brief review of nanosize electro-explosion powder properties demonstrates good prospects of their application in novel technologies and the necessity for further investigations in this field.

6. References

1. Savage S.J., Gringer O. (1992) Ultra-fine powder production methods an overview, in *Advances in Powder Metallurgy and Particulate Materials. Novel Powder Processing. Proceedings of Powder Metallurgy World Congres*, Princeton, NJ, 7, 1..
2. *Exploding Wires* (1959) in William G.Chace (eds), Air Force Cambridge Research Center, Plenum Press Inc., New York & Chapman and Hall Ltd., London.
3. *Exploding Wires* (1962), in William G.Chace (eds), Proc. of Second Conf. on the Exploding Wires, 2, Plenum Press Inc., NY.
4. *Exploding Wires* (1964), in William G.Chace (eds), Proc. of Second Conf. on the Exploding Wires, 3, Plenum Press Inc., NY.
5. Exploding Wires (1968), in William G.Chace (eds), Proc. of Second Conf. on the Exploding Wires, 4, Plenum Press Inc., NY.

6. Karioris F.G., Fish B.R. and Royster G.W. (1962) Aerosols from Exploding Wires, in WilliamG.Chace (eds), *Exploding Wires*, **2**, 299-313.
7. Jonson R.L., Siegel B. (1970) A Chemical reactor utilising successive multiple Electrical Explosions of metal Wires (Oak Ridge), *Review of Sci. Instr.*, **41**, 6, 854-859.
8. Hale R.W., Ordway D.E., Tan P. (1973) Development and Test of Technique to Create Liquid Suspensions of Sub-micron Metallic Particles, Sage Action Inc. Ithaca, NY, Office of Naval Research, Arlington, Va. Report No. SAL-RR-7301.
9. Yavorovsky N.A., Ylyin A.P., Proskurovskaiya L.T. ets. (1984) Phenomenon of the heat explosion in ultrafine powders of pure metals, *in Proc. of the Ist All-Union Symposium on Micro-Kinetics and Chemical Gasdynamics*, Department of the Institute of Chemical Physics of the Academy of Sciences of the Soviet Union, Chernogolovka. v.1, p.1, 47.
10. Yavorovsky N.A. (1996). Production of ultrafine metal powders byelectro-explosion, *Izvestya VUZov, Fizika*, **4**, Pp. 114 - 135.
11. Ivanov G.V.,Yavorovsky N.A., Kotov Y.A., and et al. (1984). Self-propagating process of the ultrafine metal powders sintering, *Dokladi AN SSSR*, **275**, 4, 873-875.
12. Ivanov G.V., Ivanov V.G., Kuznetzov V.P. (1984). Generation of the thermal waves of the meta-stable state relaxation in disperse metal media, *in Proc. of the I All-Union Symposium on Micro-Kinetics and Chemical Gasdynamics*, Department of theInstitute of the Chemical Physics of the Academy of Sciences of the Soviet Union, Chernogolovka. v. 1. p.1. -. 47.
13. Ylyin A.P., Proskurovskaiya L.T. (1990) Aluminium air oxidation in the ultra-disperse state,*The Powder Metallurgy*, 9, 32 - 35.
14. Merzhanov A.G. (1977) New elementary models of the burning of the second order,*Dokladi AN SSSR*, **233**, 6,. 764 - 769.
15. Ivanov V.G., Ivanov G.V. (1986) High-temperature oxidation and self-ignition of rare-earth metals, *Combustion, Explosion and Shock Waves*, **20**, 6,. 11- 13.
16. Ylyin A.P., Yablunovskii V.A., Gromov A.A. (1996) Effect of additives on the combustion of ultra-dispersed aluminium powder and on the chemical combining of air nitrogen,*Combustion, Explosion and Shock Waves*, **32**, 2, 11- 13.
17. Ivanov V.G., Leonov S.N., Savinov G.L., Gavrilyuk O.V., Glazkov O.V. (1994). Combustion of ultra-disperse aluminium - gel foam mixture, *Combustion, Explosion and Shock Waves*, **30**, 4, Pp. 167- 168.
18. Ivanov V.G., Ivanov G.V.,Gavriluk O.V., Glazkov O.V. (1995) Combustion of ultra-fine aluminium in fluid media, in *Chemical gas dynamics and combustion of energetic materials* (Workshop-95), Tomsk,. 40-41.

COLD GAS GENERATORS: MULTIPLE USE IN HAZARDOUS SITUATIONS

V.A.SHANDAKOV, V.F.KOMAROV, V.P.BOROCHKIN
ALTAI Federal Research & Production Centre,
Bijsk 659322, Russia

One of the most convenient for use gas sources to perform the work in pneumatic systems, to vacate or fill the space in technological equipment (release of ballast capacities of submarine, emergency filling chemical reactors) and to suppress flame in electric and electronic systems are composite solid propellants (CSP). Their main advantage over other sources is fast response and long-term reliable service without special maintenance inspections. Their major drawback is high temperature of gases that requires substitution of traditionally used structure materials. Therefore, for successful use of solid propellant gas sources it is necessary to develop CSP with a temperature of gaseous combustion products 300...330K. For their wide application in practice it is necessary to provide a certain level of ecological purity of combustion products. Let's discuss the alternatives of this problem.

The schematic drawing of temperature profile [1-2] in the burning CSP is shown in Fig.1. In the vicinity of the burning surface the chemical reactions of components decomposition and solid phase oxidation (T_0 - T_s zone) are beginning. Then the weak temperature rise occurs in the zone of diffusion mixing of gasification products (T_s - T_c zone) followed by its sharp build-up in the zone of gas-phase chemical reactions (T_c - T_m zone) to its maximum (curve 1). If some part of an oxidiser of the CSP is replaced by a substance decomposing with heat absorption and the products of its decay appear in highly oxidised form, such as

$$H_2C_2O_4 \rightarrow H_2O + CO_2 + CO,$$

then the maximal gas temperature, T_m, will be inevitably reduced (curve 2) that was realised in practice. However, in virtue of the fact that CSP combustion is a self-sustaining process, as soon as the heat coming to the propellant from the zone of chemical reactions becomes insufficient to gasify condensed material, the process is terminated. Temperature $T_m \approx 700$ K seems to be a limiting value.

V.E. Zarko et al. (eds.), Prevention of Hazardous Fires and Explosions, 341–346.

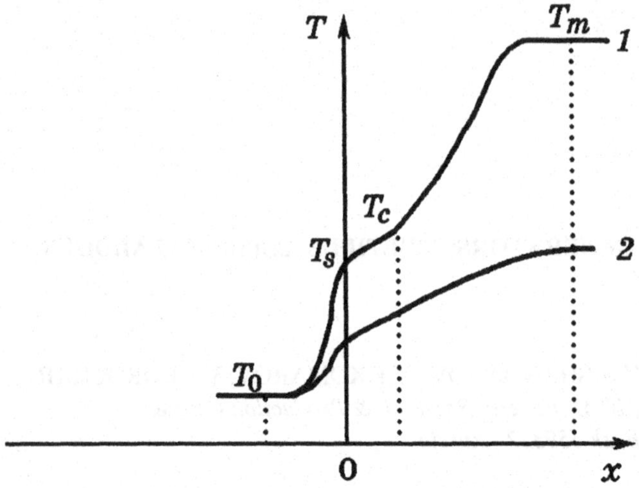

Fig.1. Temperature profile in the burning solid propellant. T_0 is the initial temperature, T_s is the surface temperature, T_m is the maximum flame temperature.

It is possible to reduce T_m further if one separates propellant and cooling agent, with gaseous combustion products of CSP being filtrated through a layer of the cooling agent. In this case it is possible to lower temperature of exhausting gases up to 400...450 K. A discrepancy between the requirement to decomposition onset temperature (T_D) of a substance-cooler (below this temperature the gases cannot be cooled) and desirable service temperature range of the device (it should be higher that T_D) is a limitation. However, such systems are suitable for application in equipment with certain limitations. Their gaseous combustion products contain water, water-soluble gases such as ammonia and, as far as cooling is going at pressure, they can be condensed, i.e. the system is unstable in time. Besides, gaseous combustion products of these systems are toxic and in some cases are fire hazardous when mixed with the air.

There exists another opportunity to reduce the temperature of gaseous combustion products. It is known the Zeldovich criterion [1] of stability of stationary CSP combustion in the form of inequality

$$\frac{E}{2RT_g^2}\cdot(T_s - T_0) \le 1 \qquad (1)$$

where T_s is the surface temperature of burning CSP;

T_g is the gas temperature of burning CSP;

T_0 is the initial temperature of CSP charge.

If CSP is manufactured as a composition which it is not capable of self-sustaining combustion without preliminary heating to a specified temperature T_0 and at the same time the heating is performed by gaseous and condensed combustion products of the propellant itself, then the combustion process, as we have already reported [3], becomes real. To do this, CSP is made as a porous material with through holes and the operation of gas generator is carried out according to the scheme contrary to the usual (Fig.2). Combustion zone is located in the dead-end side (1) of combustion chamber, it is followed by CSP charge (2), filter (3) and only after that by nozzle (4). The pressure being developed in the combustion zone pushes the gas through the body of porous

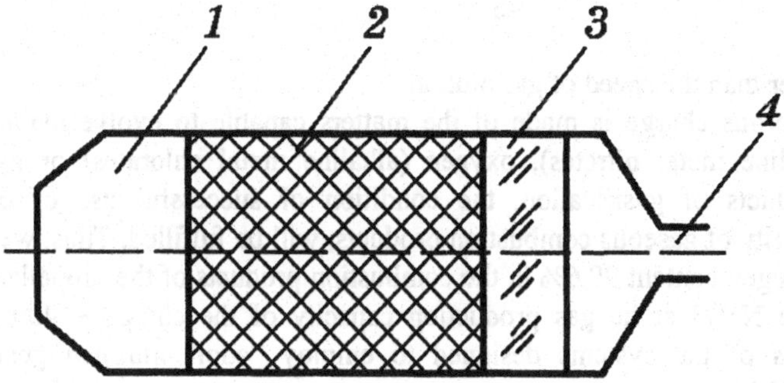

Fig.2. Scheme of porous solid propellant charge burning in the combustion chamber.

charge, heating it up to the temperature of gasification. In fact, we have the process similar to the linear pyrolysis, the combustion products of the charge work as a body heater.

Thermal balance under the ideal heat exchange between a porous medium and gas can be described by the following equation (assuming uniform temperature distribution at given cross-section of charge)

$$C_0 \frac{\partial T}{\partial \tau} = -G_H \frac{\partial T}{\partial x} \tag{2}$$

where $G_H = C_p \cdot G$;
G is the mass flow rate of gas;
C_p is the specific heat of gas;
C_0 is the volumetric heat capacity of the medium;
T is the temperature;
τ is the time.

If the gas temperature at inlet to the porous system changes in time as $T_0(\tau)$ and at the initial moment of time the medium temperature space distribution is determined by the relationship $F(x)$, the solution of equation (2) takes the form

$$T(\tau,x) = \begin{cases} F\left(\tau - \dfrac{C_0 \cdot x}{G_H}\right), & \tau > \dfrac{C_0 \cdot x}{G_H} \\ T_0\left(x - \tau \cdot \dfrac{G_H}{C_0}\right), & \tau < \dfrac{C_0 \cdot x}{G_H} \end{cases} \tag{3}$$

Hence, it follows that temperature perturbations spread along the porous body in case of ideal gas-porous material heat exchange with velocity

$$V = \frac{G_H}{C_0} \tag{4}$$

which is lower than the speed of gas motion

If porous charge is made of the matters capable to evolve mainly nitrogen (azides, alkaline metal nitrates), oxygen (alkaline metal chlorates) or hydrogen as gaseous products of gasification, the condition of successful use of the system, ecological purity of gaseous combustion products, will be fulfilled. Thus, we succeeded to obtain nitrogen content 99.6% in the combustion products of the propellant based on sodium azide NaN3 at he gas production capacity of the charge ~ 400 l/kg. The characteristics of the systems designed to employ "cool" nitrogen generators are presented in Table.

Table . Characteristics of the system using "cool" nitrogen generators

System / Characteristic	Dubl-1M	Rescue raft	Life jacket	Air bag (PSDP)	Telephone office fire protection
1. Volume of gas supply, l	600	2140	16	60	300
2. Operation time, s	20	60	4	0.04	8
3. Average rate of gas release, l/s	30	36	4	1500	38
4. Gas pressure in the sys-tem, MPa	5.6	0.11	0.11	0.11	0.11
5. Gas generator weight, kg	21	15	0.22	4.5	8
6. Probability of failure-less operation	0.98	0.99	0.99	0.999	0.97

The system of emergency change-over switch of the oil-gas pipeline to a by-pass line (Dubl M system) [4,5] is applied at the facilities of GPU ASTRAKHANGAZPROM. The system operation has been permitted for 9 years at the temperature range ±50 ° C and at atmospheric pressure 84 to 108 kPa at the upper level of relative humidity 95% at T=35 ° C. The system is actuated via power unit at voltage 4.5 ... 12V. Operation temperature range of all devices listed in Table such as a rescue raft for pilots when landing to the water surface, personal life jacket for vessel passengers, the passive system of car driver protection (PSDP) with an air bag, the system of nitrogen flame dilution in electronic cabinets of automatic telephone offices is rated to be ± 50 ° C and determined by propellant properties. The other parameters of the devices are different. First of all it refers to the average rate of gas release. Thus, for a personal life jacket it is 4 l/s and for PSDP it is 1500 l/s. This is achieved both by geometrical structure of a porous charge and by specific means of control of the burning rate. The dependencies of linear burning rate for sodium azide based charges on the volume porosity of charge and on pressure drop between the front and rear bottom of combustion chamber are shown in Fig.3.

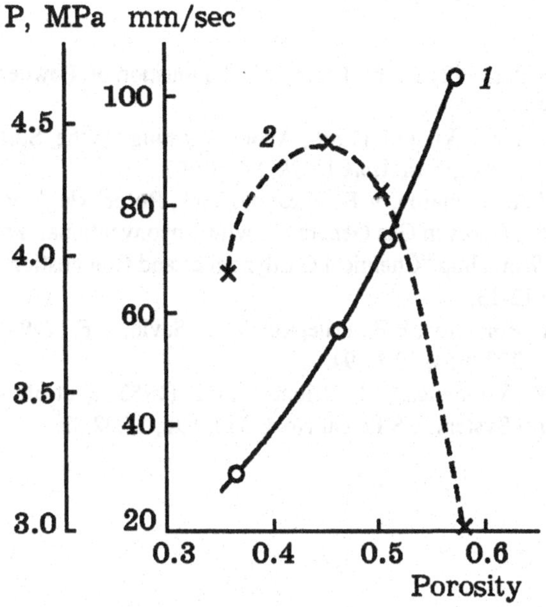

Fig.3. Burning rate (curve 1) and pressure drop ΔP (curve 2) vs volume porosity of the charge.

Along with the system of flame suppression at telephone office that works by way of air dilution with nitrogen, a lot of automatic and manual fire extinguishers have been designed and are under production that based on traditional for fire equipment principle, namely, delivering fire extinguishing powders, liquid or foam to the fire location. These are powerful installations for fire protection of warehouses and storages, automatic systems for gasoline stations, devices for spraying powder to predetermined areas, most fire hazardous for trucks, tractors and electric transport, and routine fire-extinguishers to local fire zone in working rooms. They differ from regular means only by "cool" nitrogen gas generator and work according to the above mentioned principle. They can be easily started up with the help of electric or mechanical pulse.

In the present paper attention was paid to a large number of cool gas generators used for design developments. These are pneumatic drive of Dubl 1M system, inflation of different bags, systems of nitrogen and traditional fire protection and this is only a small part of the possibilities of their successful use. The authors hope that the above examples attract attention of producers of modern techniques and technologies to use of solid propellants in their developments.

References

1. Zeldovich, Ya.B. (1942) To the Theory of Combustion of Powders and Explosives, *GETF*, **12**, p 498.
2. Aristov, Z. I., Leipunskii, O.I. (1946) About Warming Up the Surface of Burning Powder. *Doklady Akademii Nauk USSR*, **54**, p. 507.
3. Shandakov, V.A., Komarov, V.F., Puzanov, V.N., Borochkin, V.P. (1995) Method to Produce Cold Gases in Gas Generators with Unconventional Propellants. International Workshop "Chemical Gasdynamics and Combustion of Energetic Materials" , p 12-13.
4. Dubinin, V.A., Romanov, E.P., Prilepkin, V.A., Savin, V.F. (1994) Gas Generator. RF Patent No 2023956, 30.11.94.
5. Anisimov, V.M., Verkevich, V.I., Orionov, Yu.E. (1983) Shut-off Valve Actuator Remote Control System. US Patent No 4, 373, 698, 15.02.83.

FAST RESPONSE FIRE EXTINGUISHMENT SYSTEM BASED ON MILITARY EQUIPMENT

B.VETLICKY, M.KRUPKA
Energetic Materials Consulting Ltd.
University of Pardubice, Czech Republic

Some years ago a problem arouse in our Company of elaborating system for extinguishment of solid propellants in technological processes where the time of response upon the ignition beginning and subsequent quick extinguishment is highly topical.

When starting design of the system, the following considerations were taken into consideration. It is important to detect arising fire as soon as possible and it is necessary to start extinguishment immediately, before the fire expands. This philosophy is valid and self-evident, not only for extinguishment of solid propellants, but in general too.

The first problem was to find a sensing element with sufficiently quick response onto the fire beginning. After numerous trials with elements working in the visible part of spectrum (false signals), we chose elements sensitive in the infrared spectrum where

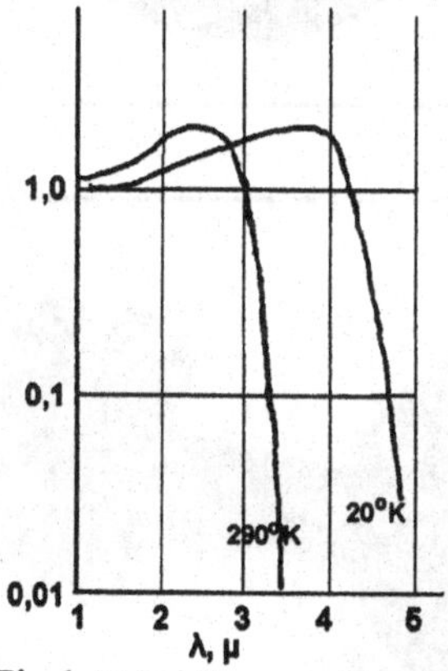

Fig.1. PbS cell spectral sensitivity.

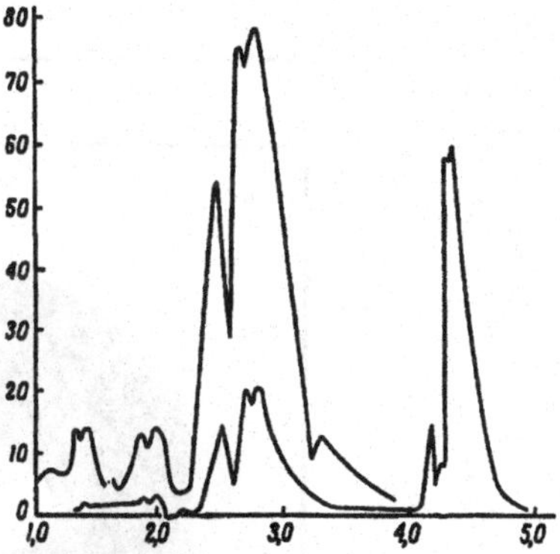

Fig.2.Emission spectrum of the flame.

three-atomic molecules, like CO_2 and H_2O, emit. These molecules are always present in the systems under consideration. In this part of spectrum, however, solid particles emit as well producing continuous spectrum known for a maximum emission-flame temperature relationship. Therefore, we chose as a sensing element a semiconductor cell based on PbS. Its spectral characteristics are shown in Fig.1. For the reference, the

V.E. Zarko et al. (eds.), Prevention of Hazardous Fires and Explosions, 347–350.

emission spectrum of the open flame of hydrocarbons (in given case, combustion of natural gas) is shown in Fig. 2.

It follows from Figs.1 and 2, that the chosen sensing element is not ideal. However, the experiments demonstrated that it is sufficiently sensitive even in the non-optimal spectral wavelength band. Moreover, this was indirectly confirmed from another point of view. These sensing elements were used in the rockets of air-air type SIDEWINDER and in analogous Russian rockets ATTOL and they produced acceptable response onto the combustion products of aircraft engines.

In the system developed the sensing element signal on the flame appearance was recorded by simple electronic device that switched on the extinguishment circuit.

Fig.3. Pyrotechnics valve before (a) and after (b) operation.

In search of a convenient and fast operating valve construction we came to the conclusion , that the only acceptable and reliable device is a pyrotechnics valve. Owing to the necessity of providing sufficient water or extinguishing liquid flow rate we chose a simple design based upon a valve structure manufactured of plexiglass (organic glass), whose operation features are well-known. The opening side of the valve was equipped with a high-speed detonator. The arrangement is illustrated in Figs. 3a and 3b.

The destroyed membrane is screened out by a conveniently dimensioned metallic grid, located downstream. The proper blasting of the membrane is carried out by an independent electric power source. The circuit of electric power is switched on by a simple electronic device reacting to the change in sensing element resistance. The amount of water or extinguishing liquid depends then only upon the pipeline diameter and the liquid pressure.

The basic problem is the sensitivity and functional speed of the whole system. In the first tests we used for the high sensitivity achievement the mirror optics (highly polished aluminium). Later on we used the optics typical of the photographic technique (i.e. not only for the infrared domain). Finally it was established that a colour filter is sufficient.

To determine the response speed by blasting of an electric fuse the time interval between electric current switching-on and the switching-on of the sensing element circuit was measured.

Results of measurement of the response time delay in the system electric fuse - infrared sensing element.

A typical diagram of the system operation is shown in Fig.4. The letters A-E stand for specified moments of time. Their description is given below.

A: Start of the response of sensing element
B: Rise time to the full signal value
C: Start of the fuse blasting
D: Resistance bridge break
E: Rise time to the full battery voltage

CA: Time delay between the measurement start and first sensing element response
CB: Time delay between the measurement start and sensing element signal amplitude reaching the maximum value (system response time)
AB: Sensing element signal rise time up to the maximum value
CD: Fuse failure time

Table. Measured values of different time delays:

Measurement No.	CA [ms]	CB[ms]	AB[ms]	CD[ms]
1	2,4	5,5	3,1	10,5
2	2,1	5,2	3,1	10,7
3	2,56	7,0	4,44	11,6
4	1,9	5,7	3,8	9,4
5	1,9	4,5	2,6	13,6
Average	2,17	5,58	3,4	11,16

It follows from these results that in the case of use of standard high-speed detonators, the total response time does not exceed 12 milliseconds and the sensing element response is sufficiently shorter. This is because the time of functioning the detonator primary and secondary explosives is of the order of few μs. In the case of use of μs - response electric fuses the system response delay would become even shorter. However, from the point of view of providing sufficient flow rate in the pipeline of the extinguishing system the resulting effect would not be very significant.

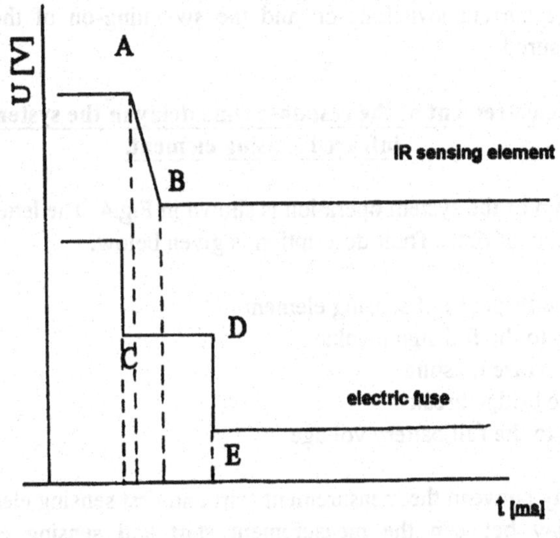

Fig.4. The time diagram of system operation.

As a summary it can be stated that the proposed and realised system has time response approximately 2 orders of magnitude faster than the ordinary commercial systems for extinguishment. At the same time the proposed system is simple and highly reliable.

ON NEAR-LIMITING MECHANISMS OF CATASTROPHIC EXPLOSIONS PENETRATING THROUGH CHANNELS

V.I.MANZHALEI
Lavrentyev Institute of Hydrodynamics of SB RAS,
630090 Novosibirsk, Russia

1. Introduction

Detonation waves and flame propagating at low velocities in smooth channels and porous medium attract significant attention of researchers at present time. Such interest is connected not only with their wave structure being investigated not enough, but also with the fact, that their existence is registered at the conditions, when propagation of classic combustion and detonation waves turns to be impossible [1-8].

From hazard point of view such regimes must be revise the combustion limits. Gas detonation processes in very thin tubes in our experiments and in mines, tubes at chemical producing and power engineering have some common determinative features. Therefore, the results of experiments in thin tubes with the help of approximate theories can be applied for channels with greater cross-section, even for mines.

In this paper the existence and parameters of low-velocity gas detonation modes were studied for smooth capillaries. There are two low-velocity modes, proper low-velocity detonation and galloping wave.

2. Structures of Low-Velocity and Galloping Detonation

The low-velocity gas detonation in circular tubes, which was observed before in stoichiometric acetylene-oxygen mixture [1-4], is the complex consisting of shock wave and flame front moving with a shock wave velocity owing to the gas mass leaving the flow core into boundary layer. Such regimes were observed under initial pressures lower then that for the spin detonation. In the coordinate system, connected with the stationary shock wave, a gas flow behind it slows down up to zero velocity (Fig. 1a). If there is flame in the flow, then it is stabilized at the tube cross-section, where gas velocity on an central axis equals to the normal flame velocity. According to the experimental data, in mixture C_2H_2+2.5O_2 this cross-section is placed at the distance 3-8 channel diameters from shock wave [2,3]. The flame is a nearly flat disk in a flow core with the oblique flame attached to it in a boundary layer. The detonation velocity varies in the range 0.45-0.60 of the CJ detonation velocity without losses, and shock heating is of such value, that a gas local volume at its motion between the shock and flame have

V.E. Zarko et al. (eds.), Prevention of Hazardous Fires and Explosions, 351–360.

not enough time for self-ignition. The structure of low-velocity detonation is shown in Fig. 1b.

It was revealed in tests with stoichiometric acetylene-oxygen mixture and fixed channel diameter [2,3] that the existence domains of usual multi-front detonation and low-velocity detonation are separated by interval of the initial pressure, where the stationary low-velocity detonation is impossible owing to the finishing of induction period before gas local volume gets to the flame, and the multi-front detonation is also impossible there. In this initial pressures range the detonation realized in the galloping mode, which is auto-oscillations between the low-velocity detonation state and multi-front detonation beyond their existence domains as stationary objects.

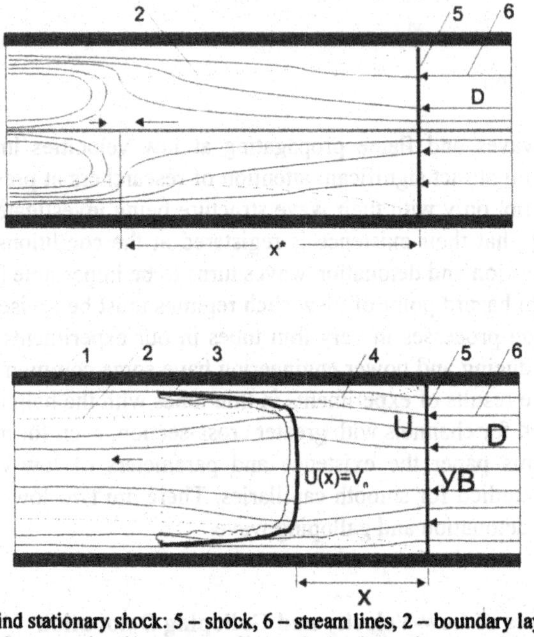

Figure 1. **a) Flow field behind stationary shock: 5 – shock, 6 – stream lines, 2 – boundary layer, X*- distance from shock till the point of flow stopping, b) structure of low-velocity detonation: 1 – tube wall, 3 – flame, 4 – sound line, X – distance from shock to flame.**

3. Experiment

Detonation was initiated by a spark in a steel tube (length 40 cm, internal diameter 20 mm), and then got into a capillary tube with diameter 1.0 mm and length 110-130 cm. The capillary was masked by dark screen with transversal slits (width 2 mm) cut through it in every 40 or 20 mm. Length of oblique flame in the boundary layer according to experimental data was 3-4 channel diameter commonly, and self radiation of gas layer beyond the flame was small as compared with a flame chemo-luminescence. This allows to register precisely the flame passing by the slits with a single photo-multiplier FEU-18A sensitive in the blue domain of visible spectrum. The photo-multiplier was placed in several meters from the capillary. Signal was recorded by a digital oscilloscope C9-16 (Fig.2).

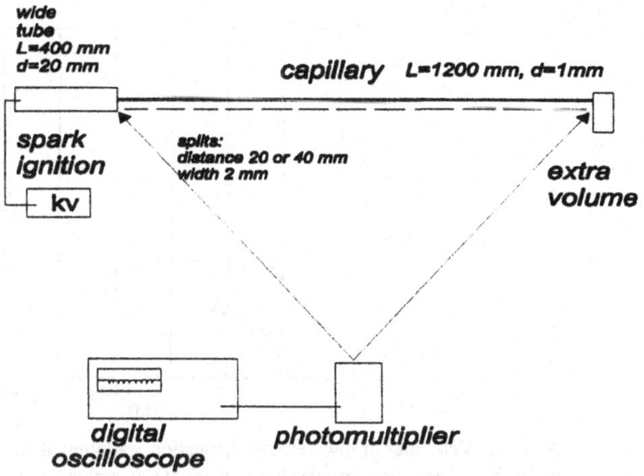

Figure 2.

Tube before each experiment was pumped out up to residual pressure less than 0.1 mm Hg, and gas rests were ousted at filling into separate volume. The using of digital oscilloscope considerably reduced the velocity measurement error on separate intervals between slits and allowed constructing a dependence of velocity versus coordinate. It differs from the data of Refs. [1-3] where only average velocity along all the capillary length, excluding the initial transitional area, was measured. An accuracy of signal digitizing was 0.5 μs, that at wave velocity 1 km/s and distance between slits 40 mm provided the registration of visible local flame velocity with accuracy better than 25 m/s. The experiments were carried out for mixtures C_2H_2+9.5 O_2, C_2H_2+5O_2, C_2H_2+2.5 O_2, C_2H_2+1.5 O_2, C_2H_2+O_2.

4. Experimental Results and Their Discussion

Boundaries of existence domains of different detonation modes are given in Table 1, and analysis of experimental and calculation results is presented in Table 2. Shown on Fig. 3, are plots of average wave velocities versus pressure including three types of detonation modes: multi-front detonation, galloping and low-velocity waves. Given in Fig.4, is the example of velocity plots for low-velocity (a) and galloping (b) detonation along a capillary. Velocities of high-velocity detonation and low-velocity waves after initial transitional area are commonly constant, though some high-frequency oscillations are observed. Such velocity dependencies on a coordinate are evidence that all three detonation modes in a capillary are self-sustaining. The description of low-velocity wave behavior in more details will be published. At present we are interested in the lowest limit of detonation modes in narrow channels and possibility of its theoretical prognostication. It was observed in the experiment that with leaning the mixture, domains of low-velocity modes become wider intensively, and with the mixture enrichment (in respect of fuel) they get narrow up to zero in C_2H_2+O_2.

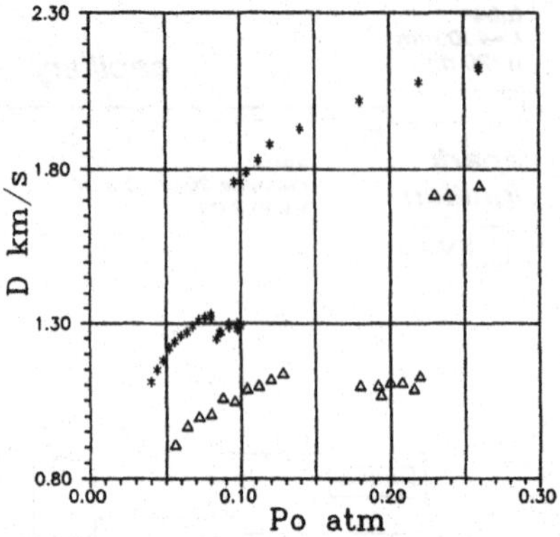

Figure 3. **Velocities of low-velocity detonation, galloping detonation and multi-front detonation for mixtures C_2H_2+2.5O_2 (stars) and C_2H_2+5O_2 (triangles) versus initial pressure. The limits of detonation modes see in Table 1.**

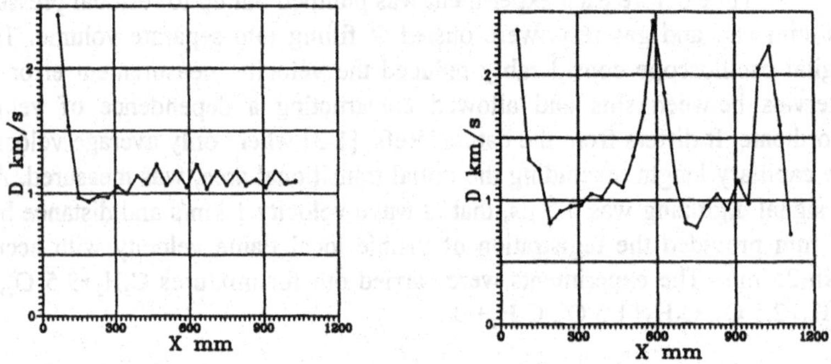

a) **b)**

Figure 4. **Velocities of low-velocity (a) detonation (P_0=0.112 atm) and galloping (b) detonation (P_0=0.208 atm) versus distance in the capillary of 1 mm initial diameter. Mixture C_2H_2+5O_2.**

TABLE 1.. Existence domains of different types of detonation waves in a capillary of diameter 1.0 mm.

Gas mixture	Lowest limit of spin detonation		Galloping detonation limits	Low-velocity detonation limits	
	Po (atm)	D km/s	Po (atm)	Po (atm)	D km/s
C_2H_2+ 9.5 O_2	0.60	1.60	0.60-0.20	0.20-0.08	0.8
C_2H_2+ 5 O_2	0.23	1.81	0.23-0.13	0.13-0.056	0.9
C_2H_2+ 2.5 O_2	0.10	1.93	0.10-0.08	0.08-0.036	1.1
C_2H_2+ 1.5 O_2	0.06	2.10	0.06-0.04	0.04-0.03	1.35
C_2H_2+O_2	0.07	2.23	About 0.07	no LV-detonation	

5. Approximate Calculation of the Lower Limit of Low-Velocity Detonation

Calculation of low-velocity detonation was carried out by the iteration method. The boundary layer parameters on the capillary wall from the shock up to flame are calculated for a given detonation velocity according to Refs. [9,10]. The Prandtl number Pr was considered to be equal to 0.7. The dependence of dynamic viscosity η on temperature T was given by the power function.

At first the transversal flow velocity component behind the shock at the boundary layer has been calculated depending on distance X from the shock was. Then the distance X^* was calculated, at which the entire gas flow passed through the shock wave is turned to be pumped out into a viscous layer. This is the stop point of non-viscous flow with respect to the shock. The boundary layer was considered to be laminar, so the distance X_{sf} from the shock up to a cross-section where flame is stabilized is related to the coordinate X^* via relationship $X_{sf} = X^* (1-V_n/U)^2$. Here U is the flow velocity immediately after the shock, V_n is the flame normal velocity that depends on gas temperature T before flame according to the power-law [2]

$$V_n = V_{n0} (T/T_0)^{3/2},$$

V_{n0} is the normal velocity of laminar flame at temperature T_0=293K. Magnitude V_{n0} = 12.16 m/sec for mixture C_2H_2+2,5O_2 was taken from Ref. [2], where it was obtained by averaging experimental data. For other mixtures V_{n0} was obtained by averaging the flame normal velocity dependencies according to data collected in Ref. [11]. The isentropic braking the flow core between a shock wave and central part of flame caused relatively small rise of temperature (2-3%) and pressure. This was not considered at the flame velocity calculation.

Flame was considered to be infinitely thin. The ratio of pressures π_f and reverse ratio of densities σ_f on flame in the central potential part of flow

$$\pi_f = 1 - q_f M^2, \qquad \sigma_f = 1 + q_f/\gamma,$$

were calculated by the Mach number of the flame front $M = U_n/C$, where the non-dimensional heat release energy q_f=$q_d(T_0/T)$ is defined as

$$q_d = (\gamma M_d^2/2)/(\gamma+1).$$

Here M_d=D_J/C_0 is the Max number of the CJ detonation velocity D_J, C_0 is the initial sound velocity in gas, γ is the heat capacity ratio, C is the sound velocity before flame front while C_f is the sound velocity behind flame - $C_f = C(\sigma_f\pi_f)^{1/2}$, flow velocity - $U_f = V_n\sigma_f$, the flow Mach number - $M_f = U_f/C_f$.

Flow pattern becomes narrow behind flame, and gas is accelerated till the local sound velocity C_1. Flow acceleration behind flame was described by formulas for isentropic flow. Non-dimensional pressure on sound surface is

$$\pi_1 = p_1/p_f = \{[2(\gamma - 1) + M_f^2]/(2\gamma - 1)\}^{\gamma/(\gamma-1)},$$

with densities ratio being $\sigma_1=\rho_1/\rho_f$. Sound velocity on sound surface can be expressed by pressure ratio $\sigma_1 = \pi_1^{1/\gamma}$, $C_1 = C_f(\pi_1/\sigma_1)^{1/2}$. The area of plane flame $S_f = \pi(d - 2\delta_f)^2/4$, where δ_f is the boundary layer thickness in cross-section of plane flame localization. Area of the capillary cross-section $S_0 = \pi d^2/4$. On the sound surface the cross-section of gas flow, passed through the plane flame, decreases till value

$$S_1 = (S_f U_f)/(\sigma_1 C_1).$$

Flow in the boundary layer behind the cross-section, where plane flame is situated, was supposed to be non-viscous. The boundary layer was divided into $n=10^3$ flow tubes. In external flow tubes the oblique flame took place. In flow tubes attached to the capillary wall, flame was not considered. Boundary between above mentioned flow tubes was determined by the Klimov criterion $K = (A/U_n^2)\cdot(dU_t/dl)$, where A is the gas temperature conductivity before the flame front element, U_n is the gas velocity component normal to flame, dU_t/dl is the derivative of tangential component of flow velocity along the dl-element of flame front.

The Klimov criterion [12] was derived to analyze a flame blow out (extinguishing) at a turbulent gas combustion. It should be taken into account that in the shear flow the flame is exposed to tension along its front. Divergence of heat and mass fluxes along the front appears, as the result the heat release zone gets narrow, and heat and diffusion fluxes normal to front increase [13-14]. At equal diffusion and thermal diffusivity coefficients, flame becomes extinguished [15], if $K > \pi/2$.

At approximate calculation the gas velocity distribution in the boundary layer and temperature distribution similar to that for velocity were taken according to Blasius (for example, see [16])

$$u_\delta = D - (D - V_n)\,\eta\,[2 - \eta^2(2 - \eta)],$$
$$T_\delta = T_0 - (T - T_0)\,\eta\,[2 - \eta^2(2 - \eta)],$$

where $\eta=y/\delta_f$ is the transversal coordinate directed from the wall to axis of capillary. An element of oblique flame was considered to be located in the beginning of flow tube. Ratio of the cross-section at the end of flow tube ΔS_e to initial cross-section

$$\Delta S_\delta = \pi\,(d - 2y)\,\Delta y$$

was obtained according to formulas that take into account the parameters change for oblique flame

$$\pi_{f\delta} = 1 - q_{f\delta}\,(V_{n\delta}/C_\delta)^2,\quad \sigma_{f\delta} = 1 + q_f/\gamma,$$

$$\Delta S_{f\delta}/\Delta S_\delta = \sigma_{f\delta} u_\delta^2/[u_\delta^2 + V_{n\delta}^2(S_{fd} - 1)]^{1/2},$$

where $q_{f\delta} = q_d\,(T_0/T_\delta)$, $V_{n\delta} = V_{n0}\,(T_\delta/T_0)^{3/2}$, $C_\delta = C_0\,(T_\delta/T_0)^{1/2}$. At the further isentropic expansion till pressure P_1 corresponding to sound surface in the central flow part, ratio of the finite cross-section of flow tube to section $\Delta S_{f\delta}$ of flow tube immediately behind the flame element is defined as

$$\Delta S_e/\Delta S_{f\delta} = (M_{f\delta}/M_e)\{[2(\gamma-1) + M_e^2]/[2(\gamma-1) + M_{f\delta}^2]^{(\gamma+1)/(2\gamma-2)},$$

where

$$M_{f\delta} = \{[u_\delta^2+V_{n\delta}^2(\sigma_{f\delta}-1)]/(\sigma_{f\delta}\pi_{f\delta}C_\delta^2)\}^{1/2}$$

is the Mach number behind the flame element that depends on transversal coordinate y, and the Mach number of flow at the end of flow tube equals

$$M_e^2 = [2/(\gamma-1)]\{[1+M_{f\delta}^2/(2\gamma-2)](\pi_1\pi_f/\pi_{f\delta})^{(1-\gamma)/\gamma} - 1\}.$$

For flow tubes, where flame was absent due to strong tension, heat release q_f was considered equal zero at calculation of the flow tubes section change. These flow tubes were attached to the capillary wall. Then the sum of finite cross-sections $S=S_1+\Sigma\Delta S_{ei}$, $i=1,...,n$ of flow tubes located in the boundary layer and central flow tube was calculated. These finite cross-sections are in the transversal section of the capillary where flow velocity in the central flow tube equals the local sound velocity. In all flow tubes in the boundary layer (with and without flame) at the above transversal capillary section the flow velocity turned to be supersonic, $M_e>1$.

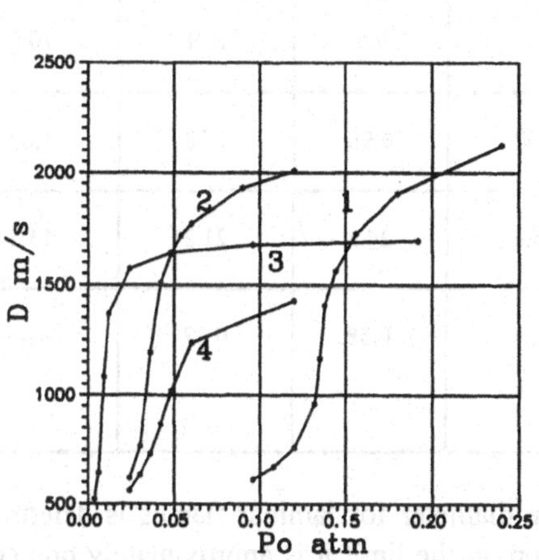

Figure 5. Results of approximate calculation of the low-velocity detonation limits.

A solution was made for by the iteration method, changing the detonation velocity D till the equality S to area S_0 of capillary cross-section was achieved. Varying the Prandtl number in a boundary layer before flame from 0.7 up to 1.0 and γ from 1.2 up to 1.4 slightly influenced the calculation results. Since the proposed model is rough enough, the main aim was to obtain the lowest limit of low-velocity detonation. At the calculations the basic variables were varied, on which a limit location depends. The calculation results are presented in Fig. 5 and Table 2. With decreasing the initial pressure, the detonation velocity turned to be of sharply decreasing magnitude and then achieved a limiting value due to progressing growth of gas portion that passes near the capillary wall without combustion. The detonation limit value was found at the inflection points of curves. Curve 2 corresponds to mixture C_2H_2+2.5O_2 with q=12.9 and V_{n0}=12.1 m/s . On curve 1 V_{n0} is two times larger, and on curve 3 it is two times less. Limit pressures were obtained to be 3.7 times higher and 4 times lower respectively. On curve 4 V_{n0}=12.1 m/sec, and q is 2 times less, as the result the limit pressure arose 1.25 times only. Thus, the limit pressure weakly depends on heat release and is approximately inversely proportional to V_{n0}^2. The calculated detonation limits turned to be close to the experimental data, which allows considering the proposed model to be applicable for calculation of the lowest detonation limit in the case when detonation is low-velocity one, and estimating the limit when it is a spin mode.

TABLE 2. Comparison of experimental limit initial pressure Po (atm) for detonation in capillary (diameter 1.0 mm) with the approximate calculation results.

Gas mixture	C_2H_2+ 9.5O_2	C_2H_2+ 5O_2	C_2H_2+ 2.5O_2	C_2H_2+ 1.5O_2	C_2H_2+ O_2
Experimental limit of detonation	0.08	0.056	0.036	0.03	0.07-gallop, spin
Calculation limit of detonation	0.103	0.06	0.035	0.044	0.092
Flame velocity in m/sec	7.0	9.3	12.1	10.0	7.0
Viscosity 10^5 kg/m sec	1.97	1.90	1.78	1.67	1.57
Criterion of thermal similarity Pe	25.6	24.3	21.2	15.2	26.0
Criterion of gas-dynamic-thermal similarity PeM_f	0.54	0.68	0.77	0.46	0.55

A critical tube diameter for laminar flames is determined by Peclet number $Pe^* \approx 60$. For detonation, at the limit it is approximately one cell at the tube diameter [17-19,21]. Cell size is proportional to ignition delay [20], which would be behind the undisturbed shock moving with spin detonation velocity. An ignition delay exponentially depends on temperature. Therefore the detonation limit for mixtures with different ratio of fuel and oxidizer for a tube with given diameter is determined by the magnitude of exp(E/RT) and does not depend on flame velocities of those gas mixtures.

A lowest limit of low-velocity laminar detonation is determined not by the Peclet number, but by magnitude inversely proportional to the Klimov criterion, applied to oblique flame near a tube wall. Therefore this limit is inversely proportional to flame velocity square in cold gas, V_{n0}^2. Namely this dependence of the low-velocity detonation limit caused its discovery in mixtures which have the highest ratio of flame velocity V_{n0} to sound velocity C_0 that provides for the existence area of low-velocity detonation a possibility to be detached from of the existence areas of detonation and laminar flame. It follows from this that the natural criterion determining the lowest limit by pressure for low-velocity detonation in smooth channel is the product of the thermal similarity criterion $Pe=PrV_{n0}\rho d/\eta$ by the criterion of gas-dynamic similarity $M_f=V_{n0}/C_0$.

Values of criterion PeM_f are given in Table 2 together with calculation results for low-velocity detonation limits and values of Pe criterion. It should be noted that the low-velocity detonation limits with respect to pressure in the experiment are determined with accuracy of 10%, approximately of the same value is an error for the laminar flame velocities V_{n0} given in the Table 2. The general error of the PeM_f criterion, caused by this, can reach approximately 25%, and the Pe criterion up to 15%. Nevertheless, values of the PeM_f criterion, given in the Table, demonstrate nearly the same relative average-quadratic deviation like values of the Pe criterion. This proves, along with theoretical conclusions, that the PeM_f criterion determines the lowest limit of low-velocity detonation more accurate, than the Peclet criterion. Average critical value $(PeM_f)^*=0.60\pm0.11$.

6. Conclusions

1) Detonation limits of acetylene-oxygen mixtures in a narrow smooth channel are investigated. The low-velocity detonation propagating with velocity 0.45-0.60 of the CJ detonation velocity without losses, is the limit of detonation. It was shown that the limit initial pressure for low-velocity detonation propagation in thin smooth channels is up to 7 times lower than the limit initial pressure for spin detonation in the same mixtures.

2) The approximate model of low-velocity detonation is proposed, which now gives overestimated values of wave velocity, but allows calculating the lowest limit of low-velocity detonation with good accuracy. In the presented calculations empirical formulae were used to give flame velocities. Because nowadays the combustion velocity calculation does not make difficulties, the closed model may be constructed.

3) It was demonstrated that the low-velocity detonation limit in narrow channels is lower then the limit of spin (high-velocity) detonation in mixtures with maximal ratio of laminar combustion velocity to sound velocity. The analogous conclusion can be derived for limits of explosion propagation in wide tubes, where a boundary layer and flame behind the shock wave are turbulent, for instance, in fuel-air mixtures.

4) It was determined the lowest value of the critical Peclet number for self-sustaining explosion propagation in smooth channel - $Pe^*=15$, and the average critical Peclet number for low-velocity detonation in acetylene-oxygen mixtures - $\langle Pe^*\rangle=22.4$.

5) For acetylene-oxygen mixtures it was shown that the low-velocity detonation limit can not be determined by the thermal similarity criterion Pe with sufficient accuracy. This limit may be determined more preferably by the gas-dynamic-thermal similarity criterion PeV_{n0}/C_0. In the experiment with acetylene-oxygen

mixtures it was discovered the lowest critical value $(PeV_{n0}/C_0)^* = 0.46$, and average value $\langle (PeV_{n0}/C_0)^* \rangle = 0.60 \pm 0.11$.

Acknowledgment

The work was partially supported by the Russian Foundation for Basic Research, grants N 93-013-17602 and N 97-01-01073.

7. References

1. Manzhalei V.I. (1991) Gaseous Detonation in Capillaries., *13 IGDERS, Nagoya, Abstracts and Information*, 49.
2. Manzhalei V.I. (1992) Low-velocity gas detonation in capillaries, *Reports of RAS*, **324**, 582-584.
3. Manzalei V.I. (1992) Detonation regimes of Gases in capillaries, *Combustion, Explosion, and Shoch Waves*, **28**, 93-100.
4. Manzhalei V.I., Mitrofanov V.V., Aksamentov S.M. (1993) Investigation of Galloping Detonation, *Proceedings of Russian-Japanese Seminar on Combustion, Chernogolovka*, 178-180.
5. Manzalei V.I., Subbotin V.A. (1996) On the possibility of measuring turbulent and laminar flame velocities at an initial high temperature, *Combustion, Explosion, and Shoch Waves*, **32**, N4, 43-46.
6. Pinaev A.V. (1996) Combustion regimes and criteria of flame propagation in jammed space,*Combustion, Explosion, and Shoch Waves*, **30**, 52-60.
7. Babkin V.S., Bunev V.A., Korzhavin A.A. (1980) Flame propagation in porous inertial media, *Combustion of Gases and Natural Fuels, Chernogolovka.*
8. Kaufman C.W., Yan Chuajan, Nichols J.A. (1982) Gaseous detonation in porous media, *19th Symp. (Intern.) on Comb., Haifa*, Pittsburgh.
9. Mairls G. (1962) Damping in a shock tube, determined by action of the non-stabilized boundary layer, *Shock Tubes*, Editors Rakhmatulin H.A. and Semionov S.S, Moscow, 286-319.
10. Mirels H. (1963) Test Time in Low-Pressure Shock Tubes, *Phys. Fluids* 6, 1201-1214.
11. Ivanov B.A. (1969) *Physics of acetylene combustion*, Moscow, Chemistry.
12. Klimov A.M. (1963) Laminar flame in turbulent flow, *PMTF N 3*, 49-58.
13. Kovasznay L.S.G. (1956) Combustion in turbulent flow, *Jet Propulsion* **26**, 485-497.
14. Karlovitz B., Denniston D.W.Jr., Knapschaefer D.H., Wells F.E. (1953) Studies of turbulent flames, *4th Symp. Combust. Baltimore*, Williams and Wilkins, 613-620.
15. Kuznetsov V.R., Sabel'nikov V.A. (1986) *Turbulence and combustion*, Moscow, Nauka.
16. Denisov Yu.N., Troshin Ya.K. (1959) Pulsing and spin detonation of gas mixtures in tubes*Reports of AS USSR* **125**, 110-113.
17. Loitsjansky L.G. (1970) *Mechanics of Liquid and Gas*, Moscow, Nauka.
18. Manzhalei V.I., Mitrofanov V.V., Subbotin V.A. (1974) Measuring of detonation front non-homogeneities in gas mixtures at increased initial pressures, *Combustion, Explosion, and Shoch Waves*, **10**, 703-710.
19. Voitsekhovskii B.V., Mitrofanov V.V., Topchian M.Ye. (1963) *Structure of detonation front in gases*, - Novosibirsk, SD AS USSR.
20. Vasiliev A.A., Nikolaev Ju.A. (1978) Closed theoretical model of a detonation cell,*Acta Astronautica* **5**, 983-996.
21. Vasiliev A.A. (1991) The limit of stationary propagation of gaseous detonation,*Fluid Mechanics and its applications* **5**. *Dynamic structure of detonation in gaseous and dispersed media*, Kluwer Academic Publishers, London, 27-49.